Inorganic and Organometallic Reaction Mechanisms

2nd Edition

Inorganic and Organometallic Reaction Mechanisms

2nd Edition

Jim D. Atwood

Jim D. Atwood
Department of Chemistry
State University of New York at Buffalo
NSM Complex
Buffalo, NY 14260-3000

This book is printed on acid-free paper. ∞

Library of Congress Cataloging-in-Publication Data
Atwood, Jim D., 1950–
    Inorganic and organometallic reaction mechanisms / Jim D. Atwood.
  —2nd ed.
        p.   cm.
    Includes bibliographical references and index.
    ISBN 1-56081-955-3
    1. Chemical reaction, Conditions and laws of.   2. Organometallic
chemistry.   I. Title.
QD501.A89   1996
541.3′9—dc20   96-12525
                                                                                   CIP

© 1997 VCH Publishers, Inc.

Printed in the United States of America

ISBN 1-56081-955-3 VCH Publishers, Inc.

Printing History:
10 9 8 7 6 5 4 3 2 1

Published jointly by

VCH Publishers, Inc.   VCH Verlagsgesellschaft mbH   VCH Publishers (UK) Ltd.
333 7th Avenue   P.O. Box 10 11 61   8 Wellington Court
New York, New York 10001   69451 Weinheim, Germany   Cambridge CB1 1HZ
                                                                   United Kingdom

Preface

The intent for this book remains as for the first edition, a textbook for a one-semester course in kinetics and mechanisms of inorganic reactions to be taught at the upper division undergraduate or at the graduate level. The coverage remains as in the first edition, but each chapter has updated references that reflect the ongoing interest in mechanistic inorganic chemistry. The blend between coordination and organometallic complexes in this book remains a unique feature. Several users have commented favorably about the referencing and problems for the first edition; these features have been strengthened in the second edition.

Chapters 1 (Chemical Kinetics), 3 (Substitution Reactions of Octahedral Werner-Type Complexes), and 7 (Stereochemical Nonrigidity) were changed only to a minor extent by adding references to new research. Chapter 2 (Ligand Substitution Reactions on Square Planar Complexes) was augmented by a section on dissociative processes which have come into clearer focus since the writing of the first edition. Chapter 6 (Homogeneous Catalysis) was augmented by expansion of the section on asymmetric hydrogenation. Chapter 8 (Oxidation-Reduction Reactions: Electron Transfer) was augmented by a large section on organometallic electron transfer. The two remaining chapters were rewritten significantly. Chapter 4 (Organometallic Substitution Reactions) has the most significant changes. The opening section on metal carbonyl complexes was tightened and reorganized to emphasize the main points, while secondary material was minimized. Significant additions were made to the section on reactions of seventeen-electron complexes. A significant portion of the metal carbonyl dimer ($M_2(CO)_{10}$) material was deleted—at the time of the first edition, this was controversial but has been resolved for 10 years. Chapter 5 (Oxidative Addition and Reductive Elimination) was essentially rewritten. New references

(34) on stereochemistry of H_2 addition, carbon–hydrogen bond activation, reductive elimination, and so on, were added.

Mary Atwood and Barbara Raff word-processed the entire manuscript; their skill and dedication were essential to this project. My mentors—T. L. Brown and E. L. Muetterties—introduced me to the field of kinetics and mechanisms; my 21 Ph.D. graduates and seven current graduate students force me to continue to learn. My thanks to all.

Jim D. Atwood
Buffalo, NY

Preface to the First Edition

This book is designed for students who have had one semester of Inorganic Chemistry after General Chemistry, typically senior or graduate-level students. It may be used in its entirety for a one-semester course in reaction mechanisms or as a portion of a more general coverage of advanced inorganic chemistry. The book is thoroughly referenced to provide an introduction to the literature of each area of inorganic and organometallic reaction mechanisms. A unique feature is the coverage of both classic inorganic reaction mechanisms (substitution and electron transfer reactions of coordination complexes) and the newer area of organometallic reaction mechanisms (substitution, oxidative-addition, reductive-elimination, fluxional behavior, and homogeneous catalysis). The former material is covered briefly, with emphasis on the basic mechanisms; the latter is discussed more thoroughly and is especially well referenced to the current literature.

Chapter 1 covers the concepts in chemical kinetics needed for a discussion of inorganic and organometallic reaction mechanisms but is not designed to be a thorough coverage of the field. Basic rate laws, kinetic terms, and kinetic techniques are discussed, with inorganic examples. Chapter 2 describes the stereochemistry, reactivity effects, and mechanisms of substitution reactions on square planar complexes. Chapter 3 provides similar coverage of substitution reactions of octahedral coordination complexes. Chapter 4 covers organometallic substitution reactions in considerable detail. Metal carbonyl complexes, including metal alkyl complexes, metal hydrides, and metal nitrosyl complexes, are discussed. This chapter is extensively referenced and especially up-to-date. Chapter 5 continues the organometallic coverage with oxidative-addition and reductive-elimination reactions. Stereochemistry and mechanisms are the primary features. Using the background of Chapters Four and Five, Chapter 6 goes into mechanisms of homogeneously catalyzed reactions. Hydrogenation, hydroformylation, acetaldehyde synthesis, hydrocyanation, olefin metathesis, polymerization, and methanol homologation are covered. Chapter 7 is an introduction to the large area of stereochemical nonrigidity. Coordination number isomerization, metal migration around a ring, and metal cluster fluxionality make up this chapter. Chapter 8 presents inorganic electron transfer through outer and inner sphere mechanisms.

This book is based on the lecture notes of the author for Chemistry 538, a graduate course in Kinetics and Mechanisms of Inorganic Reactions at the State University of New York at Buffalo. Gordon Harris used the manuscript in rough draft form for the same course and suggested many revisions based on his experience. I am also grateful to Gregory Geoffroy of Pennsylvania State University and William Jones of the University of Rochester, who reviewed the manuscript and made many helpful suggestions.

I would like to take this opportunity to acknowledge a number of other people who have aided in the preparation of this book: Mary Atwood, Ann Pierce, and Sue Slawatycki for their secretarial assistance; Jerry Keister, Melvyn Churchill, and Ken Takeuchi for proofreading and suggestions at the rough draft stage; and to the members of my research group over the three years of this project (Dave Sonnenberger, Mike Wovkulich, Ron Ruszczyk, Wayne Rees, Tom Janik, Brian Rappoli, Bih Huang, Martha Harris, Abdul Shojaie, and Karen Bernard) for help with proofreading and for keeping the research going without the full attention of the author.

Jim D. Atwood

Contents

3. Substitution Reactions of Octahedral Werner-Type Complexes   71

4. Organometallic Substitution Reactions   95

1

Chemical Kinetics

This chapter will consider aspects of kinetics that are required to differentiate between mechanistic possibilities in the remaining chapters. It is not designed to provide a comprehensive treatment of chemical kinetics; refer to References 1–5 for a more detailed treatment. We will consider simple rate laws, reaction profiles, and the most commonly used techniques for data generation.

1.1. Rate Laws

The *rate* of a chemical reaction is the rate of change of the concentration of a reactant or a product.

$$\text{rate} = \frac{-d[\text{reactant}]}{dt} = \frac{d[\text{product}]}{dt} \tag{1.1}$$

The units of rate are usually in terms of molarity per second (M/s). The *rate law* describes the rate of a reaction in terms of the concentrations of all species that affect the rate. For a simple reaction such as an isonitrile rearrangement,

$$CH_3NC \rightarrow CH_3CN \tag{1.2}$$

the rate law is expressed as

$$\text{rate} = -k[CH_3NC] \tag{1.3}$$

1

where k is the *rate constant*, which will be derived shortly. This reaction in which the rate depends linearly on the concentration of one reactant is termed a *unimolecular* reaction. A reaction where the rate depends linearly on the concentration of two reactants (Eq. 1.4) is a *bimolecular* or *second-order reaction*.

$$Br(g) + H_2(g) \rightarrow HBr(g) + H(g) \tag{1.4}$$

$$rate = -k[Br][H_2] \tag{1.5}$$

The rate law in general may be written

$$rate = -k \prod_i [A_i]^{\alpha_i}[X_j]^{\beta_j} \tag{1.6}$$

where the A_i are reactants, k is the rate constant, and X_j are other species (catalysts) that may affect the rate. The reaction order is defined as

$$reaction\ order = \sum_i \alpha_i \tag{1.7}$$

Since the rate of a reaction has units of M/s the units of a rate constant, k, will depend on the reaction order. For a first-order reaction $\left(\sum_i \alpha_i = 1 \right)$ the rate constant has units per second (s^{-1}). For a second-order reaction $\left(\sum_i \alpha_i = 2 \right)$ the units of the rate constant are $M^{-1}s^{-1}$.

1.2. Integrated Rate Expressions

1.2.1. First-Order Reactions

Most reactions are either first order or are executed under conditions that approximate first order. For a first-order reaction of this type,

$$A \rightarrow B \tag{1.8}$$

$$rate = k[A] = \frac{-d[A]}{dt} \tag{1.9}$$

which yields on rearrangement

$$\frac{-d[A]}{[A]} = k\ dt \tag{1.10}$$

Integration from $t = 0$ (A_0) to $t = t$ (A_t) gives

$$-ln[A_t] + ln[A_0] = kt \tag{1.11}$$

which is the equation of a straight line ($y = mx + b$) with $y = ln[A_t]$, $x = t$ and $b = ln[A_0]$. Thus, a plot of $ln[A_t]$ versus time for a first-order process produces a line with slope equal to the rate constant, k. This is shown in Figure 1.1. The units of k derived from the slope are s^{-1}. Examination of Equation 1.11 reveals some other properties of first-order reactions.

$$ln \frac{[A_t]}{[A_0]} = -kt \qquad (1.12)$$

$$\frac{[A_t]}{[A_0]} = e^{-kt} \qquad (1.13)$$

$$[A_t] = [A_0]e^{-kt} \qquad (1.14)$$

Thus, a plot of concentration of A will vary exponentially with time. When the reaction is half way to completion ($[A_t] = 1/2[A_0]$) the time elapsed is termed the half-life ($t_{1/2}$). Many first-order reactions are described in terms of $t_{1/2}$.

$$\frac{[A_{t_{1/2}}]}{[A_0]} = e^{-kt_{1/2}} = 0.5 \qquad (1.15)$$

This indicates that

$$-kt_{1/2} = -0.693 \qquad (1.16)$$

or

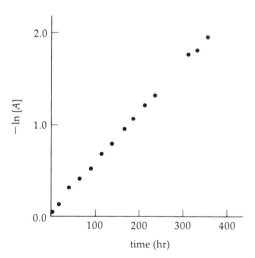

Figure 1.1. A plot of $-ln[A_t]$ versus time for the reaction of *trans*-Cr(CO)$_4$[P(OPh)$_3$]$_2$ with CO at 90°C. The reaction is followed by the decrease in the absorption of the starting complex at 1935 cm^{-1}. The rate constant derived from the slope is $1.49 \times 10^{-6}s^{-1}$.

$$t_{1/2} = \frac{0.693}{k} \tag{1.17}$$

Note that the half-life is inversely related to the rate constant.

1.2.2. Reversible First-Order Reactions

Some reactions, especially isotopic exchange reactions, are reversible under the reaction conditions.

$$A \underset{k_r}{\overset{k_f}{\rightleftarrows}} P \tag{1.18}$$

The rate law for this reaction is

$$\text{rate} = -k_f[A] + k_r[P] \tag{1.19}$$

At $t = 0$, $P_0 = 0$, and $[A_t] = [A_0]$; at any time

$$[P] = [A_0] - [A_t] \tag{1.20}$$

Substituting 1.20 into 1.19 gives

$$\frac{-d[A]}{dt} = k_f[A_t] - k_r([A_0] - [A_t]) \tag{1.21}$$

At equilibrium no net reaction occurs, thus

$$\frac{d[A]}{dt} = 0 \tag{1.22}$$

Applying Equation 1.22 to Equation 1.19 gives

$$k_f[A_e] = k_r[P_e] = k_r([A_0] - [A_e]) \tag{1.23}$$

or

$$[A_0] = \frac{k_f + k_r}{k_r} [A_e] \tag{1.24}$$

Substitution of 1.24 into 1.21 leads to

$$\frac{-d[A]}{dt} = (k_f + k_r)[A] - (k_f + k_r)[A_e] \tag{1.25}$$

Separation of the variables and integration leads to

$$ln \left(\frac{[A_0] - [A_e]}{[A_t] - [A_e]} \right) = (k_f + k_r)t \tag{1.26}$$

and a plot of $ln([A_t] - [A_e])$ versus time will give a straight line of slope $k_r + k_f$. To define either k_r or k_f uniquely one must also evaluate the equilibrium constant,

$K_{eq} = \dfrac{k_f}{k_r}$ to provide two equations with two unknowns. The most difficult problem in treating reversible first-order reactions is in accurately measuring $[A_e]$.

1.2.3. Second-Order Reactions

The only type of second-order reaction that we need to consider is a reaction that is first order in two reagents with a rate law

$$\text{rate} = k[A][B] \tag{1.27}$$

This rate law can be integrated, but the reaction is commonly run with either A or B in large excess ($[B_0] >> [A_0]$). In this case the $[B]$ will not change significantly during the reaction.

$$\text{rate} = k[B][A] = (k[B])[A] = k_{obs}[A] \tag{1.28}$$

These conditions ($[B_0] >> [A_0]$) are referred to as pseudo *first-order conditions* because the reaction may be treated as a first-order reaction. A plot of $ln[A]$ versus time will yield a straight line of slope, k_{obs}. The reaction has to be investigated for several initial concentrations of B, always maintaining the large excess of B over A. This generates a series of k_{obs} for different $[B_0]$. Referring to Equation 1.28,

$$k_{obs} = k[B] \tag{1.29}$$

such that a plot of k_{obs} versus $[B]$ will be linear with a slope of k. Such a plot is shown in Figure 1.2.[6] It should be noted that the units of k_{obs} are s^{-1}, while the units of k are $M^{-1}s^{-1}$ for a second-order reaction.

1.2.4. Consecutive First-Order Reactions

A ligand substitution reaction frequently will not stop after one substitution, but will continue to more highly substituted complexes. These would then be consecutive reactions. The plots of concentration of the substituted species as a function of time are rather distinctive, a sample plot is shown in Figure 1.3.[7] The reactions may be represented in general as

$$A \xrightarrow{k_1} B \tag{1.30}$$

$$B \xrightarrow{k_2} C \tag{1.31}$$

and

$$\frac{-d[A]}{dt} = k_1[A], \quad \frac{d[B]}{dt} = k_1[A] - k_2[B], \quad \frac{d[C]}{dt} = k_2[B]$$

Integrating, recognizing that at $t = 0$, $[A] = [A_0]$, $[B] = [C] = 0$ and at any time $[A_0] = [A] + [B] + [C]$, yields

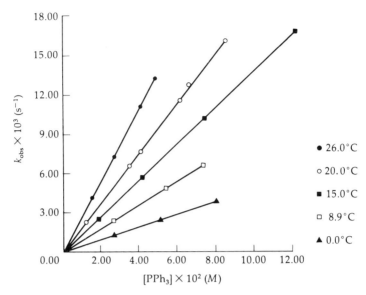

Figure 1.2. Plots of rate versus molar concentration of PPh$_3$ for substitution on V(CO)$_6$ under pseudo first order conditions. Note the intersection at k_{obs} = 20. [Reprinted with permission from Q. Shi, T. G. Richmond, W. C. Trogler, and F. Basolo, *J. Am. Chem. Soc.* **104,** 4032 (1982). Copyright 1982 American Chemical Society.]

$$[A] = [A_0]e^{-k_1t} \tag{1.32}$$

$$[B] = \frac{[A_0]k_1}{k_2 - k_1}\left(e^{-k_1t} - e^{-k_2t}\right) \tag{1.33}$$

$$[C] = [A_0]\left[1 - \frac{k_2}{k_2 - k_1}e^{-k_1t} + \frac{k_1}{k_2 - k_1}e^{-k_2t}\right] \tag{1.34}$$

A simplifying assumption called the *steady state approximation* is often utilized for consecutive reactions. The assumption is that the [B] will be small and will not change very much during the reaction ($k_1 < k_2$),

$$\frac{d[B]}{dt} \cong 0 \tag{1.35}$$

$$\frac{d[B]}{dt} = k_1[A] - k_2[B] \cong 0 \tag{1.36}$$

$$[B] = \frac{k_1}{k_2}[A] = \frac{k_1}{k_2}A_0\,e^{-kt} \tag{1.37}$$

$$[C] = [A_0] - [A] - [B] \tag{1.38}$$

$$= A_0\left[1 - e^{-k_1t} - \frac{k_1}{k_2}e^{-k_1t}\right] \tag{1.39}$$

$$\cong A_0[1 - e^{-k_1t}] \tag{1.40}$$

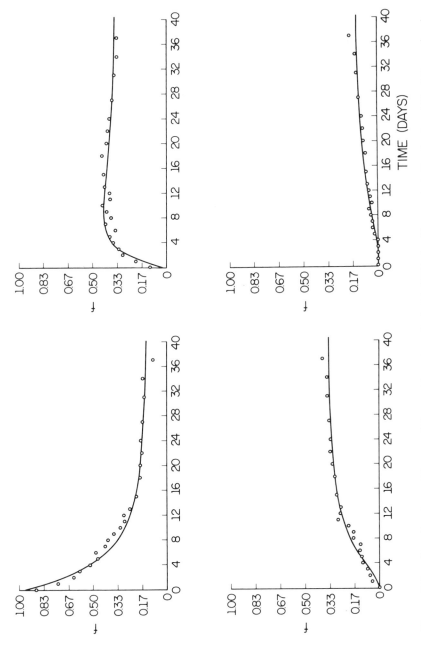

Figure 1.3. Plots for the enrichment of the position *cis* to Br in Mn(CO)$_5$Br. The *x* axis is the fraction (f) of the molecules corresponding to 0, 1, 2, and 3 ^{13}CO molecules *cis* to Br. This is the pattern expected for consecutive formation of products.

As the reactions become more numerous the analysis becomes more complicated. As long as each individual reaction is first order, then a general treatment can be made.[5,8] The following treatment is valid for any number of parallel or consecutive first-order reactions. The differential equations can be represented as:

$$\frac{dA_i}{dt} = -k_{i1}A_i - k_{i2}A_i \cdots k_{im}A_i + k_{1i}A_2 + \cdots + k_{mi}A_m \tag{1.41}$$

or

$$\frac{dA_i}{dt} = (-k_{i1} - k_{i2} \cdots - k_{im})A_i + k_{1i}A_1 + k_{2i}A_2 + \cdots + k_{mi}A_m \tag{1.42}$$

where the A_is are the concentrations of the species, the k_{ij}s are the rate constants for the reaction $A_i \rightarrow A_j$, and m is the number of species involved kinetically in the reaction. Equation 1.42 can be put into the form

$$\frac{dA_i}{dt} + \sum_{j=1}^{m} K_{ij}A_j = 0 \tag{1.43}$$

where $K_{ij} = -k_{ji}$, except when $i = j$ and then $K_{ij} = \sum_{p=1}^{m} k_{ip}$, except for $p = i$.

We will now assume a particular solution of the form usually observed

$$A_i = B_i e^{-\lambda t} \tag{1.44}$$

where the A_is are the concentrations of the species desired, the B_is are constants, and λ is a parameter that is yet to be determined. Substituting Equation 1.44 into Equation 1.43 gives

$$\lambda B_i e^{-\lambda t} + \sum_{j=1}^{m} K_{ij}B_j e^{-\lambda t} = 0 \tag{1.45}$$

which simplifies to

$$\lambda B_i + \sum_{j=1}^{m} K_{ij}B_j = 0 \tag{1.46}$$

and

$$\sum_{j=1}^{m} (K_{ij} - \delta_{ij}\lambda)B_j = 0 \text{ for } i = 1, 2, \cdots m \tag{1.47}$$

where

$$\delta_{ij} = 1 \quad \text{if} \quad i = j \quad \text{or} \quad 0 \quad \text{if} \quad i \neq j \tag{1.48}$$

Equation 1.47 is the typical expression for the general eigenvalue (λ) — eigenvector (B) problem and can be solved by digital computer methods. For each λ_r there will be a set of relative values of B_j that are solutions of Equation 1.47 that we will designate B_{jr}. These particular solutions will not usually be the solutions desired, but one can obtain a general solution as a linear combination of the particular solutions,

$$A_i = \sum_{r=1}^{m} B_{ir} Q_r e^{-\lambda rt} \tag{1.49}$$

where the Q_rs are the coefficients of the linear combinations and may be determined from the initial conditions.

A couple of examples will serve to illustrate the set-up for this analysis. For the simple example

$$A_1 \underset{k_{21}}{\overset{k_{12}}{\rightleftarrows}} A_2 \underset{k_{32}}{\overset{k_{23}}{\rightleftarrows}} A_3 \tag{1.50}$$

of two consecutive reversible reactions, the secular equation is

$$\begin{vmatrix} k_{12} - \lambda & -k_{21} & 0 \\ -k_{12} & k_{21} + k_{23} - \lambda & -k_{32} \\ 0 & -k_{23} & k_{32} - \lambda \end{vmatrix} = 0 \tag{1.51}$$

which can be readily solved. This analysis can be very useful in the analysis of data from isotope exchange reactions; $Mn(CO)_5Br$ reacting with ^{13}CO can serve as an example.[7] If we assume that the intermediate generated by CO dissociation is rigid, then the scheme shown in Figure 1.4 would give all possibilities where k_c refers to dissociation *cis* to Br and k_t refers to dissociation *trans* to Br. Defining the A_is as shown in Figure 1.5 one can establish the K_{ij} in terms of k_c and k_t. Applying the initial conditions

$$A_1^0 = .945\, A_0 \tag{1.52}$$

$$A_2^0 = .011\, A_0 \tag{1.53}$$

$$A_4^0 = .044\, A_0 \tag{1.54}$$

$$A_3^0 = A_5 = A_6 = \ldots A_{12} = 0 \tag{1.55}$$

(A_i^0 are the initial concentrations of the A_i and A_0 is the concentration of $Mn(CO)_5Br$) to equation 1.49 for $t = 0$ gives a set of 12 simultaneous linear equations that may be solved for the Q_rs. An isotopic exchange problem of this type is extremely difficult to solve without computer methods.

1.2.5. Other Commonly Observed Rate Laws

Two other rate laws are often observed when inorganic reactions are run under pseudo first-order conditions. For a number of inorganic reactions the plot of k_{obs}

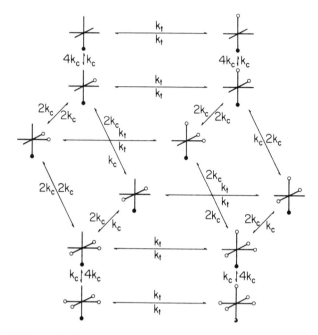

Figure 1.4.   Scheme for isotopic exchange of ^{13}CO with $Mn(CO)_5Br$ where k_c refers to dissociation of CO *cis* to Br and k_t refers to dissociation *trans* to Br.

versus concentration of the adding ligand is linear, but it does not intercept at k_{obs} = 0 for [L] = 0. A sample plot for substitution on $Ir_4(CO)_{12}$ is shown in Figure 1.6 for three different entering ligands.[9] Each ligand intercepts the Y axis at a constant value of k_{obs}. This type plot is typical for the rate law

$$\text{rate} = k_1[Ir_4(CO)_{12}] + k_2[Ir_4(CO)_{12}][L] \qquad (1.56)$$

This can be rearranged to

$$\text{rate} = (k_1 + k_2[L])[Ir_4(CO)_{12}] \qquad (1.57)$$

$$= k_{obs}[Ir_4(CO)_{12}] \qquad (1.58)$$

with

$$k_{obs} = k_1 + k_2[L] \qquad (1.59)$$

such that the plot of k_{obs} versus [L] provides k_1 as the intercept and k_2 as the slope of the linear plot. This rate law indicates competing mechanisms, one of which is ligand dependent, and the other of which is independent of [L].

   If a reaction is composed of two steps, the first of which is a rapid pre-equilibrium,

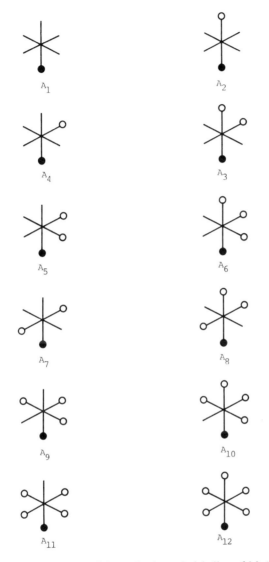

Figure 1.5. Definition of the A_i for isotopic labeling of $Mn(CO)_5Br$.

$$A + B \underset{k_1}{\overset{k_1}{\rightleftharpoons}} C \overset{k_2}{\rightleftharpoons} D \tag{1.60}$$

then the rate behavior is somewhat different. The rate of appearance of D is given by

$$\text{rate} = k_2[C] \quad \text{or} \quad k_{\text{obs}}([A] + [C]) \tag{1.61}$$

Realizing that

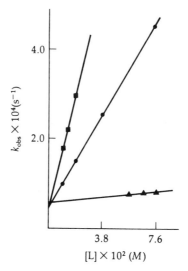

Figure 1.6. Plots of k_{obs} versus [L] for the reaction of three different ligands with $Ir_4(CO)_{12}$. The circles represent PPh_3, the triangles, $AsPh_3$, and the pluses represent $P(OPh)_3$. The values of k_2 obtained from the slope are PPh_3, $5.3 \times 10^{-3} M^{-1}s^{-1}$; $AsPh_3$, $2.7 \times 10^{-4} M^{-1}s^{-1}$; and $P(OPh)_3$ $1.3 \times 10^{-2} M^{-1}s^{-1}$. The value of k_1 obtained from the x-axis intercept is the same for each ligand, $6.0 \times 10^{-5}s^{-1}$. [Reprinted with permission from D. Sonnenberger and J. D. Atwood, *Inorg. Chem.* **20**, 3243 (1981). Copyright 1981 American Chemical Society.]

$$\frac{[C]}{[A][B]} = K \tag{1.62}$$

and substituting leads to

$$k = \frac{k_2 K[B]}{1 - K[B]} \tag{1.63}$$

or

$$\frac{-dA}{dt} = \frac{k_2 K[A][B]}{1 + K[B]} \tag{1.64}$$

One obtains a plot of k_{obs} versus [B] under pseudo first-order conditions with varying the initial concentrations of B. A sample plot is shown in Figure 1.7 for the electron transfer reaction

$$cis - Ru(NH_3)_4Cl_2^+ + Cr^{2+} \rightarrow Ru(NH_3)_4(H_2O)Cl^+ + CrCl^{2+} \tag{1.65}$$

where $A = cis - Ru(NH_3)_4Cl_2^+$ and $B = Cr^{2+}$.[10] A rate law of the form in Equation 1.64 is also observed in substitutions of classic coordination complexes with an octahedral geometry. For this rate law the dependence on [B] varies between 0 and 1 depending on the relative magnitude of $K[B]$ and 1. This rate law illustrates the care

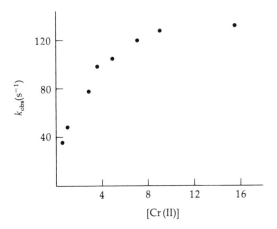

Figure 1.7. Plot of the observed rate constant versus concentration of Cr(II) in the reduction of *cis*-Ru(NH$_3$)$_4$Cl$_2^+$. This shows the rate law of the form given in Equation 1.64. [Reprinted with permission from W. G. Movius and R. G. Linck, *J. Am. Chem. Soc.* **92**, 2677 (1970). Copyright 1970 American Chemical Society.]

required to fully delineate a rate law since a partial investigation could lead to an error.

1.3. Activation Parameters

The rate law is very useful in assigning a mechanism for a reaction, but other data are also beneficial; activation parameters are usually essential. A reaction's activation parameters are derived from the temperature dependence of its rate constant. The relationship can be derived from absolute reaction rate theory,[11] but empirical relationships are used more often. The Arrhenius equation,

$$k = A\, e^{-E_a/RT} \tag{1.66}$$

where k is the rate constant for a reaction, A is a preexponential factor, E_a is the energy of activation, R is the gas constant, and T is the absolute temperature, was established a number of years ago.[12] A plot of $ln(k)$ versus $\left(\dfrac{1}{T}\right)$

$$ln(k) = \left(-\frac{E_a}{RT}\right) + ln\, A \tag{1.67}$$

gives a straight line of slope $\dfrac{-E_a}{R}$ and intercept of $ln(A)$. Energies of activation are not currently cited very often; enthalpies and entropies of activation are more useful. The rate constant is related to the free energy of activation, $\Delta G^{\ddagger}$, by the equation

$$k = \frac{k'T}{h} e^{\frac{-\Delta G^\ddagger}{RT}} \tag{1.68}$$

where k' is Boltzman's constant and h is Planck's constant. Substituting $\Delta G^\ddagger = \Delta H^\ddagger - T\Delta S^\ddagger$ gives

$$k = \frac{k'T}{h} e^{\frac{-\Delta H^\ddagger}{RT}} e^{\frac{\Delta S^\ddagger}{R}} \tag{1.69}$$

Rearranging and taking the logarithm gives

$$ln\left(\frac{k}{T}\right) = \frac{-\Delta H^\ddagger}{RT} + ln\left(\frac{k'}{h}\right) + \frac{\Delta S^\ddagger}{R} \tag{1.70}$$

and a plot of $ln\left(\frac{k}{T}\right)$ versus $\left(\frac{1}{T}\right)$ gives $\Delta H^\ddagger$ from the slope and $\Delta S^\ddagger$ from the intercept. This type of plot, which is called an *Eyring plot*, is shown in Figure 1.8.

The value of $\Delta H^\ddagger$ is not very informative regarding the mechanism. The range of values, $+10$ to $+35$ kcal/mole, cannot be readily interpreted, although the value for associate reactions is usually smaller than for dissociative reactions. For dissociative reactions, $\Delta H^\ddagger$ is used as a measure of the bond strength of the dissociating ligand. The entropy of activation is quite useful in determining the mechanism. Values that are below -10 eu indicate an associative reaction, and values of $\Delta S^\ddagger$ that are greater than $+10$ eu indicate a dissociative reaction. One must be careful in interpreting the mechanism for reactions that have $\Delta S^\ddagger$ between -10 and $+10$ eu because solvent reorganization may also contribute, especially for polar solvents and charged metal complexes.

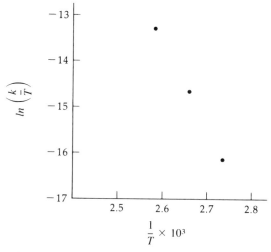

Figure 1.8. Plot of ln (k/T) versus $(1/T)$ giving the enthalpy of activation $(\Delta H^\ddagger)$ and the entropy of activation $(\Delta S^\ddagger)$ for reaction of $Cr(CO)_4PPh_3(P(OPh)_3)$ with CO. The values, $\Delta H^\ddagger = 36.6$ kcal/mole and $\Delta S^\ddagger = 24.4$ eu were obtained.

1.3.1. Volume of Activation

The volume of activation is a supplement to the enthalpy and entropy of activation obtained by measuring the variation of the rate constant with pressure

$$\frac{d(lnk)}{dP} = \frac{-\Delta V^{\ddagger}}{RT} \tag{1.71}$$

The volume of activation ($\Delta V^{\ddagger}$) is a measure of the compressibility difference between the ground state and transition state. The $\Delta V^{\ddagger}$ changes with mechanism as $\Delta S^{\ddagger}$; a positive value indicates a dissociative mechanism and a negative value indicates an associative mechanism. Unfortunately, the volume of activation incorporates both the volume changes in the reactants and volume changes in the surrounding solvent (*electrostriction* of solvent). In some cases this has led to disagreement in the interpretation of volumes of activation.[13,14]

1.3.2. Reaction Profiles

The energetics for a reaction are often described graphically with a reaction profile (e.g., Fig. 1.9). The difference in energy between the reactants and the highest energy point (the *transition state*) is the *activation barrier* as illustrated in Figure 1.9. If there is a relative minimum in the reaction profile, then an intermediate exists with the lifetime of the intermediate related to the depth of the relative minimum. There are usually insufficient data to define all of the energy parameters in a reaction profile. Any process that accelerates a reaction must lower the activation barrier by either raising the energy of the ground state or decreasing the energy of the transition state. Kinetic results are often interpreted only in terms of ground state changes. This is a confusion of kinetics and thermodynamics.

1.4. Microscopic Reversibility

The principle of *microscopic reversibility* states that any molecular process and its reverse occur with equal rates at equilibrium.[15] In mechanistic terms it states that the lowest energy path in the forward direction must also be the lowest energy path in the reverse direction. Thus, if the mechanism is known in one direction, then it is known in the other direction. The application is most straightforward for isotopic exchange of a ligand where the reverse process must be the same. Microscopic reversibility can be used to rule out certain mechanisms. It is also true for an isotopic exchange reaction that if a preference exists for dissociation from a specific site, then the entering ligand must show the same preference for that site (the low energy path in one direction must be the low energy path in the reverse direction). This has been applied to the interpretation of exchange of labeled CO on $Mn(CO)_5X$.[16] Another application will be in reversible reactions such as the oxidative-addition and reductive-elimination of a molecule such as H_2.

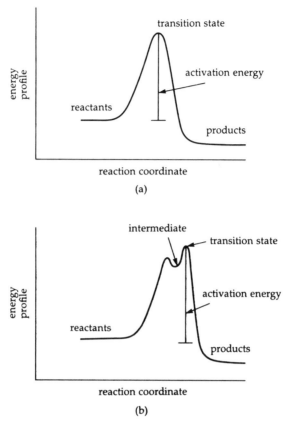

Figure 1.9. Reaction profiles illustrating the activation energy and the transition state. Profile (b) shows a reaction scheme that has an intermediate.

1.5. Linear Free Energy Relationships

While one must be careful to distinguish between kinetics and thermodynamics it is often useful in determining the mechanism for reactions of related complexes to correlate kinetic parameters with thermodynamic parameters. Such correlations are called *linear free energy relationships*. A sample plot is shown in Figure 1.10 for $Cr(CO)_5L$ where $\Delta G^{\ddagger}$ and ΔG are correlated.[17] A linear free energy relationship with a slope of one is an indication that the differences in the complexes affect the kinetic parameter (k or $\Delta G^{\ddagger}$) and the thermodynamic parameter (K or ΔG) in the same way—the difference in complexes has little effect on the transition state energy. Two examples will serve to illustrate linear free energy relationships. Reaction of $Co(CO)_3NO$ with ligands leading to replacement of CO gives the relationship shown in Figure 1.11 where the rate is correlated with the basicity of the ligand toward a proton (the half-neutralization potential, ΔHNP).[18] Several

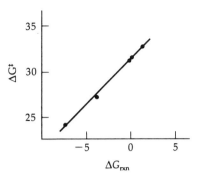

Figure 1.10. A linear free-energy plot of $\Delta G^{\ddagger}$ and ΔG_{rxn} for dissociation of L from Cr(CO)$_5$L. [Reprinted with permission from M. J. Wovkulich and J. D. Atwood, *J. Organomet. Chem.* **184,** 77 (1979). Copyright 1979 Elsevier Sequoia S.A.]

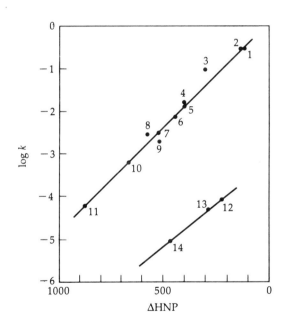

Figure 1.11. The linear free-energy relationships of the rate of reaction with the polarizability of the nucleophile for the reaction of CoNO(CO)$_3$ with (1) P(C$_2$H$_5$)$_3$; (2) P(n-C$_4$H$_9$)$_3$; (3) P(C$_6$H$_5$)(C$_2$H$_5$)$_2$; (4) P(C$_6$H$_5$)$_2$(C$_2$H$_5$); (5) P(C$_6$H$_5$)$_2$(n-C$_4$H$_9$); (6) P(p-CH$_3$OC$_6$H$_4$)$_3$; (7) P(O-n-C$_4$H$_9$)$_3$; (8) P(C$_6$H$_5$)$_3$; (9) P(OCH$_3$)$_3$; (10) P(OCH$_2$)$_3$CCH$_3$; (11) P(OC$_6$H$_5$)$_3$; (12) 4-picoline; (13) pyridine; (14) 3-chloropyridine. The ΔHNP measures basicity, which is believed to parallel polarizability for these compounds. [Reprinted with permission from E. M. Thorsteinson and F. Basolo, *J. Am. Chem. Soc.* **88,** 3929 (1966). Copyright 1966 American Chemical Society.]

features are apparent from this relationship. Nitrogen donors of comparable basicity to phosphorus donors are considerably less reactive, but linear relationships exist within a series for the same donor. The slope of the line (2.5) indicates attack at the metal with bond formation in the transition state, which is consistent with the rate law, rate = $k[Co(CO)_3NO][L]$. For substitution of CO on $Mo(CO)_4$(X-o-phen) with substituted orthophenathrolines both a ligand independent and a ligand dependent step are seen.[19] The free energy relationships for both steps as shown in Figure 1.12 are useful in assigning the mechanism.[19] For the ligand-independent (k_1) step the rate increases with the ability of the o-phenanthroline to donate electron density, which is consistent with a dissociative reaction and stabilization of the transition state. For the ligand-dependent step the rate decreases with increasing ligand basicity, which is consistent with less nucleophilic attack for a complex with more electron density. A number of special linear free energy relationships have been developed. While these have been used primarily for organic reactions, there have been some applications to inorganic systems. The Hammett relationship correlates rates involving aromatic complexes with the nature and location of the substituents (meta and para) on the aromatic ring that are measured as the Hammett constant (σ).[20]

$$\log \frac{k}{k_0} = \rho\sigma \tag{1.72}$$

The rate constants k and k_0 are for the substituted and unsubstituted aromatic molecule, respectively. The constant σ depends on the substituent, and ρ depends on the reaction.

The Hammett constant has been effectively correlated with the rate of base hydrolysis for complexes of cobalt (Co), $Co(en)_2$(X$-C_6H_4CO_2)_2$, and with the oxidation potential of substituted ferrocene complexes.[21,22] The Taft relationship is a modification of the Hammett relationship, which is appropriate for aliphatic compounds and ortho substituted aromatic species and which has been applied to the base hydrolysis of alkyl carboxylate complexes of cobalt.[23] A few other specialized free energy relationships have also been used. We will discuss the Swain-Scott relationship in Chapter 2.

1.6. Kinetic Techniques

One needs to generate a series of concentrations for the species of interest at various times for a kinetic experiment. Any technique that will generate a measure of the concentration is appropriate. In the next section we will consider a few of the most important techniques for obtaining kinetic data on inorganic systems.

1.6.1. Absorption Spectra

The most often used techniques involve measuring the absorption spectra in either infrared, visible, or ultraviolet spectral regions. Beer's law,

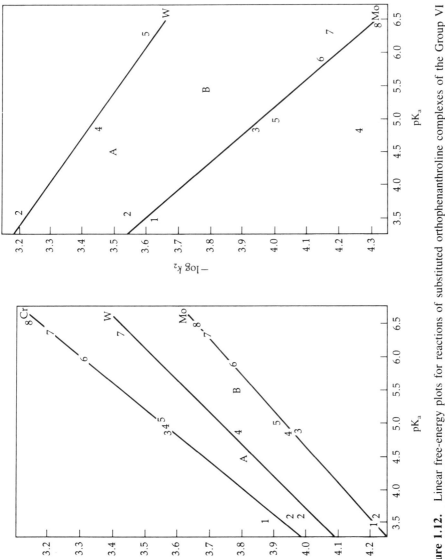

Figure 1.12. Linear free-energy plots for reactions of substituted orthophenanthroline complexes of the Group VI metals. Note that the ligand-independent and ligand-dependent steps depend on the pK_a in opposite ways. [Reprinted with permission from J. R. Graham and R. J. Angelici, *Inorg. Chem.* **6**, 994 (1967). Copyright 1967 American Chemical Society.]

$$D = \epsilon l[A] \tag{1.73}$$

relates directly the concentration of a species to its optical absorbance. D is the optical absorbance, ϵ is the molar absorptivity, and l is the path length in centimeters. Thus, measuring the absorbance of a complex gives its concentration. A spectrophotometer often measures the transmittance of light by a sample. The transmittance can be converted to absorbance by the equation,

$$D = \log \frac{I_0}{I} \tag{1.74}$$

where I_0 is the intensity of the incident light and I is the intensity of the transmitted light. These quantities are defined for an absorption spectrum in Figure 1.13. The area under an absorption curve should be used in quantitative studies; however, the change in the absorption maximum is a useful approximation for a single absorption that does not overlap with other absorptions. A Beer's law plot of the quantity that is to be used versus the concentration of the absorbing species serves as a check on the procedure. Sample Beer's law plots are shown in Figure 1.14. The plot for *trans*-$Cr(CO)_4(PBu_3)_2$ shows deviations above 2.8×10^{-4} M; data collected at higher concentrations would be difficult to interpret.[24] The Beer's law plot serves as a check on any approximations made in the spectral analysis and on the behavior of the absorbing complex. Association or dissociation of the complex would lead to nonlinear Beer's law plots because the species formed would not have identical absorption properties and the extent of association or dissociation would vary with the concentration of the complex.

Ultraviolet and visible absorption spectra usually involve broad absorptions that overlap with other species present in solution. While this sometimes makes interpretation of products more difficult, kinetic analysis can be done on overlapping absorption spectra. For the simple first-order reaction

$$A \rightarrow B \tag{1.75}$$

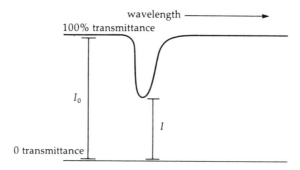

Figure 1.13. The definition of I_0 and I for a transmittance spectrum. The 100% transmittance is the base line for the spectrum.

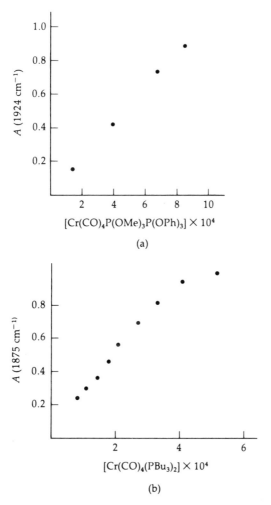

Figure 1.14. Beer's law plots of absorbance versus concentration for the most intense absorption of two chromium carbonyl complexes: (a) a normal linear Beer's law plot; (b) a similar complex showing deviation from linearity above $A = 0.7$. The ϵ value is $1.1 \times 10^3 M^{-1}$ for $Cr(CO)_4P(OMe)_3P(OPh)_3$ and 2.5×10^3 for $Cr(CO)_4(PBu_3)_2$, ignoring data above $A = 0.7$.

the absorbance at any time will be given by

$$D_t = \epsilon_A[A] + \epsilon_B[B] \qquad (1.76)$$

and at completion by

$$D_\infty = \epsilon_B([A_0] + [B_0]) \qquad (1.77)$$

The term needed for kinetic analysis

$$ln\,\frac{A_0}{A_t} = ln\left(\frac{D_0 - D_\infty}{D_t - D_\infty}\right) = kt \tag{1.78}$$

may be represented purely as absorbances. One should always keep in mind, how-ever, that inability to assign the spectra completely from which kinetic data are obtained makes the data of limited utility.

Infrared absorption spectra are limited primarily to ligands (CO, NO, NCR, CNR, etc.) that absorb in regions of the infrared that are uncluttered by common solvents; however, infrared absorptions are usually not overlapping and provide for straightforward analysis. In Chapter 4, "Organometallic Substitution Reactions," infrared monitoring will be commonly observed for carbonyl compounds. An example using the nitrosyl absorption will be illustrated here.[24]

$$[Re(CO)_2(NO)Cl_2]_2 + 2L \rightarrow 2Re(CO)_2(NO)Cl_2(L) \tag{1.79}$$

The chloride-bridged dimer has a ν_{NO} at 1803 cm^{-1} and the product has a ν_{NO} at 1780 cm^{-1}. By monitoring the decrease in the 1803 cm^{-1} absorption one can obtain kinetic plots that are shown in Figure 1.15.[24] The analysis is straightforward giving

$$rate = k[Re_2(CO)_4(NO)_2Cl_4][L] \tag{1.80}$$

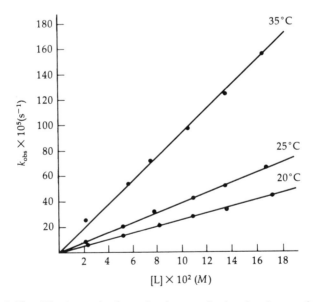

Figure 1.15. Kinetic results from the decrease in the absorbance of the nitrosyl (1803 cm^{-1}) of $Re_2(CO)_4(NO)_2Cl_4$ upon reaction with entering nucleophile. [Reprinted with per-mission from F. Zingales, A. Trovati, and P. Uguagliata, *Inorg. Chem.* **10**, 510 (1971). Copyright 1971 American Chemical Society.]

The dependence on the concentration of the entering ligand suggests nucleophilic attack of L on the Re dimer.

1.6.2. Nuclear Magnetic Resonance

In contrast to the absorption spectra just considered nuclear magnetic resonance (NMR) provides data on the lifetime of systems that are at equilibrium. The time scale accessible for a typical NMR spectrum is 10^0–10^{-6} seconds, which covers a range of rates (10^0–$10^6 s^{-1}$) that are not readily studied by other techniques.[25,26] A system that undergoes exchange on a timescale appropriate for NMR investigation will have changes in the NMR spectrum that can provide rate data in the form of lifetimes τ_A which are the inverse of the rate constants.

$$\frac{-dA}{dt} = \frac{1}{\tau_A} [A], k = \frac{1}{\tau_A} \tag{1.81}$$

A typical set of spectra obtained at different temperatures are shown in Figure 1.16 for a two-site system (exchange occurs between two equivalent sites). The *fast exchange* spectrum is the sharp single resonance seen at high temperature indicating that exchange is fast in comparison to the NMR timescale, and that the environment for the proton is averaged. As the temperature is decreased the resonance begins to broaden in the near fast exchange spectra. *Coalescence* is defined as the point just before the resonance splits into two resonances. The spectrum is said to be in the intermediate exchange region when the two resonances are distinct but overlapping. The two resonances are separate but broadened at lower temperature, and the spectrum is in the *slow exchange region*. When no exchange is occurring the two resonances are sharp and the spectrum is at the *slow exchange limit* or *stopped exchange*. Equations have been derived for the lifetime in the different regions of the spectra, such that a series of rate constants versus temperature are generated and activation parameters can be calculated. In general, the preferred method for analysis of NMR data involves computer line-shape fitting, which has been a powerful method, especially for fluxional molecules. The technique is illustrated in Figures 1.17 and 1.18 for $H_2Fe(P(OEt)_3)_4$.[27] The observed spectra are shown in Figure 1.17 with the best calculated fit, while other calculated possibilities are shown in Figure 1.18. This technique often provides the only way to differentiate mechanistic possibilities in fluxional molecules.

NMR line broadening caused by exchange can also be used to generate kinetic data. ^{13}C NMR data for electron exchange between $Os(CN)_6^{4-,3-}$ can illustrate the type of data.[28] The diamagnetic $Os(CN)_6^{4-}$ has a ^{13}C resonance at 142.9 ppm with a half-width (the width at half-height) of 4.5 kz. As shown in Figure 1.19 addition of varying amounts of $Os(CN)_6^{3-}$ in a variety of conditions causes broadening of the ^{13}C resonance for $Os(CN)_6^{4-}$.[28] Since this is in the slow-exchange region the rate constant for exchange can be obtained as the slope of a plot of the increased line width versus $[Os(CN)_6^{3-}]$. These plots are shown in Figure 1.19. Such line-broadening experiments have also been used to evaluate an exchange on $Ni(en)_3^{2+}$ (^1H, ^{13}C,

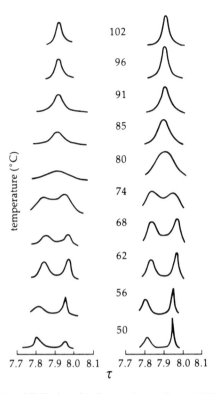

Figure 1.16. The proton NMR data for free and coordinated NH_3 from complexes with alkyl cobalt(III) complexes. This illustrates a two-site exchange. [Reprinted with permission from R. J. Guschl, R. Stewart, and T. L. Brown, *Inorg. Chem.* **13**, 959 (1974). Copyright 1974 American Chemical Society.]

and ^{14}N NMR)[29] and methyl exchange between $CpFe(Me)(CO)_2$ and $CpFe(CO)_2^-$ (^1H NMR of the Cp).[30] The range of NMR techniques is nicely illustrated by the study of self-exchange reactions of $CpM(CO)_3X$ with $CpM(CO)_3^-$ (M = Cr, Mo, W) whose rates vary from $10^{-3}\,M^{-1}s^{-1}$ to $10^4\,M^{-1}s^{-1}$.[31] For the faster reactions (X = I) line-broadening experiments were used; for reactions with intermediate rates (X = Br) magnetization transfer was used; and for the slower reactions (X = Cl) the ^1H NMR was used to determine the amount of the resonating material.[31] The magnetization transfer is a double-resonance technique in which one selectively excites a resonance of one substance and watches the transfer to the other substance.

NMR has become a very powerful technique for generating kinetic data on inorganic systems. The short coverage presented here is to indicate possibilities, but it is not comprehensive. With the continual improvement in magnets and computers the potential for kinetic studies from high field, multinuclear and variable temperature NMR is increasing.

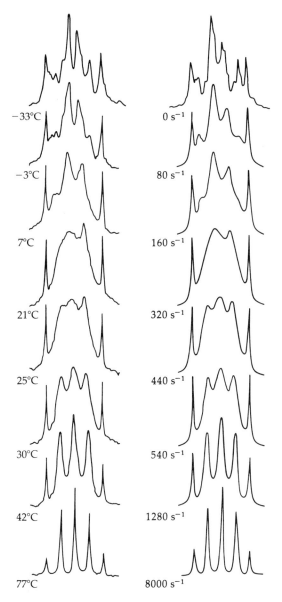

Figure 1.17. Calculated (best-fit) and observed 90 MHz proton NMR line shapes for $H_2Fe[P(OEt)_3]_4$. [Reprinted with permission from J. P. Jesson and E. L. Muetterties, *Tetrahedron* **30,** 232(9) (1974) and *J. Am. Chem. Soc.* **93,** 4701 (1975). Copyright 1974 Pergamon Press and 1975 American Chemical Society.]

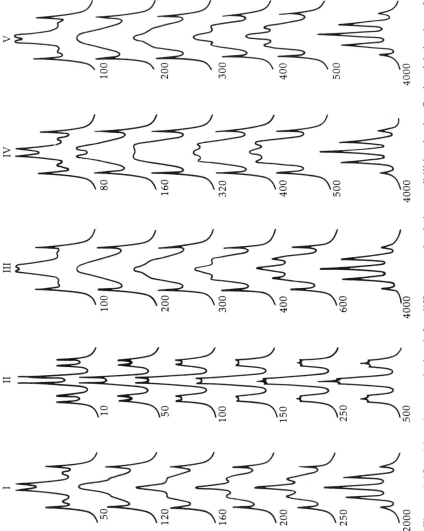

Figure 1.18. Line shapes calculated for different mechanistic possibilities in the fluxional behavior of $H_2Fe[P(OEt)_3]_4$. [Reprinted with permission from J. P. Jesson and E. L. Muetterties, *J. Am. Chem. Soc.* **93**, 4801 (1975). Copyright 1975 American Chemical Society.]

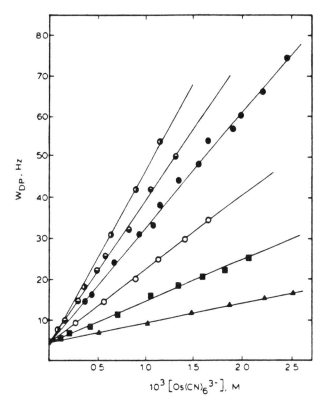

Figure 1.19. Dependence of the $Os(CN)_6^{4-}$ ^{13}C NMR line width (W_{DP}) on the concentration of $Os(CN)_6^{3-}$ in aqueous solution at 25°C, $I = 0.2$ M (▲), 0.5 M (■), and 1.0 M (●) with $NaClO_4$ and $I = 1.0$ M with $LiClO_4$ (◖), KCl (◐), and NH_4Cl (◖). [D. H. Macartney, *Inorg. Chem.* **30**, 3337 (1991).]

1.6.3. Gas Uptake

A number of the reactants of interest to inorganic chemists are gases such as CO, H_2, C_2H_4, and so on. Reactions where gases are given off or taken up may be monitored by studying the amount of gas versus time. This is an especially important technique for homogeneously catalyzed reactions where a reactant is frequently a gas. A constant pressure device is necessary since the reaction may be dependent on the pressure. A typical model has been described and is shown in Figure 1.20.[32] The reaction of $RuCl_2$ with CO in dimethylacetamide was followed by gas uptake.[33]

$$Ru(II) + CO \rightarrow Ru(II)(CO) \qquad (1.82)$$

The plot of absorbed CO versus time is shown in Figure 1.21.[33] The CO uptake data were used to deduce the [Ru(II)] by the relationship,

$$[Ru(II)] + [Ru(II)(CO)] = [Ru(II)]_0 \qquad (1.83)$$

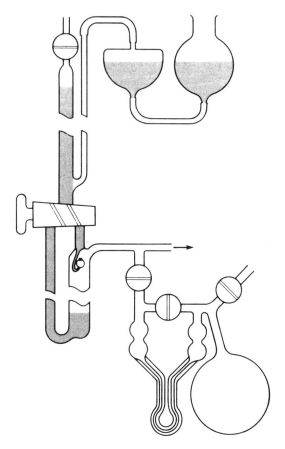

Figure 1.20. An apparatus for measuring the uptake of gases at constant pressure. [Reprinted with permission from J. E. Taylor, *J. Chem. Ed.* **42,** 618 (1965). Copyright 1965 Journal of Chemical Education.]

giving

$$\frac{-d[CO]}{dt} = \frac{d[Ru(II)(CO)]}{dt} = k_l[Ru(II)] \tag{1.84}$$

The value of k_l was evaluated from plots such as those shown in Figure 1.22.[33] Thus, once the concentration data were obtained the kinetic analysis was the same.

1.6.4. Stopped-Flow Techniques

Many reactions occur too rapidly for standard absorption spectroscopy. In the time required for sampling the reaction would be completed. A simple technique that

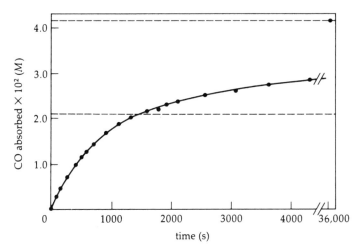

Figure 1.21. A plot of CO uptake versus time for reaction with Ru(II) chloride. [Reprinted with permission from B. C. Hui and B. R. James, *Can. J. Chem.* **48,** 3613 (1970). Copyright 1970 National Research Council of Canada.]

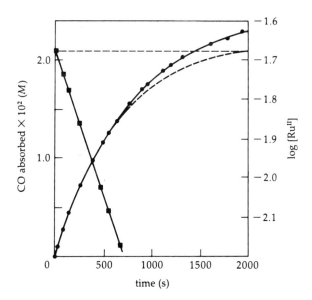

Figure 1.22. Kinetic plots to evaluate the rate constant from gas uptake experiments. The squares represent log [Ru(II)] versus time. [Reprinted with permission from B. C. Hui and B. R. James, *Can. J. Chem.* **48,** 3613 (1970). Copyright 1970 National Research Council of Canada.]

allows one to investigate rapid reactions is the stopped-flow technique. A schematic of the apparatus is shown in Figure 1.23. The reactants are separately brought to a constant temperature, then mixed immediately prior to obtaining the spectra. The spectrometer is set on one wavenumber and the absorption versus time spectrum is recorded. The data would then be analyzed normally. The stopped-flow technique has primarily been used with ultraviolet (UV)-visible spectrophotometers, although some experiments have been done with infrared radiation.[34,35] Data for the reaction of $Co_2(CO)_8$ with $AsPh_3$ are shown in Figure 1.24.[36]

1.6.5. Other Techniques

To reiterate, any measurement that gives the amount of material as a function of time can be used to generate kinetic data. Fast time-resolved infrared detection has been used for several organometallic complexes. In this experiment a reactive

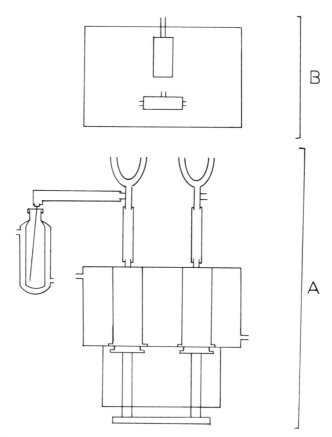

Figure 1.23. Schematic of a single stopped-flow apparatus. In (a), equilibrated solutions of the reactants are pulled into the syringes; the bar connecting the plungers of both syringes is pushed. In (b) the two solutions pass through a mixer into a spectrophotometer, where the absorbance spectrum is recorded.

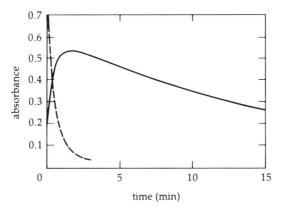

Figure 1.24. Typical absorbance plot obtained from a stopped-flow apparatus for reaction of $Co_2(CO)_8$ with $AsPh_3$. This plot shows the decrease in $Co_2(CO)_8$ (dashed line, 2042 cm^{-1}) and the growth and slow decrease of $Co_2(CO)_7AsPh_3$ (solid line, 1996 cm^{-1}). [Reprinted with permission from M. Absi-Halabi, J. D. Atwood, N. P. Forbus, and T. L. Brown, *J. Am. Chem. Soc.* **102,** 6248 (1980). Copyright 1980 American Chemical Society.]

intermediate is generated by UV-visible flash photolysis followed by detection with rapid infrared spectroscopy.[37,38] Flash photolysis followed by UV-visible detection has also been used. A sample is shown in Figure 1.25 for oxidation of $CpMo(CO)_3\bullet$ by an organic radical.[39] The rate constant evaluated, $(2.09 \pm 0.03) \times 10^9$ $s^{-1}M^{-1}$ indicates that rapid rates can be examined by such techniques.[39]

While all of the kinetic techniques discussed thus far are isothermal (constant temperature), it is possible to obtain kinetic data by nonisothermal techniques that involve a continuous change in temperature during the reaction. The potential advantages and limitations have been summarized.[40] The nonisothermal technique

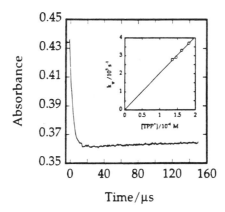

Figure 1.25. The change in absorbance for reaction of $CpMo(CO)_3\bullet$ with an organic radical (triphenylpyrylium radical) versus time. [Won and Espenson, *Organometallics* **14,** 4275 (1995).]

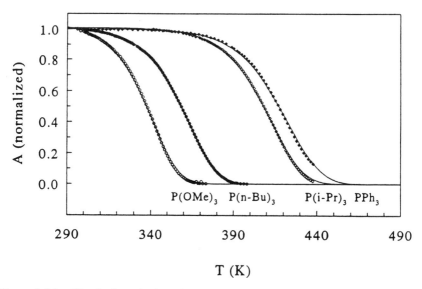

Figure 1.26. Kinetic data obtained for CpRh(CO)$_2$ reaction with P-donor ligands while constantly varying the temperature. [Zhang and Brown, *Inorg. Chim. Acta,* **240,** 427 (1995).]

allows $\Delta H^{\ddagger}$ and $\Delta S^{\ddagger}$ to be determined in one experiment. The type of data obtainable are shown in Figure 1.26 for reaction of CpRh(CO)$_2$ with phosphorus ligands.[40] The determined activation parameters for PBu$_3$, $\Delta H^{\ddagger} = 13.4$ kcal/mole and $\Delta S^{\ddagger} = -33.9$ eu were in reasonable agreement with those determined by standard techniques ($\Delta H^{\ddagger} = 13.5$ kcal/mole and $\Delta S^{\ddagger} = -38$ eu.[41]

1.7. Classification of Mechanisms

A number of inorganic reactions may be classified according to simple mechanistic schemes.[42] For many reactions the term *dissociative,* for a reaction which shows no dependence on the species reacting with the metal complex or *associative,* for a reaction whose rate depends on the nature and the concentration of the species reacting, are sufficient. In some cases, however, a more precise separation of mechanistic types is useful. The three basic mechanistic types are D (dissociative), A (associative), and I (interchange); each is illustrated as follows:

$$M - X \rightleftarrows \overset{Y}{[M + X]} \to M - Y \qquad D \qquad (1.85)$$

$$M - X + Y \rightleftarrows \left[M \overset{X}{\underset{Y}{\diagdown}} \right] \to M - Y + X \qquad A \qquad (1.86)$$

$$MX + Y \rightleftarrows M - X \cdots Y \rightleftarrows M - Y \cdots X \to M - Y + X \qquad I \qquad (1.87)$$

For a D mechanism an intermediate of reduced coordination number is formed, whereas for an A mechanism an intermediate of increased coordination number is formed, and for an I mechanism no intermediate of altered coordination number is formed. Activation profiles for the D and A mechanisms are shown in Figure 1.27. The I mechanism is a concerted exchange of X and Y between the inner and outer coordination spheres of the metal. I reactions may have a variety of transition states, but we shall distinguish two possibilities: I_a where the transition state will display substantial bonding of the metal to both the entering and leaving groups and there would be some dependence on the entering group and I_d where the transition state has only very weak bonding to the entering and leaving group in the transition state with

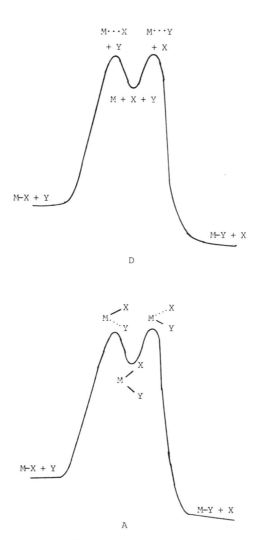

Figure 1.27. Energy profiles for (a) dissociative and (b) associative reactions.

a very small dependence of the entering group on the rate. These are usually referred to as *associative interchange* (I_a) and *dissociative interchange* (I_d).

The four possibilities can be separated as follows. If the rate shows a dependence on the entering group (nature and concentration), then it is either A or I_a; if there is no dependence on the entering group, then the mechanism is D or I_d. D and I_d mechanisms may be separated depending on whether an intermediate of reduced coordination number exists. This may be a difficult distinction to make since the primary difference is in the depth of the minimum in the energy profile (Fig. 1.27). Competition experiments are most often accomplished to test the selectivity toward entering groups. Similar considerations may be used to differentiate I_a and A mechanisms.

1.8. References

1. R. G. Wilkins, *The Study of Kinetics and Mechanism of Reactions of Transition Metal Complexes* (Allyn and Bacon, Boston, 1974).

2. I. Amdur and G. G. Hammes, *Chemical Kinetics* (Academic Press, New York, 1966).

3. S. W. Benson, *The Foundation of Chemical Kinetics* (McGraw-Hill, New York, 1960).

4. W. C. Gardiner, Jr., *Rates and Mechanisms of Chemical Reactions* (W. A. Benjamin, Inc., New York, 1969).

5. A. A. Frost and R. G. Pearson, *Kinetics and Mechanisms* (John Wiley and Sons, New York, 1961).

6. Q. Shi, T. G. Richmond, W. C. Trogler, and F. Basolo, *J. Am. Chem. Soc.* **104**, 4032 (1982).

7. J. D. Atwood and T. L. Brown, *J. Am. Chem. Soc.* **98**, 3155 (1976).

8. F. A. Matsen and J. L. Franklin, *J. Am. Chem. Soc.* **72**, 3337 (1950).

9. D. Sonnenberger and J. D. Atwood, *Inorg. Chem.* **20**, 3243 (1981).

10. W. G. Movius and R. H. Linck, *J. Am. Chem. Soc.* **92**, 2677 (1970).

11. L. P. Hammett, *Physical Organic Chemistry*, 2nd ed. (McGraw-Hill, New York, 1970).

12. S. Arrhenius, *Z. Phys. Chem.* **4**, 226 (1889).

13. C. H. Langford, *Inorg. Chem.* **18**, 3288 (1979).

14. T. W. Swaddle, *Inorg. Chem.* **19**, 3203 (1980).

15. R. L. Burwell, Jr., and R. G. Pearson, *J. Phys. Chem.* **70**, 300 (1966).

16. T. L. Brown, *Inorg. Chem.* **7**, 2673 (1968).

17. M. J. Wovkulich and J. D. Atwood, *J. Organomet. Chem.* **184**, 77 (1979).

18. E. M. Thorsteinson and F. Basolo, *J. Am. Chem. Soc.* **88**, 3929 (1966).

19. J. R. Graham and R. J. Angelici, *Inorg. Chem.* **6**, 992 (1967).

20. L. P. Hammett, *J. Am. Chem. Soc.* **59**, 96 (1937).

21. F. Aprile, V. Cagliotti, and G. Illuminati, *J. Inorg. Nucl. Chem.* **21**, 325 (1961).

22. W. F. Little, C. N. Reilley, J. D. Johnson, and A. P. Sanders, *J. Am. Chem. Soc.* **86**, 1382 (1964).

23. W. E. Jones and J.D.R. Thomas, *J. Chem. Soc. (A)*, 1481 (1966).

24. F. Zingales, A. Trovati, and P. Uguagliati, *Inorg. Chem.* **10**, 510 (1971).

25. K. Vrieze and P.W.N.M. van Leeuwen, *Prog. Inorg. Chem.* **14**, 1 (1971).

26. R. S. Drago, *Physical Methods in Chemistry* (W. B. Saunders Company, Philadelphia, 1977).

27. J. P. Jesson and E. L. Muetterties, *J. Am. Chem. Soc.* **93**, 4701 (1971).

28. D. H. Macartney, *Inorg. Chem.* **30**, 3337 (1991).

29. S. Soyama, M. Ishii, S. Funakashi, and M. Tanaka, *Inorg. Chem.* **31**, 536 (1992).

30. P. Wang and J. D. Atwood, *J. Am. Chem. Soc.* **114**, 6424 (1992).

31. C. L. Schwarz, R. M. Bullock, and C. Creutz, *J. Am. Chem. Soc.* **113**, 1225 (1991).

32. J. E. Taylor, *J. Chem. Ed.* **42**, 618 (1965).

33. B. C. Hui and B. R. James, *Can. J. Chem.* **48**, 3613 (1970).

34. S. E. Rudolph, L. F. Charbonneau, and S. G. Smith, *J. Am. Chem. Soc.* **95**, 7083 (1973).

35. M. S. Corraine and J. D. Atwood, *Inorg. Chem.* **28**, 3781 (1989).

36. M. Absi Halabi, J. D. Atwood, N. P. Forbes, and T. L. Brown, *J. Am. Chem. Soc.* **102**, 6247 (1980).

37. A. J. Dixon, M. A. Healy, P. M. Hodges, B. D. Moore, M. Poliakoff, M. B. Simpson, J. J. Turner, and M. A. West, *J. Chem. Soc. Faraday Trans.* **82**, 2083 (1986).

38. M. Poliakoff and E. Weitz, *Adv. Organomet. Chem.* **25**, 277 (1986).

39. T.-J. Won and J. H. Espenson, *Organometallics* **14**, 4275 (1995).

40. S. Zhang and T. L. Brown, *Inorg. Chim. Acta* **240**, 427 (1995).

41. M. E. Rerek and F. Basolo, *Organometallics* **2**, 372 (1983).

42. C. H. Langford and H. B. Gray, *Ligand Substitution Processes* (W. A. Benjamin, Inc., New York, 1965).

1.9. Problems

1.1. Derive the integrated rate equation for a reversible first-order process.

1.2. From the following data calculate E_a, $\Delta H^{\ddagger}$ and $\Delta S^{\ddagger}$.

T(°C)	$k(s^{-1})$
30.1	6.7×10^{-5}
40.2	3.3×10^{-4}
47.8	1.0×10^{-3}

1.3. For the substitution reaction $ML_5X + Y \rightarrow ML_5Y + X$, the rate was studied under conditions of excess ligand Y, and the reaction goes to completion.

   A. How can one determine the order of the reaction in ML_5X from a single kinetics run?

   B. Assuming the order in ML_5X is 1, the following plot is obtained.

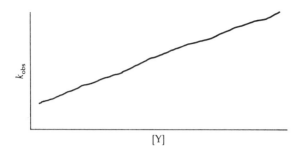

Write the general form of the rate law for substitution and express the constant(s) in terms of the slope and/or intercept.

1.4. The following data were obtained for the reaction $Cr(CO)_4(PPh_3)_2 + CO \rightarrow Cr(CO)_5PPh_3$.

k	$T(°C)$
1.96×10^{-4}	50
4.29×10^{-4}	55
8.73×10^{-4}	60

Derive the enthalpy and entropy of activation ($\Delta H^{\ddagger} = 31.3$ kcal/mole, $\Delta S^{\ddagger} = 21.2$ eu). Calculate the rate constant for the reaction at 40°C and 130°C. Which value will be more accurate?

1.5. Given the absorbance versus time data calculate the rate constant for reaction of $Cr(CO)_4(P(OPh)_3)AsPh_3$ at 70°C.

A (1913 cm^{-1})	t (min)
0.374	2.0
0.312	7.0
0.259	14.0
0.203	22.0
0.155	32.0
0.108	43.0
0.078	54.0
0.050	67.0

1.6. Given the following data calculate all the rate parameters for the reaction [A. J. Hart-Davis, C. White, and R. H. Mawby, *Inorg. Chim. Acta* **4**, 441 (1970)], of (η^5-C_9H_7)Mo(CO)$_3$Br with L in THF.

T(°C)	Ligand	Ligand Concentration (M)	$10^4 \, k_{obs}$ (sec^{-1})
5.0	P(OMe)$_3$	0.065	2.62
		0.149	4.22
		0.222	5.55
		0.286	6.64
		0.391	8.52
		0.422	8.98
8.8	PPh$_3$	0.046	3.25
		0.093	3.25
		0.140	3.40
		0.287	3.96
		0.298	4.03
10.0	P(OMe)$_3$	0.094	6.61
		0.208	9.26
		0.248	11.2
		0.410	15.6
15.0	PPh$_3$	0.097	8.04
		0.146	8.45
		0.230	8.82
		0.299	9.08
		0.388	9.68
	PBu$_3^n$	0.094	16.3
		0.135	21.2
		0.195	28.0
		0.248	32.3
		0.317	40.3
	P(OPh)$_3$	0.105	7.87
		0.223	8.12
		0.296	8.22
	P(OMe)$_3$	0.098	12.6
		0.133	14.0
		0.193	16.4
		0.231	17.7
		0.297	20.4
19.5	P(OMe)$_3$	0.099	20.6
		0.153	23.7
		0.192	26.4
		0.231	28.3
		0.300	33.0
20.0	PPh$_3$	0.075	15.2
		0.152	15.8
		0.256	16.6
		0.366	16.7
		0.389	18.3
25.7	PPh$_3$	0.041	34.9
		0.082	35.1
		0.216	37.4
		0.294	38.1

1.7. Given the following absorption spectra calculate the rate constant for deoxygenation of $[(NH_3)en_2Co(\mu\text{-}O_2)Co(NH_3)en_2]^{+4}$. [Y. Sasaki, K. Z. Suzuki, A. Matsumato, and K. Saito, *Inorg. Chem.* **21**, 1825 (1982).]

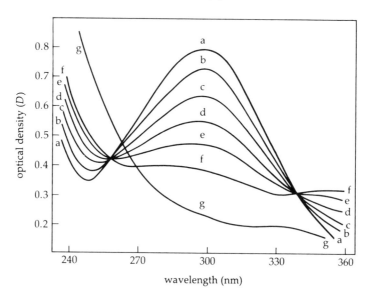

Change in absorption spectra of 8×10^{-5} M $[(en)_2(NH_3)Co(III)(\mu\text{-}O_2^{2-})Co(III)(NH_3)\text{-}(en)_2]^{4+}$ in 0.01 M NaClO$_4$ solution at ca. 17°C: (a) immediately after dissolution; (b) after 10 min; (c) after 20 min; (d) after 35 min; (e) after 55 min; (f) after 2 h; (g) after 2 days.

1.8. Given the following kinetic data for substitution on CpRh(CO)$_2$ derive the rate law and suggest a mechanism for substitution.

L	[CpRh(CO)$_2$], M	[L], M	$10^5 k_{obs}$, sec^{-1}	$10^4 k$, M^{-1} sec^{-1}
P(n-C$_4$H$_9$)$_3$	0.004	0.053	17.5	33.0
P(n-C$_4$H$_9$)$_3$	0.008	0.122	38.3	32.0^a
P(n-C$_4$H$_9$)$_3$	0.012	0.188	66.0	35.0
P(n-C$_4$H$_9$)$_3$	0.002	0.245	86.0	34.0
P(n-C$_4$H$_9$)$_3$	0.008	0.410	130.0	32.0^a
P(n-OC$_4$H$_9$)$_3$	0.0024	0.088	3.55	4.0
P(n-OC$_4$H$_9$)$_3$	0.0013	0.147	5.6	3.8
P(n-OC$_4$H$_9$)$_3$	0.0024	0.234	8.0	3.4
P(n-OC$_4$H$_9$)$_3$	0.0012	0.660	25.3	3.8

^{a} The reaction was followed by the rate of CO evolution. Other data were obtained by the infrared method.

1.9. Set up the differential equations and the secular equation for the following scheme.

$$A \underset{k_{-1}}{\overset{k_1}{\rightleftharpoons}} B$$

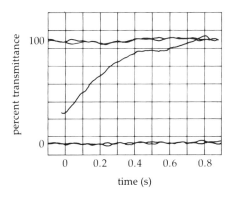

1.10. Given the data below for the reaction

$$[Co(NH_3)_5OH_2]^{3+} + X^- \rightleftharpoons [Co(NH_3)_5X]^{2+} + H_2O$$

where k_a is for the aquation, derive a linear free energy relationship and discuss the implications [A. Haim, *Inorg. Chem.* **9**, 426 (1970).]

X^{-1}	$k_a(s^{-1})$	$K_{eq}(M^{-1})$
NCS$^-$	4.1×10^{-10}	2.7×10^3
N$_3^-$	2.1×10^{-9}	8.3×10^2
F$^-$	8.6×10^{-8}	25
Cl$^-$	1.7×10^{-6}	1.11
Br$^-$	6.5×10^{-6}	0.35
NO$_3^-$	2.7×10^{-5}	0.08

1.11. The plot of the infrared spectrum at 1658 cm^{-1} that resulted from a stopped-flow investigation of the reaction of a 0.02M solution of a ketone with 0.24M cyclopentylmagnesium bromide is shown. Calculate the rate constant. [S. E. Rudolph, L. F. Charbonneau, and S. G. Smith, *J. Am. Chem. Soc.* **95**, 7083 (1973).]

1.12. Given the following data for O$_2$ uptake by Ir(CO)(PPh$_3$)$_2$Cl derive the form of the rate law. [P. B. Chock and J. Halpern, *J. Am. Chem. Soc.* **88**, 3511 (1966).]

Partial Pressure O$_2$(mm)	$k_{obs}(s^{-1})$
242	1.21×10^{-4}
371	1.93×10^{-4}
473	2.48×10^{-4}
630	3.25×10^{-4}

1.13. What is wrong with the scheme shown below for isotopic enrichment in $Mn(CO)_5X$?

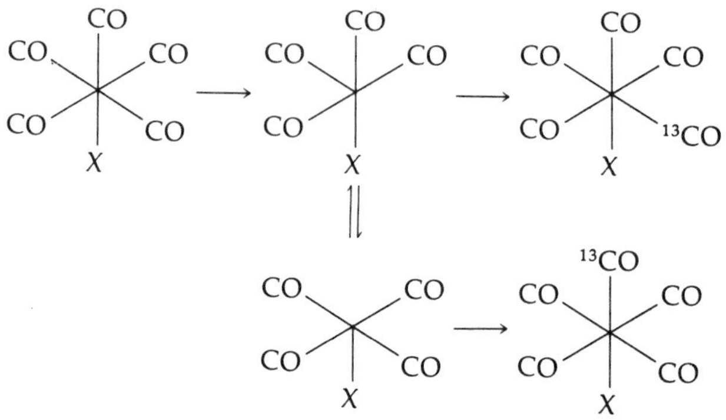

1.14. The competition ratio is the ratio of rates for two separate species with one reactant. This is usually used to test the ability of an intermediate to discriminate between two possible reactants. Values of the competition ratio between the ligand L and piperidine for $Mo(CO)_5$ are shown.

L	Competition Ratio
$P(OMe)_3$	0.74
PPh_3	0.85
$AsPh_3$	1.0

What information on the nature of the intermediate does this provide?

1.15. Falk and Halpern have studied the kinetic isotope effect in the reaction *trans*-$Pt(PR_3)_2LCl + Py \rightarrow Pt(PR_3)_2PyL^+ + Cl^-$

$$L = H^- \text{ or } D^-$$

and found $k_H/k_D = 1.4$ (*J. Am. Chem. Soc.* **87**, 3003 (1965)). The Pt–H stretching frequency is 2183 cm^{-1} in the ground state; make some reasonable assumptions and calculate the Pt–H stretching frequency in the transition state during substitution.

1.16. The $\Delta G^{\ddagger}$ values for the kinetic process involving rate-determining loss of ligand L from a complex CL,

$$CL \xrightarrow{k} C + L$$

are as follows:

L	$\Delta G^{\ddagger}$ (kcal/mole)	L	$\Delta G^{\ddagger}$ (kcal/mole)
CH_3OH	13.0	Py	18.7
CH_3CN	14.7	CH_3NC	20.1
$S(CH_3)_2$	17.4	$P(OCH_3)_3$	20.2
$N(CH_3)_3$	18.3	imidazole	21.1
PPh_3	18.5	$P(n\text{-}C_4H_9)_3$	23.5

Using these values derive a set of ligand dissociation parameters δ_L, employing linear free energy concepts. In the reaction $CoNO(CO)_3 + L \rightarrow CoNO(CO)_2L + CO$ with the rate law, Rate $= k_2[Co(NO)(CO)_3][L]$. The rates at comparable conditions for some L are

Ligand	$k(M^{-1}sec^{-1})$
P(n-Bu)$_3$	9.2×10^{-2}
P(OCH$_3$)$_3$	1.8×10^{-3}
PPh$_3$	1.0×10^{-3}
Py	3.9×10^{-5}

Do these data correlate with the derived δ_L values? Discuss the reasons for the observed relationship or the lack of it.

1.17. Given the following sets of absorption data determine the reaction order in the absorbing species. Discuss any difficulties or uncertainty you have in interpretation of the data.

(a)	seconds	absorption
	0.0	0.150
	1.13	0.083
	2.25	0.045
	3.38	0.026
	4.50	0.014
	5.63	0.007
	6.75	0.003

(b)	seconds	absorption
	0.0	0.0
	1.5	1.30
	3.0	2.26
	4.5	2.97
	6.0	3.50
	7.5	3.88
	9.0	4.17
	10.5	4.39

1.18. Given the following data calculate $\Delta H^{\ddagger}$ and $\Delta S^{\ddagger}$ for reaction of $(\eta^6\text{-}C_6H_6)Cr(CO)_3 + 3 PBu_3 \rightarrow Cr(CO)_3(PBu_3)_3 + C_6H_6$.

T(°C)	$10^4 k_2 (M^{-1}s^{-1})$
112	0.0800
135	0.410
150	1.29
165	3.81

What does this indicate about the mechanism?

1.19. Discuss the implications of the following data for exchange reactions of two Ni(II) complexes.

	k	$\Delta H^{\ddagger}$(kcal/mole)	$\Delta S^{\ddagger}$(eu)	$\Delta V^{\ddagger}$
$Ni(en)_3^{2+}$	20	17	2	11
$Ni(H_2O)_6^{2+}$	4×10^4	12	2	7

[S. Soyama, M. Ishii, S. Fumakashi, and M. Tanaka, *Inorg. Chem.* **31,** 536 (1992).]

2

Ligand Substitution Reactions on Square-Planar Complexes

Square-planar complexes are often formed when the metal ion contains eight electrons in its d orbitals. This includes complexes of Pt(II), Pd(II), Au(III), Rh(I), and Ir(I). Such square-planar complexes are invariably spin paired (diamagnetic). Substitution reactions of square-planar complexes of Rh(I) and Ir(I) are evidently quite rapid and have not been studied sufficiently. Square-planar complexes also undergo oxidative addition reactions that will be considered in Chapter 5. For substitution reactions the most often studied complexes are of Pt(II), and we will discuss these in some detail in this chapter. Reactions of complexes of Pd(II), Au(III), and Ni(II) will be considered briefly later in this chapter. Several good general reviews of square-planar substitution reactions have appeared.[1-4]

2.1. Substitution Reactions

The reaction that we are considering is substitution of one ligand for another on $trans$-Pt(T)(L)$_2$X, where X is the leaving group, T is the ligand $trans$ to the leaving group, and L are ligands cis to the leaving group.

$$Pt(T)L_2X + Y \rightarrow Pt(T)L_2Y + X \tag{2.1}$$

The substitution reaction proceeds with retention of the configuration at the metal as shown in Figure 2.1. Reactions 2.2 and 2.3 provide examples that show the stereochemistry of substitutions on square-planar complexes.[5]

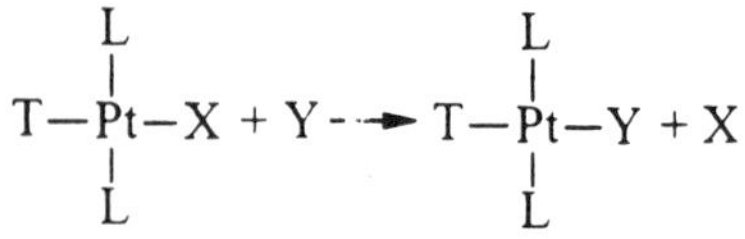

$$T-\underset{\underset{L}{|}}{\overset{\overset{L}{|}}{Pt}}-X + Y \longrightarrow T-\underset{\underset{L}{|}}{\overset{\overset{L}{|}}{Pt}}-Y + X$$

Figure 2.1. Substitution reactions of square-planar complexes proceed with retention of stereochemistry.

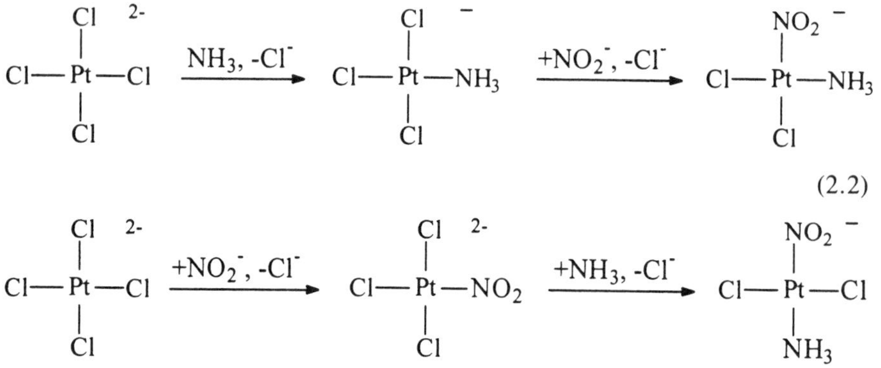

$$\underset{\underset{Cl}{|}}{\overset{\overset{Cl}{|}}{Cl-Pt-Cl}}^{2-} \xrightarrow{NH_3, -Cl^-} \underset{\underset{Cl}{|}}{\overset{\overset{Cl}{|}}{Cl-Pt-NH_3}}^{-} \xrightarrow{+NO_2^-, -Cl^-} \underset{\underset{Cl}{|}}{\overset{\overset{NO_2}{|}}{Cl-Pt-NH_3}}^{-}$$

$$(2.2)$$

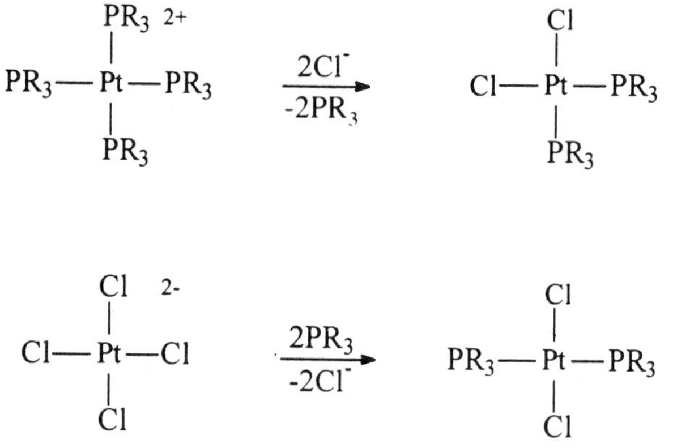

$$\underset{\underset{Cl}{|}}{\overset{\overset{Cl}{|}}{Cl-Pt-Cl}}^{2-} \xrightarrow{+NO_2^-, -Cl^-} \underset{\underset{Cl}{|}}{\overset{\overset{Cl}{|}}{Cl-Pt-NO_2}}^{2-} \xrightarrow{+NH_3, -Cl^-} \underset{\underset{NH_3}{|}}{\overset{\overset{NO_2}{|}}{Cl-Pt-Cl}}^{-}$$

$$\underset{\underset{PR_3}{|}}{\overset{\overset{PR_3}{|}}{PR_3-Pt-PR_3}}^{2+} \xrightarrow[-2PR_3]{2Cl^-} \underset{\underset{PR_3}{|}}{\overset{\overset{Cl}{|}}{Cl-Pt-PR_3}}$$

$$(2.3)$$

$$\underset{\underset{Cl}{|}}{\overset{\overset{Cl}{|}}{Cl-Pt-Cl}}^{2-} \xrightarrow[-2Cl^-]{2PR_3} \underset{\underset{Cl}{|}}{\overset{\overset{Cl}{|}}{PR_3-Pt-PR_3}}$$

These reactions indicate that the order of addition of reagents can affect the product geometry. The following substitution reactions show the preparation of the three isomers of $Pt(Py)(NH_3)(Cl)(Br)$. A combination of the effect of the *trans* ligand

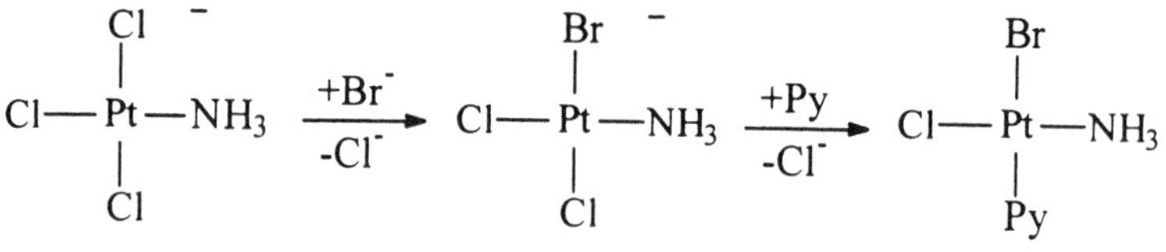

$$\underset{\underset{Cl}{|}}{\overset{\overset{Cl}{|}}{Cl-Pt-NH_3}}^{-} \xrightarrow[-Cl^-]{+Br^-} \underset{\underset{Cl}{|}}{\overset{\overset{Br}{|}}{Cl-Pt-NH_3}}^{-} \xrightarrow[-Cl^-]{+Py} \underset{\underset{Py}{|}}{\overset{\overset{Br}{|}}{Cl-Pt-NH_3}}$$

$$(2.4)$$

$$
\begin{array}{c}
\text{Cl} \\
| \\
\text{Cl}\!-\!\text{Pt}\!-\!\text{Py} \\
| \\
\text{Cl}
\end{array}
\;^{-}
\xrightarrow[-\text{Cl}^-]{+\text{Br}^-}
\begin{array}{c}
\text{Br} \\
| \\
\text{Cl}\!-\!\text{Pt}\!-\!\text{Py} \\
| \\
\text{Cl}
\end{array}
\;^{-}
\xrightarrow[-\text{Cl}^-]{+\text{NH}_3}
\begin{array}{c}
\text{Br} \\
| \\
\text{Cl}\!-\!\text{Pt}\!-\!\text{Py} \\
| \\
\text{NH}_3
\end{array}
\qquad (2.5)
$$

$$
\begin{array}{c}
\text{Cl} \\
| \\
\text{Cl}\!-\!\text{Pt}\!-\!\text{Py} \\
| \\
\text{Py}
\end{array}
\xrightarrow[-\text{Cl}^-]{+\text{NH}_3}
\begin{array}{c}
\text{NH}_3 \\
| \\
\text{Cl}\!-\!\text{Pt}\!-\!\text{Py} \\
| \\
\text{Py}
\end{array}
\;^{+}
\xrightarrow[-\text{Py}]{+\text{Br}^-}
\begin{array}{c}
\text{NH}_3 \\
| \\
\text{Cl}\!-\!\text{Pt}\!-\!\text{Br} \\
| \\
\text{Py}
\end{array}
\qquad (2.6)
$$

(*trans* effect) and the group that is being substituted (leaving group effects) lead to the observed products in these reactions. Each of these will be discussed later in this chapter.

Studies of the kinetics of the reaction show a dependence on the concentration of the entering ligand as shown in Table 2.1 and Figure 2.2 for reaction of $Pt(Py)_2(NO_2)Cl$ with Py.[6]

$$Pt(NO_2)(Py)_2Cl + Py \rightarrow Pt(NO_2)(Py)_3^+ + Cl^- \qquad (2.7)$$

Under pseudo first-order conditions plots of *ln* [metal complex] versus time are linear with a slope of k_{obs}. Varying the initial concentration of the ligand (always under pseudo first-order conditions) gives plots such as those shown in Figure 2.2 from which k_1 is derived as the intercept and k_2 as the slope. Thus, the rate law for the reaction

$$trans\text{-}Pt(T)L_2X + Y \rightarrow trans\text{-}Pt(T)L_2Y + X \qquad (2.8)$$

has a ligand-independent as well as a ligand-dependent term. The ligand-dependent term (k_2) can be readily understood in terms of nucleophilic attack by the entering ligand. A square-planar complex is coordinatively unsaturated and hence has a site

Table 2.1. DEPENDENCE ON THE [Py] OF THE RATE OF SUBSTITUTION OF Cl^- ON $Pt(NO_2)(Py)_2Cl$ IN NITROMETHANE AT 25°C[6]

[Py]M	$k_{obs}(\times\ 10^3 s^{-1})$
0.02	0.703
0.05	0.832
0.10	1.35
0.20	1.86
0.30	2.20
0.40	2.96
0.50	3.60

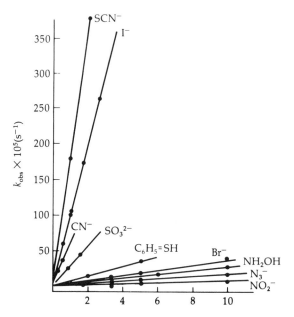

Figure 2.2. Rates of reaction of *trans*-[Pt(Py)$_2$Cl$_2$] with various nucleophiles in methanol at 30°C. [Reprinted with permission from U. Belluco, L. Cattalini, F. Basolo, R. G. Pearson, and A. Turco, *J. Am. Chem. Soc.* **87,** 241 (1965). Copyright 1965 American Chemical Society.]

for attack and an orbital available along the *z* axis. Addition of the two-electron ligand Y to the square-planar species leads to a coordinatively saturated 18-electron intermediate. Data that will be presented in a subsequent section will show the dependence of the rate of the substitution reaction on the nucleophilicity of the entering ligand.

The ligand-independent pathway could have several possible explanations; the one that is generally accepted involves attack of solvent on the square-planar complex. In this reaction the solvent becomes the entering nucleophile in the rate-determining step. The scheme showing both first- and second-order pathways is illustrated in Figure 2.3. The similarity of the two pathways is a strong feature in support of this mechanism. This is especially true since the solvents often used (i.e., H$_2$O, CH$_3$OH, etc.) have electron pairs and can be nucleophiles. The large excess in concentration of the solvent leads to the observed first-order kinetics and makes up for the generally lower nucleophilicity of the solvent. The solvent dependence of the first-order rate constant for Reaction 2.7 shows that the first-order step is dependent on the nature of the solvent and is not dissociation of a ligand. These data are shown in Table 2.2.[6] This strong dependence of the first-order rate constant on the nature of the solvent is consistent with direct nucleophilic attack by the solvent on the square-planar complex. As we proceed through our discussion we will find that all evidence supports this mechanism.

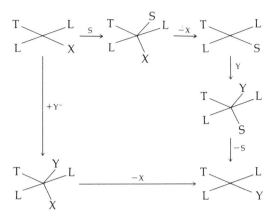

Figure 2.3. Associative mechanism for ligand attack (Y^-) and for solvent attack (S) on a square-planar complex.

2.2. *Trans* Effect

The overriding consideration in studies of substitution reactions on Pt(II) was the effect on the rate of the ligand *trans* to the substitution site.[5] The *trans effect* was recognized early (circa 1900) and played an important role in preparations of different isomers, as shown in Reactions 2.1–2.6.

The effect of a series of *trans* ligands on substitution of pyridine for chloride on *trans*-Pt(PEt$_3$)$_2$LCl was determined.[7]

$$trans\text{-Pt(PEt}_3)_2\text{LCl} + \text{Py} \rightarrow trans\text{-Pt(PEt}_3)_2\text{L(Py)} + \text{Cl}^- \tag{2.9}$$

(In many reactions that we will use the charge on the complex will depend on the nature of L as it will in Reaction 2.9. In each case the Pt is in a +2 oxidation state and the charge on the complex will be that necessary with standard charges on ligands.) The results for Reaction 2.9 are given in Table 2.3.[7] As the *trans* ligand changes through the series shown in going down the table, the rates of substitution

Table 2.2. EFFECT OF SOLVENT ON
THE RATE CONSTANT FOR THE FIRST-
ORDER STEP (REACTION 2.7) AT 30°C[6]

Solvent	$k_1(\times 10^4 \text{s}^{-1})$
dichloromethane	0.06
methanol	0.7
nitromethane	8.0
acetone	3100[a]

[a] At 20°C.

Table 2.3. EFFECT OF THE *trans* LIGAND ON RATE
IN SUBSTITUTIONS ON *trans*-Pt(PEt$_3$)$_2$L(Cl)[7]

L	$k_1(s^{-1})$	$k_2(M^{-1}s^{-1})$
PEt$_3$	1.7×10^{-2}	3.8
H$^-$	1.8×10^{-2}	4.2
CH$_3^-$	1.7×10^{-4}	6.7×10^{-2}
C$_6$H$_5^-$	3.3×10^{-5}	1.6×10^{-2}
p-ClC$_6$H$_4^-$	3.3×10^{-5}	1.6×10^{-2}
p-CH$_3$OC$_6$H$_4^-$	2.8×10^{-5}	1.3×10^{-2}
Cl$^-$	1.0×10^{-6}	4.0×10^{-4}

dramatically decrease. For this series of complexes the only change is in the *trans* ligand and the 10^4 difference in rate can be assigned to the *trans* effect. An important feature that supports the interpretation of the mechanism is that k_1 and k_2 are both affected in the same way by the change in *trans* ligand, indicating the mechanistic similarity of the two steps. From the data presented in Table 2.3, and many other reactions, a *trans* effect order has been developed as follows[8]:

$$CO, CN^-, C_2H_4 > PR_3, H^- > CH_3^- > C_6H_5^-, NO_2^-, I^-, SCN^-$$
$$> Br^-, Cl^- > Py, NH_3, OH^-, H_2O$$

This order spans a factor of 10^6 in rate and holds for all square-planar platinum complexes examined thus far.

A rate enhancement can arise by either ground-state energy differences or by transition-state energy differences. As shown in Figure 2.4 an increase in rate indicates either a ground state destabilization (increase in energy) or a transition state stabilization (decrease in energy). A ligand that exerts a strong *trans* effect will weaken the bond to the *trans* ligand, stabilize the transition state, or affect both the ground state and the transition state.

Before discussing the origin of the *trans* effect it is necessary to distinguish between the *trans* effect and the *trans influence* of a ligand. The *trans* effect concerns the effect of the ligand on the *rate* of substitution of the ligand that is *trans*. The *trans* influence is the effect of the *trans* ligand on ground state properties such as bond lengths or infrared stretching frequencies. The difference is similar to the difference between kinetics and thermodynamics. *Thermodynamics* concern ground state measurements, whereas *kinetics* concern the difference in energy between the ground state and the transition state. Similarly, the *trans* influence provides information on the ground state influence of a *trans* ligand, whereas the *trans* effect will involve the effect of the *trans* ligand on both the ground state and the transition state. In the absence of transition state differences the *trans* effect order and the *trans* influence order would be similar. As we examine the reasons for the *trans* effect, differences in the *trans* effect order and the *trans* influence order will be ascribed to transition state energy differences.

A number of explanations have been offered for the *trans* effect. The best explanation considers both the σ-donating and π-accepting capabilities of the li-

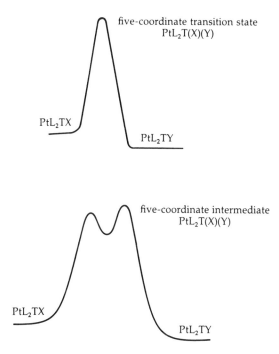

Figure 2.4. Plots of the reaction coordinate for square-planar substitution reactions show-ing the difference between a five-coordinate transition state and a five-coordinate inter-mediate.

gand.[1,3,9] The *trans* ligands in a square-planar complex share a p orbital, as shown in Figure 2.5. A ligand that is a strong σ-donor ligand contributes more electron density to the shared p orbital (p_x) and weakens the bond to the leaving group X. A pictorial demonstration of this is shown in the lower part of Figure 2.5. In the five-coordinate intermediate the T group and X group do not directly share the same p orbital so that the effect of a strong σ donor on the transition state is much smaller. Thus, the effect of a σ donor is primarily a ground-state destabilization of the Pt–X bond. As primarily a ground-state effect the *trans* effect arising from σ donation should parallel the order of *trans* influence. The expected order for the *trans* effect from σ donation is:

$$H^- > PR_3 > SCN^- > I^-, \ CH_3^-, \ CO, \ CN^- > Br^- > Cl^-$$
$$> NH_3 > OH^-$$

This order is similar to the general order of *trans* effect ligands with some very notable exceptions (CO and CN^- are badly out of order). As we discussed earlier, discrepancies between the *trans* effect order and the *trans* influence order may be attributed to transition state effects. The ligands that have a substantially larger *trans* effect than would be predicted from their *trans* influence are ligands that have π-bonding capabilities. Thus, a stabilization of the transition state by π-accepting

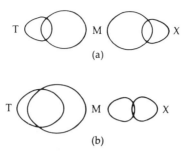

Figure 2.5. Schematic diagram showing the difference in M–X bond strengths for a weak (a) and a strong (b) σ-donor ligand. In (b), the strong σ donor weakens the bond to the *trans* ligand.

ligands is indicated. It was further shown that the σ-only explanation is not sufficient by an examination of the electronic effect of *trans* ligands in several *p*-substituted aniline complexes.[9]

$$\text{trans-(styrene)}(p\text{-}XC_6H_4NH_2)PtCl_2 + 1\text{-pentene} \qquad (2.10)$$
$$\rightarrow \text{trans-(1-pentene)}(p\text{-}XC_6H_4NH_2)PtCl_2 + \text{styrene}$$

These reactions proceed very rapidly and show a dependence on the X group as shown in Table 2.4. Since the rate of substitution decreases with increasing base strength of the *trans* aniline ligand, opposite to that expected for the σ *trans* effect, it is further indication for π-stabilization of the intermediate. This can be understood since the transition state involves ligand attack and a net increase of electrons at the metal center. A ligand with π acceptance capabilities can remove electron density from the metal and thus stabilize the transition state. The order of π-bonding ability of the ligands

$$C_2H_4, CO > CN^- > NO_2^- > SCN^- > I^- > Br^- > Cl^-$$
$$> NH_3 > OH^-$$

accounts for the ligands that are not properly ordered by considering σ donation only. The complete *trans* effect order can be composed of the expected orders for σ

Table 2.4. THE EFFECT OF CHANGING AMINE BASICITY ON RATE OF REPLACEMENT OF THE *trans* LIGAND[9]

X	$k(M^{-1}, s^{-1})$	pKa
Cl	9750	4.1
H	7030	4.63
CH_3	5000	5.08
OCH_3	3960	5.34

donation and π acceptance, and it supports the idea of the *trans* effect arising from these different ligand capabilities.

2.3. Other Effects on the Rate

While the *trans* group has a large and very significant effect on the rate of square-planar substitution reactions, other factors are also important. In this section we will consider the effect of the *cis* ligand (steric effects), the effect of the leaving group, the effect of the entering nucleophile, and the effect of the solvent.

2.3.1. *Cis* Effect

The electronic effect of the *cis* ligands is relatively small.[1,10] The reactions of *cis*-Pt(PEt$_3$)$_2$LCl illustrate the magnitude expected.

$$cis\text{-Pt(PEt}_3)_2\text{LCl} + \text{Py} \rightarrow cis\text{-Pt(PEt}_3)_2\text{L(py)} + \text{Cl}^- \tag{2.11}$$

The data are shown in Table 2.5 and may be compared with those in Table 2.3.[7] For these ligands the *cis* effect is only a factor of 3 in rate while the *trans* effect is greater than two orders of magnitude.[7] In general, the *cis* effect is small and the order is variable, unless steric factors are involved. Steric effects provide one of the best means of differentiating a dissociative mechanism from an associative mechanism. For a dissociative reaction one usually observes an increase in reaction rate with increasing steric interactions, a *steric acceleration*. For an associative reaction that involves an increased coordination number, increasing steric size would decrease the rate of reaction. Data shown in Table 2.6 provide excellent confirmation of the associative nature of substitution reactions of Pt(II) complexes.[1,7] This table illustrates several features of interest. Increasing the size of the ligands dramatically slows the rate of substitution reactions. The orientation of the aromatic ring of the aryl ligand is perpendicular to the plane of the molecule such that the substituents lie above and below the plane of the molecule and effectively block the site of attack. This is illustrated in Figure 2.6. Changing the size of the *trans* ligand has less effect on the rate of substitution as shown in Table 2.5. This can be understood by considering the trigonal bipyramidal intermediate formed. The *trans* group shares the equatorial plane and is away from the entering group and leaving group by an

Table 2.5.   THE EFFECT OF L
ON THE RATE OF REACTION OF
cis-Pt(PEt$_3$)$_2$LCl WITH Py[7]

L	$k_1(10^2\text{s}^{-1})$
Cl$^-$	1.7
C$_6$H$_6^-$	3.8
CH$_3^-$	6.0

Table 2.6. STERIC EFFECTS ON THE
RATES OF SQUARE-PLANAR
SUBSTITUTION REACTIONS AT 25°C[1,7]

Complex	k_{obs}, s^{-1}
cis-Pt(PEt$_3$)$_2$LCl	
L = phenyl[a]	8.0×10^{-2}
L = *o*-tolyl[a]	2.0×10^{-4}
L = mesityl	1.0×10^{-6}
trans-Pt(PEt$_3$)$_2$LCl	
L = phenyl	1.2×10^{-4}
L = *o*-tolyl	1.7×10^{-5}
L = mesityl	3.4×10^{-6}

[a] At 0°C.

angle of 120°. The *cis* ligands occupy axial sites in the intermediate and interact with the entering and leaving groups at a 90° angle. Thus, there is an increased effect on rate by steric interactions from the *cis* ligand. The importance of steric interactions is further shown by a comparison of the substitution reactions of Pt(dien)Cl$^+$ and the tetraethyl-substituted derivative, Pt(Et$_4$dien)Cl$^+$ (dien = H$_2$NCH$_2$CH$_2$NHCH$_2$CH$_2$NH$_2$; Et$_4$dien = Et$_2$NCH$_2$CH$_2$NHCH$_2$CH$_2$NEt$_2$; each is a tridentate ligand). The Pt(dien)Cl$^+$ undergoes substitution of Py for Cl$^-$ at room temperature. The tetraethyl substituted derivative is so sterically hindered that reaction does not occur until 80°C, and proceeds only by a first-order reaction pathway that appears to be dissociative in nature.[1]

Steric effects are further illustrated for a series of amine nucleophiles in the following reactions.[11]

$$\text{\textit{trans}-(Pr}_3\text{P)PtCl}_2\text{(NH[}^{14}\text{C]Et}_2) + \text{amine} \qquad\qquad (2.12)$$
$$\rightleftarrows \text{\textit{trans}-(Pr}_3\text{P)PtCl}_2\text{(amine)} + \text{NH[}^{14}\text{C]Et}_2$$

The plots shown in Figure 2.7 show the dramatic effect of the size of the incoming nucleophile.

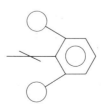

Figure 2.6. Geometry of aryl square-planar complexes showing the *ortho* substituents blocking the site of attack.

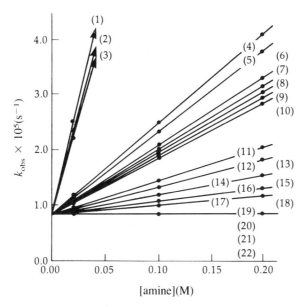

Figure 2.7. Reaction rates for *trans*-(Pr₃P)PtCl₂, (NHEt₂)* + amine in methanol at 25°C. Amine reagents (those marked with asterisk react to give new compounds): (1) 4-methylpyridine*; (2) 3-methylpyridine*; (3) pyridine; (4) methylamine; (5) aziridine*; (6) ethylamine; (7) isobutylamine*; (8) pyrrolidine*; (9) *n*-butylamine*; (10) ammonia; (11) isopropylamine*; (12) *s*-butylamine*; (13) 2-methylpyridine*; (14) piperidine; (15) dimethylamine; (16) 2,4,6-trimethylpyridine*; (17) *t*-butylamine*; (18) diethylamine; (19) di-isobutylamines*; (20) (21) di-isobutylamine*; (22) 2,4,6-trimethylpiperidine.* [Reprinted with permission from A. L. Odell and H. A. Raethel, *J. Chem. Soc., Chem. Commun.* 1323 (1968). Copyright 1968 Royal Society of Chemistry.]

2.3.2. Leaving Group Effects

One must keep the *cis* and *trans* groups constant to investigate the effect of the leaving group (and other effects) on the rate of substitution reactions. The Pt(dien)X⁺ complexes have provided excellent information regarding leaving group effects. The reaction of this complex with an entering nucleophile, Y, is shown in Figure 2.8. In dissociative reactions one expects a large dependence on the nature of the leaving group since the bond from the metal to the leaving group is broken in the transition state. For associative reactions the effect of the leaving group depends on the extent of bond breaking in the transition state. Data for the reaction

$$\text{Pt(dien)X}^+ + \text{Py} \rightarrow \text{Pt(dien)Py}^{+2} + \text{X}^- \tag{2.13}$$

are presented in Table 2.7.[12,13] The magnitude of the leaving group effect, shown in Table 2.7, indicates a substantial amount of bond breaking in the transition state. The amount of bond breaking existing in the transition state depends on the specific reaction. It was concluded from volume of activation measurements that replace-

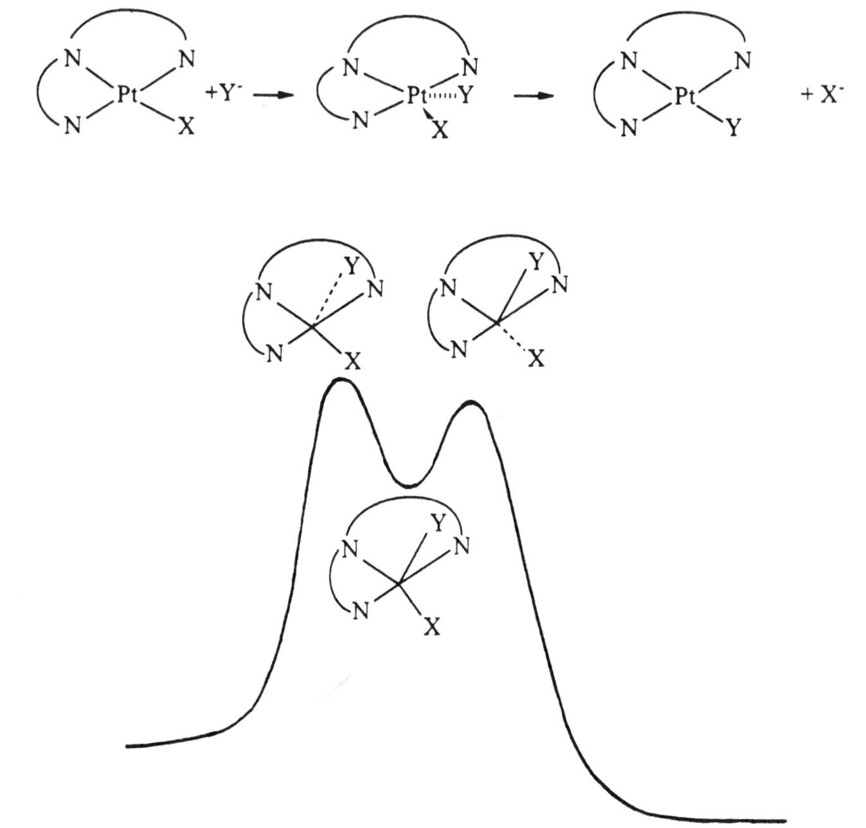

Figure 2.8. Reaction of Pt(dien)X with Y^- showing entering- and leaving-group effects.

Table 2.7. THE EFFECT OF THE
LEAVING GROUP, X, ON RATE OF
SUBSTITUTION OF Pt(dien)X^+
WITH Py[12,13]

X	k_{obs}
NO_3^-	very fast
H_2O	1900
Cl^-	35
Br^-	23
I^-	10
N_3^-	0.8
SCN^-	0.3
NO_2^-	0.05
CN^-	0.02

ment of Cl^- by Br^- on *trans*-$Pt(PEt_3)_2Cl_2$ leads to a transition state with virtually no bond breaking of the Pt-Cl bond.[14] The order of leaving group dependence for Reaction 2.13

$$NO_3^- > H_2O > Cl^- > Br^- > I^- > N_3^- > SCN^-$$
$$> NO_2^- > CN^-$$

is very similar to the inverse of the *trans* effect order. This reflects the fact that the *trans* effect depends on the strength of bonding (through either π or σ capability) and the more strongly bound ligands dissociate more slowly from the five-coordinate intermediate. In Section 2.4 (activation parameters) we will obtain some additional evidence on leaving group effects.

2.3.3. Effect of the Entering Nucleophile

For a reaction that involves attack of the entering ligand we should certainly expect a dependence on the nucleophilicity of the entering ligand. To discuss the nucleophilicity of a ligand we must define the species that will interact with the nucleophile. The order observed for Pt(II) substitution reactions[1-5]

$$PR_3 > I^-, SCN^-, N_3^- > NO_2^- > Br^- > Py > NH_3, Cl^-$$
$$> H_2O > OH^-$$

is certainly different than the order of basicities toward a proton. The *polarizability* or *softness* of a nucleophile is an important consideration. Soft nucleophiles prefer soft substrates and *hard* (*nonpolarizable*) ligands prefer hard substrates.[15] The observation that larger donors are effective nucleophiles toward Pt(II) indicates that Pt(II) is a soft center. In general, if one wants to describe the nature of a nucleophile, then one correlates the reactivity with other properties of the ligand. This type of correlation is a linear free energy relationship (LFER). Data that have been collected for displacement of chloride from *trans*-PtL_2Cl_2,[16-18]

$$trans\text{-}PtL_2Cl_2 + Y \rightarrow trans\text{-}PtL_2ClY + Cl^- \qquad (2.14)$$

are presented in Table 2.8 and shown in Figure 2.9. The ligand dependence is clearly not on the basicity toward a proton. The parameter, n_{Pt}°, defined for each ligand, is called the *nucleophilic reactivity constant*. This constant is defined as

$$\log \frac{k_Y}{k_S} = n_{Pt}^{\circ} \qquad (2.15)$$

where k_Y is the rate constant for reaction of an entering nucleophile; Y, with $Pt(Py)_2Cl_2$ and k_S, is the rate constant for attack by solvent (CH_3OH) on the complex.[17] This constant measures the reactivity of a nucleophile toward Pt and is useful for correlation with other ligand properties. In addition, plots of $\log k_Y$ for other Pt(II) complexes versus n_{Pt}° are observed to be linear. Such plots are shown in Figure 2.9. An LFER is given by

Table 2.8. EFFECT OF THE ENTERING
NUCLEOPHILE ON REACTION AT $trans$-$PtL_2Cl_2{}^a$

Y	L = Py[b]	L = PEt₃[c]	$n^o_{Pt}{}^b$
Cl^-	0.45	0.029	3.04
NH_3	0.47	—	3.07
NO_2^-	0.68	0.027	3.22
N_3^-	1.55	0.2	3.58
Br^-	3.7	0.93	4.18
I^-	107	236	5.46
SCN^-	180	371	6.65
PPh_3	249,000	—	8.93

^{a} Entries are $10^3 k$ in units of $M^{-1}s^{-1}$. ^{b} Reference 16.
^{c} Reference 17,18.

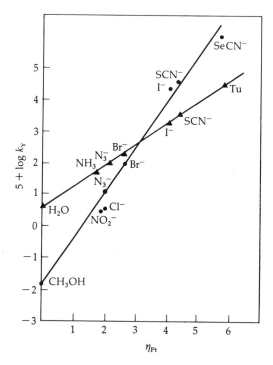

Figure 2.9. Correlation of the rates of reaction of Pt(II) complexes with the standard $trans$-$[Pt(Py)_2Cl_2]$ for different nucleophiles: ● = $trans$-$[Pt(PEt_3)_2Cl_2]$ in methanol at 30°C; ▲ = $[Pt(en)Cl_2]$ in water at 35°C. (Tu = thiourea, s = $C(NH_2)_2$.) [Reprinted with permission from U. Belluco, L. Cattalini, F. Basolo, R. G. Pearson, and A. Turco, $J.$ $Am.$ $Chem.$ $Soc.$ **87,** 241 (1965). Copyright 1965 American Chemical Society.]

$$\log k_Y = sn^o_{Pt} + \log k_S \tag{2.16}$$

and holds for a variety of different Pt(II) complexes and solvents. The constant s, which is called the *nucleophilic discrimination factor*, is a measure of the sensitivity of the metal center to the nucleophilicity of the entering ligand. A large value indicates that the rate is very sensitive to changes in nucleophilic character. Values of s for several complexes are shown in Table 2.9.[17] The comparison of the Pt(dien)X$^+$ complexes (X = Br$^-$, Cl$^-$, H$_2$O) offers confirmation of an often-stated postulate. The complex that is most reactive (best leaving group) is least discriminating in its reactions (low value of s). This is a feature that we will return to in catalytic reactions, the balance between reactivity and selectivity.

The order of the ligand nucleophilicities, shown in Table 2.8, is similar to the *trans* effect order. This can be understood in terms of the position of the Y group and the *trans* group in the transition state/intermediate. As shown in Figure 2.2, the entering nucleophile and *trans* ligand occupy approximately the same position and should have similar effects on the rate.

The entering group effects and leaving group effects are closely related since they occupy equivalent positions in the intermediate. The magnitude depends on the relative amounts of bond formation or bond weakening in the transition state. The similarity in the two is shown by the rate constants in Table 2.10 for the reaction.[19]

$$
\begin{array}{ccc}
\underset{Cl}{\overset{am}{\diagdown}}Pt\underset{Cl}{\overset{DMSO}{\diagup}} & + \ am \ \underset{k_r}{\overset{k_f}{\rightleftarrows}} & \underset{am}{\overset{am}{\diagdown}}Pt\underset{Cl}{\overset{DMSO}{\diagup}}
\end{array}
\tag{2.17}
$$

In the reverse reaction the *cis* complex is always formed because the *trans* effect of the S-bonded DMSO is much larger than that of Cl$^-$. A small *cis*-effect may exist for these complexes. This is significant when basicity differences are small as for cyclopentylamine through cyclooctylamine.[19]

Table 2.9. NUCLEOPHILIC DISCRIMINATION FACTORS FOR DIFFERENT Pt(II) COMPLEXES[17]

Complex	s
trans-Pt(PEt$_3$)$_2$Cl$_2$	1.43
trans-Pt(AsEt$_3$)$_2$Cl$_2$	1.25
trans-Pt(Py)$_2$Cl$_2$	1.0
trans-Pt(piper)$_2$Cl$_2$	0.91
Pt(en)Cl$_2$	0.64
Pt(dien)Br$^+$	0.75
Pt(dien)Cl$^+$	0.65
Pt(dien)H$_2$O^{2+}	0.44

Table 2.10. RATE CONSTANTS ILLUSTRATING
THE ENTERING AND LEAVING-GROUP EFFECTS
OF A SERIES OF AMINES FOR REACTION 2.17[19]

Amine	k_f	k_r	pK_a
cyclopropylamine	4.82	14.8	8.66
cyclobutylamine	3.60	3.2	9.34
cyclopentylamine	1.96	1.22	9.95
cyclohexylamine	1.73	1.16	9.82
cycloheptylamine	1.38	1.37	9.99
cyclooxtylamine	1.22	1.37	10.01

2.3.4. Solvent Effects

The rate law for square-planar substitution reactions includes a term that is independent of the entering nucleophile. This is ascribed to solvent attack; therefore, it is not surprising that solvent effects are important in substitution reactions on square-planar complexes. Table 2.2 presents some pertinent data. Data for the chloride exchange reaction

$$\textit{trans-}Pt(Py)_2Cl_2 + {}^{36}Cl^- \rightarrow \textit{trans-}Pt(Py)_2Cl({}^{36}Cl) + Cl^- \qquad (2.18)$$

are shown in Table 2.11.[20] Depending on the relative nucleophilicity of the solvent and the entering ligand, the observed rate law may be dependent or independent of the ligand. This is shown in Table 2.11, where the upper part of the table contains solvents of good coordinating ability that react by a ligand independent path, and the lower half of the table, which contains solvents of low coordination ability with the reactions proceeding purely by attack of the entering ligand on the square-planar

Table 2.11. EFFECT OF SOLVENT ON
THE CHLORIDE EXCHANGE REACTION[20]

solvent	$k(10^{-5}s^{-1})$
DMSO	380
H_2O	3.5
EtOH	1.4
PrOH	0.4

solvent	$k,\ M^{-1}s^{-1}$
CCl_4	10^4
C_6H_6	10^2
i-BuOH	10^{-1}
$Me_2C(O)$	10^{-2}
DMF	10^{-3}

complex. For solvents that have the capability to coordinate to the metal the rates show a direct dependence on the nucleophilicity of the solvent. For the poorly coordinating solvents the larger rates occur in nonpolar solvents where the chloride would not be solvated and would be quite reactive.

The extent of solvation in general is uncertain; however, it has been determined for cis-Pt(PBu$_3$)$_2$Cl$_2$ in several solvents as shown in Table 2.12.[21] The number n is the number of solvent molecules coordinated to the complex.[21] These data support the concept of solvent attack through a five-coordinate intermediate for the first-order step.

2.4. Activation Parameters

Activation parameters have been determined for a number of square-planar substitution reactions. A few examples are shown in Table 2.13. These activation parameters are consistent with the mechanistic scheme that we have developed. The uniformly negative entropies and volumes of activation are especially significant. Negative values of both parameters indicate an associative reaction (see the discussion in Chapter 1). The other significant feature is that the first-order path and the second-order path have very similar activation parameters, which again suggests the mechanistic similarity in both paths.

2.5. Dissociative Processes

Considerable data indicate that dissociative processes are possible for square-planar, 16-electron, organometallic complexes.[22] Ligand substitution reactions of cis-PtR$_2$S$_2$ (R = Me, Ph; S = SMe$_2$, S(O)Me$_2$) use the strong labilizing effects of Me and Ph groups to proceed through 14 e$^-$ intermediates[22]:

$$cis\text{-PtR}_2\text{S}_2 + 2\ S' \rightarrow cis\text{-PtR}_2\text{S}_2' + 2\ S \qquad (2.19)$$

Instead of the more typical associative process the primary evidence cited for S dissociation has been: (1) positive values for the volume of activation; (2) indepen-

Table 2.12.   EXTENT OF SOLVATION OF cis-Pt(PBu$_3$)$_2$Cl$_2$ IN SEVERAL SOLVENTS AND EQUILIBRIUM CONSTANTS FOR SOLVENT ASSOCIATION[21]

Solvent	n	K
CH$_3$NO$_2$	1.00	21.0
CH$_3$CN	1.00	17.0
(CH$_3$CH$_2$)$_2$O	1.07	3.0
CHCl$_3$	0.99	0.3
CH$_3$OH	2.04	110.0

Table 2.13. ACTIVATION PARAMETERS FOR SQUARE-PLANAR SUBSTITUTION REACTIONS

Reaction	Solvent	k_1			k_2			Reference
		$\Delta H^{\ddagger}$	$\Delta S^{\ddagger}$	$\Delta V^{\ddagger}$	$\Delta H^{\ddagger}$	$\Delta S^{\ddagger}$	$\Delta V^{\ddagger}$	
trans-Pt(Py)$_2$Cl(NO$_2$) + Py	CH$_3$OH	—	—	—	12	−24	−9	6
cis-Pt(Py)$_2$Cl(NO$_2$) + Py	CH$_3$OH	—	—	—	12	−32	−6	6
trans-Pt(PEt$_3$)$_2$Cl$_2$ + Py	CH$_3$OH	—	—	—	14	−25	−14	6
trans-Pt(PEt$_3$)$_2$Cl$_2$ + N$_3^-$	CH$_3$OH	—	—	—	15	−24	—	27
trans-Pt(PEt$_3$)$_2$Br$_2$ + N$_3^-$	CH$_3$OH	—	—	—	13	−24	—	27
trans-Pt(PEt$_3$)$_2$I$_2$ + N$_3^-$	CH$_3$OH	—	—	—	12	−16	—	27
trans-Pt(PEt$_3$)$_2$(2,4,6-Me$_3$C$_6$H$_2$)Br + SC(NH$_2$)$_2$	CH$_3$OH	17	−20	−11	11	−33	−13	28
cis-Pt(PEt$_3$)$_2$(2,4,6-Me$_3$C$_6$H$_2$)Br + I$^-$	CH$_3$OH	20	−14	−16	15	−29	−15	28
cis-Pt(PEt$_3$)$_2$(2,4,6-Me$_3$C$_6$H$_2$)Br + SC(NH$_2$)$_2$	CH$_3$OH	19	−17	−17	14	−29	−13	28
Pt(dien)Cl$^+$ + SCN$^-$	H$_2$O	20	−18	—	10	−25	—	27
Pt(dien)Br$^+$ + SCN$^-$	H$_2$O	19	−17	—	9	−27	—	27
Pt(dien)I$^+$ + SCN$^-$	H$_2$O	—	—	—	10	−25	—	27
Pt(dien)N$_3^+$ + SCN$^-$	H$_2$O	—	—	—	14	−22	—	27
Au(dien)Cl^{+2} + Br$^-$	CH$_3$OH	—	—	—	13	−4	—	27

dence of the rate on the nature of the entering group; (3) saturation kinetics; (4) identical rates for substitution and solvent exchange; and (5) positive values for the entropy of activation.[22]

Dissociative reactions for square-planar organometallic complexes may be more important than for square-planar coordination complexes because a strong donor ligand is important to facilitate dissociation. The characteristics of dissociative processes are quite different from associative reactions. Lack of dependence on the nature of the incoming ligand is shown in the replacement of DMSO from *cis*-Pt(Ph)$_2$(DMSO)$_2$[23]:

$$cis\text{-Pt(DMSO)}_2(Ph)_2 + L\text{-}L \rightarrow cis\text{-Pt(L-L)(Ph)}_2 + 2 \text{ DMSO} \qquad (2.20)$$

$$L\text{-}L = \text{dppe or } o\text{-phen}$$

The rate constant is 2.0 s^{-1} for both dppe and *o*-phen. Such dissociative processes also have a very small solvent dependence.[23]

Dissociation of PPh$_3$ from Pd(Br)(p-tolyl)(PPh$_3$)$_2$ prior to reactions with organotin reagents was indicated by kinetic studies.[24] An inverse first-order dependence on added PPh$_3$ was observed. However, reactions of the Pd complex with the organotin reagents occur at 110°C at rates considerably less than expected for substitution reactions.

Dissociation of ligands from square-planar complexes is also important in ligand redistribution reactions of square-planar complexes[25,26]:

$$\text{Ir(CO)(Cl)L}_2 + \text{Ir(CO)(Me)L}'_2 \rightleftharpoons \text{Ir(CO)(Cl)LL}' \qquad (2.21)$$
$$+ \text{Ir(CO)(Cl)L}'_2 + \text{Ir(CO)(Me)LL}' + \text{Ir(CO)(Me)L}_2$$

$$L, L' = \text{phosphine ligands}$$

Reactions such as Reaction 2.21 appear to be relatively common for square-planar complexes. Phosphine ligand dissociation offers the most reasonable explanation for such redistribution reactions.

2.6. Other Metal Centers

While most of the studies of square-planar substitution reactions have been of Pt(II) complexes a few studies have been accomplished on Pd(II), Ni(II), Au(III), Rh(I), and Ir(I). A study of the CN$^-$ exchange reactions for Pd(CN)$_4^{-2}$, Pt(CN)$_4^{-2}$, Ni(CN)$_4^{-2}$, and Au(CN)$_4^-$ by ^{13}C NMR allows direct comparison of these metal centers.[29]

$$\text{M}(^{13}\text{CN})_4^{-n} + {}^{13}\text{CN} \rightarrow \text{exchange measured by line-broadening}$$

The Ni(II) complex exchanged too rapidly for full analysis; the kinetic data for the other complexes are shown in Table 2.14. As an entering ligand, CN$^-$ is so strong that the rate law showed no k_1 term.

$$\text{rate} = k_2[\text{M(CN)}_4^{-n}][\text{CN}^-] \qquad (2.23)$$

Table 2.14. RATE CONSTANTS AND ACTIVATION PARAMETERS FOR CN^- EXCHANGE WITH $M(CN)_4^{-n}$ AT 24°C[29]

Complex	$k_2(M^{-1}s^{-1})$	$\Delta H^{\ddagger}$ (kcal/mole)	$\Delta S^{\ddagger}$ (eu)
$Ni(CN)_4^{2-}$	$>5 \times 10^5$	—	—
$Pd(CN)_4^{2-}$	120	4	-45
$Pt(CN)_4^{2-}$	26	6	-37
$Au(CN)_4^-$	3900	7	-24

The order of reactivity is (Ni(II) > Au(III) > Pd(II) > Pt(II)), although the difference between Pt(II) and Pd(II) and Au(III) is not as large as it is for other reactions,[29] which will be examined in the following sections.

2.6.1. Pd(II) Complexes

Analogous Pd(II) complexes react approximately 10^5 times more rapidly than do Pt(II) complexes. Data are presented in Tables 2.15 and 2.16. The greater reactivity of the Pd(II) complexes has previously been ascribed to a weaker Pd–X bond. It is generally considered that a higher effective nuclear charge leads to stronger metal ligand bonds (for ligands that do not *accept* electron density from the metal) as one proceeds from first row to second row to third row transition metals. The overall mechanism appears to be of the same type with both a ligand attack and a solvent attack term, although the solvent attack term appears to be more important for Pd(II) complexes than it is for Pt(II).[30]

Substitution on $Pd(acac)_2$ shows the similarity of entering group effects for Pd(II) and Pt(II).[31]

$$Pd(acac)_2 + 4X^- + 2H^+ \rightarrow PdX_4^{2-} + 2H(acac) \tag{2.24}$$

The rates for different X^- groups are shown in Table 2.17. The mechanism suggested for the Pd dien complexes is very similar to that suggested for Pt dien complexes (see Fig. 2.8).[31] A similar *trans* effect also appears operative in Pd(II) reactions as shown by reaction of $Pd(acac)X_2^-$.

$$Pd(acac)X_2^- + 2X^- + H^+ \rightarrow PdX_4^{2-} + H(acac) \tag{2.25}$$

Table 2.15. COMPARISON OF THE REACTIVITY OF Pd(II) AND Pt(II) COMPLEXES

Complex	k_{obs}, M = Pd(II)	k_{obs}, M = Pt(II)	Reference
$M(dien)Cl^+$	fast	3.5×10^{-5}	20
$M(dien)I^+$	3.2×10^{-2}	1.0×10^{-5}	20
$M(dien)SCN^+$	4.3×10^{-2}	3.0×10^{-7}	20
$M(dien)NO_2^+$	3.3×10^{-2}	2.5×10^{-7}	20
$trans\text{-}M(PEt_3)_2(o\text{-tolyl})Cl$	5.8×10^{-1}	6.7×10^{-6}	7

Table 2.16. RATE CONSTANTS FOR THE REACTION
OF THE TETRAHALIDES OF Pt AND Pd WITH
ACETYLACETONATE[30]

Complex	$k_1(s^{-1})$	$k_2(M^{-1}s^{-1})$	$T(°C)$
$PdCl_4^{2-}$	71	2.2	25
$PtCl_4^{2-}$	1.2×10^{-2}	1.2×10^{-3}	55
$PdBr_4^{2-}$	210	1.9	25
$PtBr_4^{2-}$	3.0×10^{-1}	1.1×10^{-3}	55

In comparing the rate constants for the solvolysis path, the reaction for $X = Br^-$ was an order of magnitude more rapid than for $X = Cl^-$.[31]

While *cis–trans* isomerization of Pt(II) complexes is a relatively slow process, often requiring elevated temperatures, many Pd complexes readily establish a *cis–trans* equilibrium in solution. A series of complexes, PdL_2X_2, have been investigated by NMR (L = PPh_nMe_{3-n}; X = Cl^-, N_3^-, CN^-).[32–34] Equilibration was achieved in a few minutes at room temperature, with the *cis* isomers generally thermodynamically more stable than the *trans* isomers. The mechanism apparently involves solvent attack on the palladium complex and is greatly catalyzed by the presence of a nucleophile.[32–34]

$$cis\text{-}PdL_2X_2 + L' \rightleftarrows trans\text{-}PdL_2X_2 + L' \tag{2.26}$$

L' = solvent or other nucleophile

Square-planar complexes of palladium show the same characteristics as platinum complexes in their substitution reactions: associative reactions, either attack by solvent or by incoming nucleophile; the *trans* ligand, the leaving group, the entering nucleophile, and the solvent all affect the rate of substitution. The significant difference is that Pd complexes undergo substitution 10^5 times more rapidly than do platinum complexes.

Table 2.17. ENTERING GROUP EFFECTS
ON THE RATE OF SUBSTITUTION OF
$Pd(acac)_2$[30]

X	$k(\times 10^2 s^{-1})$
OH^-	3.2
Cl^-	8.9
Br^-	32
I^-	fast
SCN^-	very fast

2.6.2. Ni(II) Complexes

Many Ni(II) complexes are octahedral. Those that are square planar react very rapidly—Ni(CN)$_4^{2-}$ exchanges CN$^-$ in seconds, which is significant since Ni(CN)$_5^{3-}$ is a known compound.[35] Reaction of *trans*-Ni(PEt$_3$)$_2$(*o*-tolyl)Cl with pyridine showed a two-term rate law.[6]

$$\text{\textit{trans}-Ni(PEt}_3\text{)}_2\text{(\textit{o}-tolyl)Cl + Py} \qquad\qquad (2.27)$$
$$\rightarrow \text{\textit{trans}-Ni(PEt}_3\text{)}_2\text{(\textit{o}-tolyl)Py + Cl}^-$$

rate = k_1[Ni(II)] + k_2[Ni(II)][Py]

The relative rates for the Ni(II), Pd(II) and Pt(II) complexes are 5×10^6, 1×10^5, and 1, respectively.[7] The most comprehensive data on Ni(II) square-planar complexes is for a series of dithiolate ligands using stopped-flow techniques, showing entering group, leaving group and *trans* effect similar to the Pt(II) complexes.[36]

2.6.3. Au(III) Complexes

Although Au(III) is isoelectronic with Pt(II), studies of Au(III) are much less common. The primary complication is the ease of reduction of Au(III) to Au(I) or Au(0). Table 2.18 shows a comparison of rates of Au(III) complexes to Pt(II) complexes.[36] The form of the rate law is the same as for other square-planar complexes, although the solvent attack term is less important. A plot of the observed pseudo first-order rate constants versus concentration of bromide ion is shown in Figure 2.10. This can be compared with Figure 2.7 showing that the solvent attack term (the intercept of these plots) provides a small contribution to the rate for Au(III).[37] Gold complexes react approximately 10^4 times more rapidly than do isoelectronic Pt(II) complexes. This is attributable to the larger positive charge on Au(III), which favors formation of the electron-rich, five-coordinate intermediate.[37]

2.7. Summary

Substitution reactions of square-planar complexes proceed by nucleophilic attack of an incoming ligand on the square-planar complex. A two-term rate law is observed

Table 2.18. COMPARISON OF THE RATES OF SUBSTITUTION REACTIONS OF Au(III) AND Pt(II) COMPLEXES[32]

Reaction	k_1(s^{-1})	k_2(M^{-1}s^{-1})
AuCl$_4^-$ + *Cl$^-$	0.006	1.47
PtCl$_4^{2-}$ + *Cl$^-$	3.8×10^{-5}	0
Au(dien)Cl^{2+} + Br$^-$	<0.5	154
Pt(dien)Cl$^+$ + Br$^-$	8.0×10^{-5}	5.3×10^{-3}

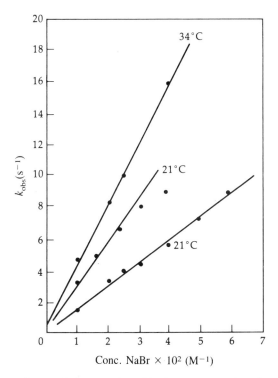

Figure 2.10. Dependence of k_{obs} on bromide ion concentration for the reaction of $[\text{Au(dien)Cl}]^{2+}$ with this reagent. [Reprinted with permission from W. H. Baddley and F. Basolo, *Inorg. Chem.* **3**, 1087 (1964). Copyright 1964 American Chemical Society.]

that includes a term that is dependent on the incoming nucleophile and a term that is independent of the incoming nucleophile. These terms correspond to attack on the complex by the entering ligand and the solvent, respectively. The rate of reaction shows a dependence on the ligand *trans* to the substitution site, on the leaving group, and on the entering nucleophile. The importance of these effects depends on the amount of bond formation or bond breaking in the transition state. One can consider the reaction as a double humped potential curve as shown in Figure 2.11.[10] These considerations were covered in a molecular orbital study.[10] If the first maximum is the transition state, then bond formation will be most important (a very sharp dependence on the entering nucleophile). If the second maximum is the transition state, then bond breaking would be most important (leaving group effects would be very large). Support for the energy profile shown in Figure 2.11, where the five-coordinate, trigonal bipyramidal species is an intermediate, has come from studies showing that this intermediate exists for a sufficient time to undergo pseudorotation.[38,39] The lifetime for $\text{PtCl}_2(\text{I})(\text{PEt}_3)_2^-$ formed by I^- addition to $\text{PtCl}_2(\text{PEt}_3)_2$ was shown to be greater than 10^{-5}s.[40] The data on square-planar substitution

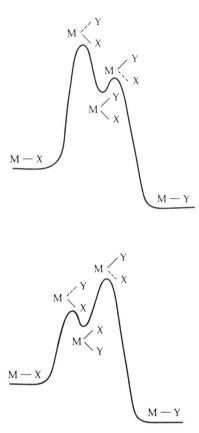

Figure 2.11. Diagrams of the reaction coordinate showing that for a specific reaction the dependence on entering and leaving groups may be more or less important.

reactions are consistent with the mechanistic scheme shown in Figure 2.3, except for the data presented in the section on dissociation from square-planar complexes.

2.8. References

1. F. Basolo and R. G. Pearson, *Mechanisms of Inorganic Reactions* (John Wiley and Sons, Inc., New York, 1968).

2. R. G. Wilkins, *The Study of Kinetics and Mechanism of Reactions of Transition Metal Complexes* (Allyn and Bacon, Inc., Boston, 1974).

3. C. H. Langford and H. B. Gray, *Ligand Substitution Processes* (W. A. Benjamin, Inc., New York, 1965).

4. M. L. Tobe, *Inorganic Reaction Mechanisms* (Nelson, London, 1972).

5. F. Basolo and R. G. Pearson, *Prog. Inorg. Chem.* **4**, 381 (1964).

6. M. Kotowski, D. A. Palmer, and H. Kelm, *Inorg. Chem.* **18,** 2555 (1979).

7. F. Basolo, J. Chatt, H. B. Gray, R. G. Pearson, and B. L. Shaw, *J. Chem. Soc.*, 2207 (1961).

8. F. Basolo, *Adv. Chem. Ser.* **49,** 81 (1965).

9. S. S. Hupp and G. Dahlgren, *Inorg. Chem.* **15,** 2349 (1976).

10. J. K. Burdett, *Inorg. Chem.* **16,** 3013 (1977).

11. A. L. Odell and H. A. Raethel, *J. Chem. Soc., Chem. Commun.,* 1323 (1968).

12. H. B. Gray and R. J. Olcott, *Inorg. Chem.* **1,** 481 (1962).

13. F. Basolo, H. B. Gray, and R. G. Pearson, *J. Am. Chem. Soc.* **82,** 4200 (1960).

14. T. Taylor and L. R. Hathaway, *Inorg. Chem.* **8,** 2135 (1969).

15. R. G. Pearson, *J. Am. Chem. Soc.* **85,** 3533 (1963).

16. R. G. Pearson, H. Sobel, and J. Songstad, *J. Am. Chem. Soc.* **90,** 319 (1968).

17. U. Belluco, L. Cattallini, F. Basolo, R. G. Pearson, and A. Turco, *J. Am. Chem. Soc.* **87,** 241 (1965).

18. U. Belluco, M. Martelli, and A. Orio, *Inorg. Chem.* **5,** 582 (1966).

19. P. D. Braddock, R. Romeo, and M. L. Tobe, *Inorg. Chem.* **13,** 1170 (1974).

20. R. G. Pearson, H. B. Gray, and F. Basolo, *J. Am. Chem. Soc.* **82,** 787 (1960).

21. P. Haake and R. M. Pfeiffer, *Inorg. Chem.* **9,** 5243 (1970).

22. (a) G. Alibrandi, L. M. Scolaro, and R. Romeo, *Inorg. Chem.* **30,** 4007 (1991). (b) R. Romeo, *Comments Inrog. Chem.* **11,** 21 (1990).

23. S. Lanza, D. Minniti, P. Moore, J. Sachinidis, R. Romeo, and M. L. Tobe, *Inorg. Chem.* **23,** 4428 (1984).

24. J. Louie and J. F. Hartwig, *J. Am. Chem. Soc.* **117,** 11,598 (1995).

25. R. L. Rominger, J. M. McFarland, J. R. Jeitler, J. S. Thompson, and J. D. Atwood, *J. Coord. Chem.* **31,** 7 (1994).

26. P. E. Garrou, *Adv. Organomet. Chem.* **23,** 95 (1984).

27. U. Belluco, R. Ettorre, F. Basolo, R. G. Pearson, and A. Turco, *Inorg. Chem.* **5,** 591 (1966).

28. R. van Eldik, D. A. Palmer, and H. Kelm, *Inorg. Chem.* **18,** 572 (1979).

29. J. J. Pesek and W. R. Mason, *Inorg. Chem.* **22,** 2958 (1983).

30. J. V. Rund, *Inorg. Chem.* **13,** 738 (1974).

31. R. G. Pearson and D. A. Johnson, *J. Am. Chem. Soc.* **86,** 3983 (1964).

32. D. A. Redfield, L. W. Cary, and J. H. Nelson, *Inorg. Chem.* **14,** 50 (1975).

33. A. W. Verstuyft and J. H. Nelson, *Inorg. Chem.* **14,** 1501 (1975).

34. A. W. Verstuyft, L. W. Cary, and J. H. Nelson, *Inorg. Chem.* **15,** 3161 (1976).

35. A. W. Adamson, J. P. Welker, and M. Volpe, *J. Am. Chem. Soc.* **72,** 4030 (1950).

36. R. G. Pearson and D. A. Sweigart, *Inorg. Chem.* **9,** 1167 (1970).

37. W. H. Baddley and F. Basolo, *Inorg. Chem.* **3,** 1087 (1964).

38. W. J. Louw, *Inorg. Chem.* **16,** 2147 (1977).

39. M. K. Cooper and J. M. Dounes, *J. Chem. Soc., Chem. Commun.*, 381 (1981).

40. A. Turco, A. Morvillo, U. Vettori, and P. Traldi, *Inorg. Chem.* **24**, 1125 (1985).

2.9. Problems

2.1. Predict the product of the following ligand substitution reactions

a. $Pt(NO_2)Cl_3^{2-} + NH_3 \rightarrow$
b. $Pt(PR_3)_4^{2+} + 2Cl^- \rightarrow$
c. $PtCl_4^{2-} + 2PR_3 \rightarrow$
d. cis-$Pt(Py)_2(NH_3)_2^{2+} + 2Cl^- \rightarrow$
e. $trans$-$Pt(Py)_2(NH_3)_2^{2+} + 2Cl^- \rightarrow$

2.2. Design the two-step syntheses of *cis*- and *trans*-$Pt(NO_2)(NH_3)Cl_2^-$ beginning with $PtCl_4^{2-}$.

2.3. The following data have been obtained for reaction of a 0.0018 M solution of $Pt(dien)Cl^+$ with Br^-.

$[Br^-](M)$	$k_{obs}(10^4 s^{-1})$
0.005	1.04
0.015	1.60
0.020	1.87
0.025	2.12

[H. B. Gray, *J. Am. Chem. Soc.* **84**, 1548 (1962).] Derive k_1 and k_2 for this reaction.

2.4. Arrange the following ligands in order of *trans*-labilizing power in square-planar substitutions.

$$I^-, Br^-, CN^-, C_6H_5^-, H_2O, PR_3$$

2.5. Present rational syntheses of the following three isomers.

2.6. How do each of the following affect the rate of a square-planar substitution reaction?

a. Changing *trans* ligand from H^- to Cl^-.
b. Changing leaving group from I^- to SCN^-.
c. Changing charge on the metal.
d. Making *cis* ligand larger.
e. Changing from H_2O to n-C_3H_7OH as solvent.
f. Changing from Pd(II) to Ni(II).

2.7. A number of Pt(II) complexes are effective antitumor agents [B. Rosenberg, L. Van Camp, J. E. Trosco, and V. H. Mansour, *Nature* **222**, 385 (1969).] Among these are *cis*-Pt(NH$_3$)$_2$Cl$_2$ and Pt(en)Cl$_2$. Assume that the binding to the protein requires two positions and that the Pt-protein complex should be stable and explain why these complexes are reasonable choices for antitumor behavior.

2.8. There has been considerable discussion about the mechanism of *cis–trans* isomerization of square-planar Pt(II) complexes, with both rate-determining dissociative and associative (solvent attack) mechanisms suggested. Given the following data [R. van Eldik, D. A. Palmer, and H. Kelm, *Inorg. Chem.* **18**, 572 (1979)], decide between these possibilities for Pt(PEt$_3$)$_2$(2,4,6-Me$_3$C$_6$H$_2$)Br in CH$_3$OH.

Reaction	$k_1 \times 10^4$	$\Delta H^{\neq}$	$\Delta S^{\neq}$
isomerization	1.90	20	−9.8
cis-Pt(PEt$_3$)$_2$RBr + I$^-$	1.86	18	−15

2.9. Order the rates of substitution reactions of the following complexes:

trans-Pt(NH$_3$)$_2$(H)Cl, *trans*-Pt(NH$_3$)$_2$(NO$_2$)Cl, *trans*-Pt(NH$_3$)$_2$(CN)Cl, Pt(dien)Cl$^+$, Pt(dien)Br$^+$, Pt(dien)I$^+$, Pd(dien)I$^+$, Au(dien)I^{2+}

2.10. Interpret the following data relating the intrinsic reactivity of a complex (log k_s) with the nucleophilic discrimination factor(s) [U. Belluco, L. Cattalini, F. Basolo, R. G. Pearson, and A. Turco, *J. Am. Chem. Soc.* **87**, 241 (1965).]

Complex	log k_S	s
trans-Pt(PEt$_3$)$_2$Cl$_2$	−6.83	1.43
trans-Pt(AsEt$_3$)$_2$Cl$_2$	−5.75	1.25
trans-Pt(pip)$_2$Cl$_2$	−4.56	0.91
Pt(en)Cl$_2$	−4.33	0.64
Pt(dien)Cl$^+$	−3.61	0.65

2.11. A competition study was carried out by reacting Pt(dien)Br$^+$ with Y$^-$ and OH$^-$ in comparable concentrations, where Y$^-$ = Cl$^-$, Br$^-$, I$^-$. The initial product formed in all these reactions is Pt(dien)Y$^+$. No Pt(dien)OH$^+$ is seen initially. Pt(dien)Y$^+$ then converts to Pt(dien)OH$^+$. (Dien is a tridentate amine.) What is the significance of the absence of Pt(dien)OH$^+$ as the initial product? Write a plausible mechanistic scheme in which Pt(dien)OH$^+$ *would* be an initial product (Langford and Gray, p. 37).

2.12. The complex [HPt(PEt$_3$)$_3$]$^+$BPh$_4^-$ was studied in acetone-d^6 in the presence of excess PEt$_3$ using ^{1}H NMR at various temperatures. In the absence of added excess ligand the spectrum for the hydridic proton looks like

Addition of PEt$_3$ causes the triplet to collapse. The collapsing of the spectra is a function of the amount of added ligand. Explain the origin of the spectrum seen for the complex, and account for the effects of added PEt$_3$ in terms of a detailed mechanism.

2.13. The five-coordinate complex $PtCl_2(I)(PEt_3)_2^-$, formed by addition of I^- to $PtCl_2(PEt_3)_2$ was shown to have a lifetime of more than 10^{-5}s (A. Turco, U. Morvillo, U. Vettori, and P. Traldi, *Inorg. Chem.* **24**, 1125 (1985)). How long would the lifetime have to be to affect the kinetics or products?

2.14. Water exchange with $Pt(H_2O)_4^{2+}$ has been open to discussion. Based on the following activation parameters [L. Helm, L. I. Elding, and A. E. Merbach, *Inorg. Chem.* **24**, 1719 (1985)], what is the most likely mechanism? Explain.

$\Delta H^{\ddagger} = 22.5 \pm 0.6$ kcal/mole
$\Delta S^{\ddagger} = -2 \pm 2$ kcal/mole
$\Delta V^{\ddagger} = -4.6 \pm 0.2$ cm^3/mole

2.15. Discuss the following sets of data for NH_3 exchange with the tetraamine complexes of Pd^{2+}, Pt^{2+}, and Au^{3+}.

	$Pd(NH_3)_4^{2+}$	$Pt(NH_3)_4^{2+}$	$Au(NH_3)_4^{3+}$
$k(s^{-1}M^{-1})$	1.6×10^{-2}	9.5×10^{-10}	1.7×10^{-1}
$\Delta H^{\ddagger}$(kcal/mole)	17	31	15
$\Delta S^{\ddagger}$(eu)	-13	4	-13

[B. Brønnum, H. S. Johansen, and L. H. Skibsted, *Inorg. Chem.* **31**, 3023 (1992).]

3

Substitution Reactions of Octahedral Werner-Type Complexes

Complexes of transition metal ions (coordination complexes) were extensively investigated in the early 1900s to test bonding concepts. The data collected on the complexes confirmed Werner's theory of two types of valence; a primary valence representing the charge on the metal (oxidation state) and a secondary valence representing ligands held tightly to the metal ion (coordination number).[1] For coordination number 6, the geometry postulated by Werner, subsequently confirmed, is octahedral. Structural studies of the hexaaquo salts of the first transition series show the octahedral coordination geometry and illustrate periodic trends in M–O bond distance.[2] Figure 3.1 illustrates the trends and emphasizes the Jahn-Teller distortion of $Cr^{2+}(d^4)$ and $Cu^{2+}(d^9)$.[2]

Isomers played an important role in the original interpretation of coordination complexes. A general interpretation of possible isomers can be complicated, but if we allow only two different ligands, then the possibilities can be readily constructed.[3,4] MA_5B has only isomer, MA_4B_2 has two possibilities—*cis* and *trans* isomers—and MA_3B_3 has two possibilities—facial and meridional. These are shown in Figure 3.2. Optical isomers also exist, usually with bidentate ligands. Isomers have had considerable impact on the mechanistic interpretation of substitution reactions of Werner-type complexes.[5-7]

The presence of six electron-donor ligands around the transition metal creates an octahedral electronic field, raising the energy of the d-orbitals by electron repulsion. The ligands lie along the x, y, and z axes, and they have a larger effect on the orbitals that lie along the axes (d_{z^2} and $d_{x^2-y^2}$) than they do on the orbitals that lie between the axes (d_{xy}, d_{xz}, d_{yz}). This *crystal field splitting* is shown in Figure 3.3. The strength of the ligand field influences the magnitude of the splitting of the

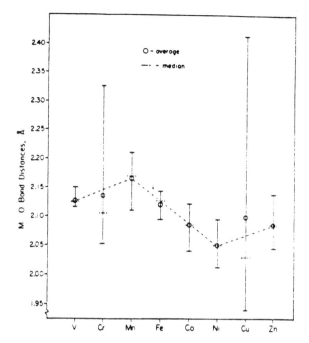

Figure 3.1. M–O bond distances for the hexaaquo complexes of the first-row transition metals. The dashed line gives the median M–O bond distance with the range shown by the vertical lines. [F. A. Cotton et al. *Inorg. Chem.* **32,** 4861 (1993).]

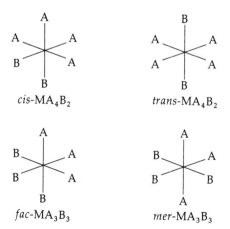

Figure 3.2. Possible stereoisomers of octahedral complexes containing only two different ligands.

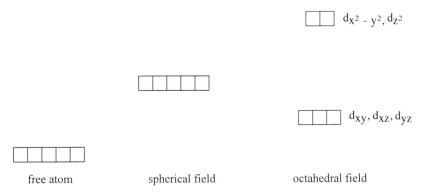

Figure 3.3.   Splitting of the d-orbitals in an octahedral crystal field.

d-orbitals. By measuring the electronic absorption spectra of a number of complexes a spectrochemical series has been developed that is a measure of the ability of the ligand to split the *d*-orbitals. We will call this *ligand strength*.[3,6]

$$I^- < Br^- < Cl^- < F^- < OH^- < acetate < oxalate < H_2O$$
$$< pyridine \sim NH_3 < en < bipy \sim o\text{-phen} < NO_2^- < CN^-$$

Both ligand strength and electronic configuration of the metal are important in determining reactivity of coordination complexes.

Most reactions of Werner-type complexes are studied in water, which may serve as a ligand.

$$M(H_2O)_6^{m+} + L \rightleftarrows M(H_2O)_5L^{m+} + H_2O \qquad (3.1)$$

$$M(H_2O)_5L^{m+} + L \rightleftarrows M(H_2O)_4L_2^{m+} + H_2O$$

$$\vdots \qquad\qquad \vdots$$

$$M(H_2O)L_5^{m+} + L \rightleftarrows ML_6^{m+} + H_2O$$

The forward reactions are referred to as *formation* reactions or, if *L* is an anion, as *anation* reactions; the reverse reactions in which water replaces a ligand are called *aquation* reactions. Reactions in either direction are ligand substitutions and are pertinent for this chapter. The rates of substitution reactions on transition metal complexes range from diffusion controlled to no reaction under ambient conditions. Complexes that react rapidly (half-life of a few seconds) are termed *labile* while those that react slowly (half-life of several minutes) are referred to as *inert*.

3.1. Kinetics

A range of kinetic behavior can be seen for substitution reactions on octrahedral transition metal complexes, depending on the metal ion and ligands.[5-7] The reactions cannot be as comprehensively defined as are substitutions on square planar complexes. The description that follows pertains to most reactions under neutral conditions. Under pseudo first-order conditions, the rate is dependent on the concentration of the entering ligand at low ligand concentration; it is independent of the concentration of L at high concentration of L. This behavior is consistent with the following two reactions (3.2 and 3.3) and the rate law derived from them (3.4).

$$M-H_2O \underset{k_{-1}}{\overset{k_1}{\rightleftarrows}} M + H_2O \tag{3.2}$$

$$M + L \overset{k_2}{\rightleftarrows} M - L \tag{3.3}$$

$$\text{rate} = \frac{k_1 k_2 [M-H_2O][L]}{k_{-1}[H_2O] + k_2[L]} \tag{3.4}$$

Since these reactions are studied in H_2O, the term $k_{-1}[H_2O]$ is a constant. Figure 3.4 shows typical plots of k_{obs} versus [L].[8] There is a dependence on the nature of L that is not surprising since Reactions 3.2 and 3.3 can be considered a competition between H_2O and L for the intermediate M. The limiting rate constant (intercept) is the same for different L. The mechanism described by Equations 3.2 and 3.3 is the dissociative or D mechanism. An I_d (interchange) mechanism (Equations 3.5 and 3.6) can also lead to a rate law,

$$M-H_2O + L \overset{K}{\rightleftarrows} M-H_2O, L \tag{3.5}$$

$$M-H_2O, L \overset{k}{\rightleftarrows} M-L, H_2O \tag{3.6}$$

$$\text{rate} = \frac{kK[M-H_2O][L]}{1 + K[L]} \tag{3.7}$$

which is indistinguishable from that for the D mechanism. To simplify the interpretation of kinetic data, most reactions are examined at one of the limits, where the reaction is expected to be first order (independent of [L]) or second order (showing first-order dependence on [L]). The metal complexes that have been studied in the most detail are those of first-row metals that are relatively inert (Cr(III), Co(III), and Ni(II)).

Rate constants for the first-order path are relatively independent of the nature of the incoming ligand, as shown in Table 3.1 for reactions of $Ni(H_2O)_6^{2+}$.

$$Ni(H_2O)_6^{2+} + L \rightarrow Ni(H_2O)_5L^{2+} + H_2O \tag{3.8}$$

This very small dependence can be compared with that in associative reactions such as substitution on square planar complexes as shown in Table 2.8 where the ligand dependence of the rate constant spans four orders of magnitude. Second-order rate constants for octahedral substitutions show a dependence on the nature of the incoming ligand as shown by the data in Table 3.2. The dependence on the nature of the ligand for the second-order rate constant is not large enough to involve nucleo-

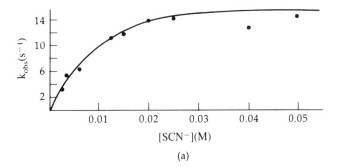

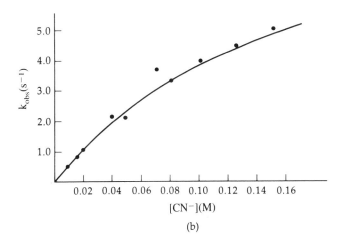

Figure 3.4. The observed rate constant as a function of [L] for (a) SCN^- and (b) CN^- in reaction with a Co(III) hematoporphyrin. [Reprinted with permission from E. B. Fleischer, S. Jacobs, and L. Mestichelli, *J. Am. Chem. Soc.* **90,** 2527 (1968). Copyright 1968 American Chemical Society.]

Table 3.1. RATE CONSTANTS FOR THE REACTION OF $Ni(H_2O)_6^{2+}$ WITH DIFFERENT NUCLEOPHILES[9]

L	$k \times (10^4 \text{ s}^{-1})^*$
SO_4^{2-}	1.5
CH_3COO^-	3
SCN^-	0.6
F^-	0.8
H_2O	3.0
NH_3	3
Py	3

* Defined in Equation 3.7.

Table 3.2. SECOND-ORDER RATE
CONSTANTS FOR ANATION REACTIONS
OF COBALIMIN [Co(III) CORRINOID]
WITH DIFFERENT LIGANDS[10]

L	$k(10^2 M^{-1} s^{-1})$
SCN^-	23
I^-	14
Br^-	10
HSO_3^-	1.7

philic attack of L on the metal complex in an associative reaction (compare with Table 2.8), but is consistent with the ability of the ligand to compete with H_2O for the five-coordinate intermediate in a D mechanism or to interact with the complex in an I_d mechanism.

3.2. Mechanisms

Substantial evidence can be cited that favors a dissociative (either D or I_d) mechanism over an associative mechanism (A or I_a) in some reactions[5,6]: (1) There is often little or no dependence on the entering ligand, as discussed earlier. (2) An increase in charge on the complex leads to a decrease in rate of substitution. An increase in charge would be expected to accelerate an associative reaction with increased attraction for the incoming ligand and more facile acceptance of the negative charge build up on the metal. A dissociative reaction would be slowed by an increase in charge since the dissociating ligand would be held more tightly. As an example, the rate constant for substitution at $Cr(en)(H_2O)_2Cl_2^+$ is $3.1 \times 10^{-5} s^{-1}$ at 25°C and at $Cr(en)(H_2O)_3Cl^{2+}$ is $3 \times 10^{-7} s^{-1}$ at 25°C.[11] (3) Steric crowding increases the rate of substitution reactions. For associative reactions an increase in the crowding around the metal inhibits approach of the incoming ligand and would decrease the rate of reaction. Steric crowding would be relieved by dissociation to yield a five-coordinate intermediate and thus lead to an increase in the rate of dissociation. As an example, the rate of Cl^- replacement on $Co(NH_3)_5Cl^{2+}$ occurs at a rate of $1.7 \times 10^{-6} s^{-1}$ at 25°C, while the rate of Cl^- removal from $Co(NMeH_2)_5Cl^{2+}$ occurs at a rate of 3.7×10^{-4}.[12] Slower rates of aquation for $Cr(NH_2Me)_5Cl^{2+}$ over $Cr(NH_3)_5Cl^{2+}$ could be interpreted as evidence for an associative reaction. However, care must be exercised in interpretation of steric factors (or any other kinetic data!). The Cr–Cl bond is significantly shorter for $Cr(NH_2Me)_5Cl^{2+}$ despite the steric interactions and a dissociative (I_d) mechanism is consistent with the data for $Cr(NH_3)_5Cl^{2+}$ and $Cr(NH_2Me)_5Cl^{2+}$.[13] (4) The rates for different leaving groups correlate with the bond strengths.[14] Data for aquation of $Co(NH_3)_5X^{2+}$ are shown in Table 3.3 and Figure 3.5. (5) The activation parameters are frequently consistent with a dissociative process. The activation parameters have been the focus of

Table 3.3.   RATE AND EQUILIBRIUM
CONSTANTS FOR THE AQUATION REACTION
OF $Co(NH_3)_5X^{2+}$ [14]

X	$k(s^{-1})$	$K_{eq}(M^{-1})$
NCS^-	4.1×10^{-10}	3.7×10^{-4}
N_3^-	2.1×10^{-9}	1.2×10^{-3}
F^-	8.6×10^{-8}	.040
Cl^-	1.7×10^{-6}	0.90
Br^-	6.5×10^{-6}	2.9
I^-	8.3×10^{-6}	8.3
NO_3^-	2.7×10^{-5}	12.0

considerable discussion regarding the distinction between I_d and I_a mechanisms. As shown by the data in Table 3.4, the entropies and volumes of activation are near zero (between ± 10, which, as discussed in Chapter 1, makes interpretation difficult), which is indicative of an interchange mechanism. The early suggestions that transition metal complexes react by dissociative reactions (D) are not supported by the majority of the activation data, although some examples of D mechanisms have been established (e.g., $Co(CN)_5H_2O^{2-}$).[23] The body of evidence for Co(III) complexes is consistent with I_d or D mechanisms. The situation for Cr(III) complexes is more complicated and is perhaps more indicative of the situation for transition metal complexes in general. The kinetic parameters for the anation reactions of $CrL_5H_2O^{3+}$,

$$CrL_5H_2O^{+3} + X^- \rightarrow CrL_5X^{2+} + H_2O \tag{3.9}$$

$$L = H_2O, NH_3$$

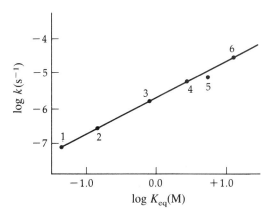

Figure 3.5.   Linear free-energy relationship in the aquation of $Co(NH_3)_5X^{2+}$ at 25°C. Points are designated: (1) $X^- = F^-$; (2) $X^- = H_2PO_4^-$; (3) $X^- = Cl^-$; (4) $X^- = Br^-$; (5) $X^- = I^-$; and (6) $X^- = NO_3^-$. [Reprinted with permission from C. H. Langford, *Inorg. Chem.* **4**, 265 (1965). Copyright 1965 American Chemical Society.]

Table 3.4. ACTIVATION PARAMETERS FOR SUBSTITUTION REACTIONS OF OCTAHEDRAL COORDINATION COMPLEXES

Complex	$\Delta H^{\ddagger}$	$\Delta S^{\ddagger}$	$\Delta V^{\ddagger}$	Ref.	Product
$Cr(H_2O)_5NO_2^{2+}$	19.8	9	—	15	$Cr(H_2O)_6^{3+}$
$Cr(H_2O)_5SO_4^+$	26.5	−1	—	15	$Cr(H_2O)_6^{3+}$
$Cr(H_2O)_5N_3^{2+}$	23.2	−8	—	15	$Cr(H_2O)_6^{3+}$
$Cr(H_2O)_5CN^{2+}$	20.2	−6	—	15	$Cr(H_2O)_6^{3+}$
$Cr(H_2O)_6^{3+}$	30.3	9	—	16	$Cr(H_2O)_5Cl^{2+}$
$Cr(en)(H_2O)_4^{3+}$	27.1	−3	—	17	$Cr(H_2O)_6^{3+}$
$Mn(H_2O)_6^{2+}$	8.1	3	−1.2	18, 19	$Mn(H_2O)_6^{2+}$
$Fe(H_2O)_6^{2+}$	7.7	−3	−4.5	18, 19	$Fe(H_2O)_6^{2+}$
$Co(H_2O)_6^{2+}$	10.4	5	6–10	18, 20	$Co(H_2O)_6^{2+}$
$Ni(H_2O)_6^{2+}$	13.9	9	7	18, 21	$Ni(H_2O)_6^{2+}$
$Co(NH_3)_5(H_2O)^{3+}$	26.6	6.7	1.2	22	$Co(NH_3)_5(H_2O)^{3+}$
$Rh(NH_3)_5(H_2O)^{3+}$	24.6	0.8	−4.1	22	$Rh(NH_3)_5(H_2O)^{3+}$
$Ir(NH_3)_5(H_2O)^{3+}$	28.1	2.7	−3.2	22	$Ir(NH_3)_5(H_2O)^{3+}$
$Co(NH_3)_5Cl^{2+}$	23.3	−6.8	−10.6	22	$Co(NH_3)_5(H_2O)^{3+}$
$Co(NH_3)_5Br^{2+}$	23.2	−3.8	−9.2	22	$Co(NH_3)_5(H_2O)^{3+}$
$Co(NH_3)_5(NCS)^{2+}$	30.1	−0.8	−4.0	22	$Co(NH_3)_5(H_2O)^{3+}$
$Co(NH_3)_5(NO_3)^{2+}$	24.3	1.9	−6.3	22	$Co(NH_3)_5(H_2O)^{3+}$
$Co(NH_3)_5(N_3)^{2+}$	33.2	13.1	16.8	22	$Co(NH_3)_5(H_2O)^{3+}$

are shown in Table 3.5. The dependence on the entering ligand for $L = H_2O$ indicates an I_a mechanism, which is supported by volume of activation measurements on $Cr(H_2O)_6^{3+}$. This large dependence on the entering ligand for $Cr(H_2O)_6^{3+}$ may be contrasted to the very small dependence of the entering ligand in the anation of $Cr(NH_3)_5H_2O^{3+}$. The anation of $Cr(NH_3)_5H_2O^{3+}$ was thus suggested as an I_d mechanism. The data in Table 3.5 indicate that the ligand environment may change the mechanism for substitution at a Cr(III) center. The increased electron density at the chromium for $Cr(NH_3)_5H_2O^{3+}$ relative to $Cr(H_2O)_6^{3+}$ favors dissociative interchange over associative interchange. The volumes of activation for both $Cr(H_2O)_6^{3+}$

Table 3.5. KINETIC PARMAMETERS FOR THE ANATION OF $CrL_5H_2O^{3+}$, $L = H_2O$, NH_3 [24,25]

X	L = H₂O			L = NH₃
	$k(10^8 M^{-1}s^{-1})$	$\Delta H^{\ddagger}$	$\Delta S^{\ddagger}$	$k(10^4 M^{-1}s^{-1})$
NCS^-	180	25	1	4.2
NO_3^-	73	26	1	—
Cl^-	2.9	30	9	0.7
Br^-	1.0	29	2	3.7
I^-	0.08	31	6	—
$CF_3CO_2^-$	—	—	—	1.4

and $Cr(NH_3)_5H_2O^{3+}$ are negative. An explanation has been offered in terms of a dissociative mechanism for entropies and volumes of activation that are near zero or even slightly negative,[21] although active discussion persists on this question.[20] The chemical evidence shown in Table 3.5 is relatively clear cut and provides evidence that I_a and I_d mechanisms are both operative in Cr(III) complexes. The substitution reactions of other Cr(III) complexes have been interpreted in terms of an I_d mechanism.[26]

Volume of activation data were cited to suggest a change in mechanism as one proceeds from early to late first-row transition metals.[19] The volumes of activation for $M(H_2O)_6^{2+}$ are shown in Table 3.6. The $Ti(H_2O)_6^{2+}$ was suggested to exchange with $^{17}OH_2$ by a purely associative mechanism (A),[27] $V(H_2O)_6^{2+}$ and $Mn(H_2O)_6^{2+}$ by associative interchange, $Fe(H_2O)_6^{2+}$ by a pure interchange mechanism and $Co(H_2O)_6^{2+}$ and $Ni(H_2O)_6^{2+}$ by a dissociative interchange mechanism. However, these conclusions have been challenged by a theoretical study that interprets all first-row $M(H_2O)_6^{2+}$ water exchange reactions as proceeding by H_2O dissociation through a $M(H_2O)_5^{2+}$ intermediate.[28]

While this discussion is ongoing, considerable evidence supports dissociative processes for substitutions on some octahedral coordination complexes. To differentiate between a D- and an I_d-type mechanism, one needs to determine whether or not an intermediate of reduced coordination number (five-coordinate) exists in the reaction coordinate.[29] The two possibilities are shown in the following scheme:

$$M(H_2O)_6^{3+} \underset{-X^-}{\overset{+X^-}{\rightleftharpoons}} [M(H_2O)_6 \cdots X^-]^{2+}$$

$$-H_2O \Updownarrow +H_2O \qquad -H_2O \Updownarrow +H_2O \qquad (3.10)$$

$$[M(H_2O)_5^{3+}] \underset{-X^-}{\overset{+X^-}{\rightleftharpoons}} M(H_2O)_5X^{2+}$$

Outer sphere complexes such as $M(H_2O)_6 \cdots X^-$ are established by flow methods, temperature jump data, and ultrasonic absorption,[30-32] while less evidence exists for five-coordinate intermediates. The question of whether these species are intermediates along the reaction coordinate is difficult to answer.[33,34] Evidence for

Table 3.6. VOLUMES OF ACTIVATION OF HEXAAQUO IONS, $M(H_2O)_6^{2+}$ [19,27]

M	$V^{\ddagger}(cm^3/mol)$
Ti^{2+}	−12.4
V^{2+}	−4.1
Mn^{2+}	−5.4
Fe^{2+}	3.8
Co^{2+}	6.1
Ni^{2+}	7.2

$Co(NH_3)_5^{3+}$ as an intermediate in the aquation of $Co(NH_3)_5N_3^{2+}$ was later questioned.[33,35] There was no evidence for $Cr(H_2O)_5^{3+}$ in the water exchange reaction, but there was a good possibility for its formation in aquation of $Cr(H_2O)_5I^{2+}$.[34]

The diversity of transition metal coordination complexes makes unambiguous conclusions about their reactivity difficult. Most complexes apparently react by an I_d mechanism, although D and I_a mechanisms may also be possible for a given complex.

3.3. Leaving Group, Chelate, and Ligand Effects

As expected for a reaction that proceeds by a dissociative mechanism, the rate is dependent on the nature of the leaving group. Data for aquation of $Co(NH_3)_5X^{2+}$ is shown in Table 3.7.[36] The order of leaving group effects

$$NO_3^- > I^- > Br^-, H_2O > Cl^-, SO_4^{2-} > CH_3CO_2^- \qquad (3.11)$$
$$> N_3^-, NCS^-, NH_3 > NO_2^-, OH^-$$

represents the order of complex stability as shown earlier in Figure 3.5.[14] Thus, the leaving group order represents the bond strength order for the ligands. The relationship between leaving group effects and ligand strength is also shown by the similarity between the spectrochemical series and the leaving group order.

Bidentate ligands are substituted more slowly than are monodentate ligands, and complexes with chelate ligands are more stable than are comparable complexes of monodentate ligands; this is termed the *chelate effect*.[36] Kinetics for substitution of bidentate ligands are presented for comparison to the monodentate analogues in Table 3.8.[9] The chelate ligands are more slowly removed. Removing a bidentate ligand can be viewed as a two step process:

Table 3.7. RATES OF AQUATION OF $Co(NH_3)_5X^{2+}$ AT 25°C[36]

X	$k(s^{-1})$
NO_3^-	2.7×10^{-5}
I^-	8.3×10^{-6}
Br^-	6.3×10^{-6}
H_2O	5.8×10^{-6}
Cl^-	1.7×10^{-6}
SO_4^{2-}	1.2×10^{-6}
$CH_3CO_2^-$	1.2×10^{-7}
NH_3	10^{-10}
NCS^-	5×10^{-10}
N_3^-	2.1×10^{-9}
NO_2^-	slow
OH^-	slow

Table 3.8. CHELATE EFFECT ON
REACTIONS OF NICKEL COMPLEXES

Complex	$k_1(s^{-1})$
$Ni(Py)_2$	38.5
$Ni(bipy)$	3.8×10^{-4}
$Ni(NH_3)_2$	5.8
$Ni(en)$	0.27

$$M\big\langle{L \atop L}\big\rangle \underset{k_{-1}}{\overset{k_1}{\rightleftharpoons}} M—L—L \tag{3.12}$$

$$M—L—L \underset{k_{-2}}{\overset{k_2}{\rightleftharpoons}} M + L—L \tag{3.13}$$

A number of explanations have been offered for the chelate effect.[5,6,36] The simplest is that after dissociation of one end of the chelating ligand there is a very high effective concentration of that end near the metal since the chelating ligand is still bound to the metal. Thus, the k_{-1} step is much larger for a chelate than it is for a related monodentate ligand. This simple explanation has been augmented by more detailed consideration for ethylenediamine, and Reaction 3.12 has been expanded to the following scheme[37]:

$$M\big\langle{N \atop N}\big\rangle \rightleftharpoons M\big\langle{N \atop N}\big\rangle \rightleftharpoons M\big\langle{N—{} \atop OH_2}\big| {} \atop N} \tag{3.14}$$

The initial bond-breaking step is slower than it is for dissociation of a unidentate ligand because angular expansion of the chelate is required to lengthen the M–N distance. This leads to a larger value of $\Delta H^{\ddagger}$ for chelate removal, as commonly observed. The water would be prevented from entering the coordination sphere by steric interactions until the nitrogen rotates away from the metal, as shown in the second step. Thus, reversal of the bond breaking (first step) would be facile. More complete analyses have been offered.[36]

Ligand effects for octahedral complexes are not as significant as they are for square-planar complexes, as described in Chapter 2, and they have not been as systematically investigated. The data provided in Table 3.9 show that a small ligand effect exists.[38] Sequential replacement of H_2O by NH_3 gives increasing rates of exchange for the remaining water ligands. The increase is less than a factor of 10 for each NH_3. Ethylenediamine has an effect similar to two NH_3 ligands.[38] In this case the NH_3 ligands may be increasing electron density at the Ni^{2+} center, weakening the $Ni–OH_2$ bonds.

Table 3.9. EFFECT OF SUBSTITUTION OF NH_3 FOR H_2O ON WATER-EXCHANGE RATES[38]

Complex	$k_1(s^{-1})$
$Ni(H_2O)_6^{2+}$	3.6×10^4
$Ni(H_2O)_5(NH_3)^{2+}$	2.5×10^5
$Ni(H_2O)_4(NH_3)_2^{2+}$	6.1×10^5
$Ni(H_2O)_3(NH_3)_3^{2+}$	2.5×10^6
$Ni(H_2O)_4(en)^{2+}$	4.4×10^5
$Ni(H_2O)_2(en)_2^{2+}$	5.4×10^5

3.4. Effect of the Metal

Some metal complexes are labile regardless of the ligand environment, whereas complexes of other metals are inert. This indication of influence of the metal center on reactivity is shown further by the data in Table 3.10 for the rates of water exchange in the hexaaquo complexes.[6] The rates vary enormously with the nature of the metal center. The bond involves an attraction of the pair of electrons on the ligand to the positive charge on the metal. An increase in the charge on the metal would lead to an increased attraction to the ligand and a decreased rate. Thus, the $+3$ ions dissociate H_2O more slowly than $+2$ ions. For the $+2$ hexaaquo ions of the first-row transition metals another explanation is required since the charge would steadily increase across the row and the rate clearly does not correlate with the increasing charge. The $Mn(H_2O)_6^{2+}$, $Fe(H_2O)_6^{2+}$, and $Co(H_2O)_6^{2+}$ complexes illustrate the effect expected from increased effective nuclear charge. Using these com-

Table 3.10. KINETIC PARAMETERS FOR WATER EXCHANGE AT 298 K[6,27,39]

	$k'(s)$	$\Delta H^{\ddagger}(kJ/mole)$	$\Delta S^{\ddagger}(J/Kmol)$	$\Delta V^{\ddagger}(cm^3/mol)$	d-Electron Configuration
$V(H_2O)_6^{3+}$	87	62 ± 1	0 ± 2	-4.1 ± 0.1	t_{2g}^3
$Cr(H_2O)_6^{2+}$	7×10^9				$t_{2g}^4 e_g^3$
$Mn(H_2O)_6^{2+}$	2.1×10^7	33 ± 1	6 ± 5	-5.4 ± 0.1	$t_{2g}^3 e_g^2$
$Fe(H_2O)_6^{2+}$	4.4×10^6	41 ± 1	21 ± 5	3.8 ± 0.2	$t_{2g}^4 e_g^2$
$Co(H_2O)_6^{2+}$	3.2×10^6	47 ± 1	37 ± 4	6.1 ± 0.2	$t_{2g}^5 e_g^2$
$Ni(H_2O)_6^{2+}$	3.2×10^4	57 ± 1	32 ± 3	7.2 ± 0.3	$t_{2g}^6 e_g^2$
$Cu(H_2O)_6^{2+}$	8×10^9				$t_{2g}^6 e_g^3$
$Cr(H_2O)_6^{3+}$	2.8×10^{-6}	110 ± 1	16 ± 4	-9.3 ± 0.3	t_{2g}^3
$Ti(H_2O)_6^{3+}$	1.8×10^5	43.4 ± 0.7	1 ± 2	-12.4	t_{2g}^1
$Fe(H_2O)_6^{3+}$	1.6×10^2	64 ± 2.5	12 ± 7	-5.4 ± 0.4	$t_{2g}^4 e_g^2$
$Ru(H_2O)_6^{3+}$	3.5×10^{-6}	90	-48	-8.3	t_{2g}^5
$Rh(H_2O)_6^{3+}$	4×10^{-8}				t_{2g}^6

plexes as the basis, $V(H_2O)_6^{2+}$ and $Ni(H_2O)_6^{2+}$ are less reactive than expected, and $Cr(H_2O)_6^{2+}$ and $Cu(H_2O)_6^{2+}$ are more reactive than expected. The extra reactivity of $Cr(H_2O)_6^{2+}$ and $Cu(H_2O)_6^{2+}$ is attributed to Jahn-Teller distortion, which results in elongated M–OH$_2$ bonds. The lowered reactivity of $V(H_2O)_6^{2+}$ and $Ni(H_2O)_6^{2+}$ requires further analysis. One would anticipate that electrons in the e_g orbitals that point at the ligands would cause an extra destabilization of the metal–ligand bond. This concept has been semi-quantitatively applied by considering Crystal Field Activation Energies (C.F.A.E.).[5]

A nonspherical ligand geometry splits the d-orbitals, stabilizing some orbitals and destabilizing others.[2,3] The magnitude of the stabilization or destabilization, which is called the *crystal field stabilization energy*, has been derived for several geometries. Since octahedral complexes react, primarily, by dissociative processes, we only need to consider C.F.S.E. for octahedral and square pyramidal geometries. [Trigonalbipyramidal geometry is also possible for the intermediate, but the results are similar to square pyramidal. In each case where there is a difference, square pyramidal is in better agreement with the experimental evidence.[5] Similar results are also obtained for associative processes through seven-coordinate transition states (see Problem 3.3).] The values for different *d*-electron configurations are given in Table 3.11.[6] The C.F.A.E. is derived by subtracting the stabilization in the square pyramidal transition state from the stabilization for the octahedral ground state. Negative values have no significance for reactivity and should be considered as zero.[6] For the hexaaquo complexes that are weak field complexes, the correlation between C.F.A.E. and rate of H$_2$O exchange (Table 3.8) is quite good. It is necessary to realize that the contribution from C.F.A.E. is only one part of the total activation process and can only be applied for complexes with the same ligands and

Table 3.11. CRYSTAL FIELD ACTIVATION ENERGIES FOR SUBSTITUTION REACTIONS ON OCTAHEDRAL COORDINATION COMPLEXES[6]

Electron Configuration	Octahedral CFSE	Square Pyramidal CFSE	CFAE
d^0	0	0	0
d^1	4	4.6	−0.6
d^2	8	9.1	−1.1
d^3	12	10.0	2
d^4 (strong field)	16	14.6	1.4
d^4 (weak field)	6	9.1	−3.1
d^5 (strong field)	20	19.1	0.9
d^5 (weak field)	0	0	0
d^6 (strong field)	24	20.0	4
d^6 (weak field)	4	4.6	−0.6
d^7 (strong field)	18	19.1	−1.1
d^7 (weak field)	8	9.1	−1.1
d^8	12	10	2.0
d^9	6	9.1	−3.1
d^{10}	0	0	0

the same charge. From the C.F.A.E. values one would anticipate that the inert complexes would be of d^3, d^8, and strong field d^6 configurations (i.e., Cr^{3+}, Ni^{2+}, Co^{3+}). This is the experimental observation.

Activation parameters for the *tris*-orthophenanthroline (*o*-phen) complexes that are strong field are also in the order expected from C.F.A.E. Values are shown in Table 3.12.[6] A series of strong field complexes, $M(CN)_6^{3-}$, has also been investigated.[6] The reaction order is

$$V(CN)_6^{3-} > Mn(CN)_6^{3-} >> Cr(CN)_6^{3-} > Fe(CN)_6^{3-} \sim Co(CN)_6^{3-}$$
$$\quad d^2 \qquad\qquad d^4 \qquad\qquad\quad d^3 \qquad\qquad d^5 \qquad\qquad d^6$$

which is very close to the order expected from C.F.A.E. values.

The crystal field activation model correctly predicts the reactivity of metal complexes in both weak field and strong field complexes. There are many approximations in the crystal field activation that can be questioned, but the utility in correlations with reactivity makes it a useful concept.

I have chosen to present the crystal field approach for the preceding analysis because it is relatively simple to remember. Alternate approaches provide a more theoretically accurate description. For example, energies for H_2O loss from $M(H_2O)_6^{2+}$ (Table 3.13) have been calculated by ab initio SCF methods.[40] This approach provides similar information to the crystal field approach.

To summarize the effect of metal on reactivity: (1) High-oxidation state complexes are less reactive than are low-oxidation state transition-metal complexes. (2) The reactivity decreases going down a column for analogous complexes. (3) A larger C.F.A.E. leads to slower reactions.

3.5. Acid and Base Catalysis

Substitution reactions on octahedral coordination complexes have been primarily investigated in aqueous solution, where the pH may vary. Thus, it is important to understand the effect of acids and bases on substitution reactions. In general, acids and bases catalyze substitution reactions of inert complexes.

Table 3.12.   REACTIONS OF STRONG FIELD COMPLEXES

Complex	No. Electrons	E_a (kcal/mole)
$Mn(o\text{-phen})_3^{2+}$	d^5	10.4
$Fe(o\text{-phen})_3^{2+}$	d^6	12.8
$V(o\text{-phen})_3^{2+}$	d^3	21.3
$Ni(o\text{-phen})_3^{2+}$	d^8	25.2

Table 3.13. ENERGIES FOR H_2O
LOSS FROM $M(H_2O)_6^{2+}$ FROM
ab initio SCF CALCULATIONS[40]

M	E(kcal/mole)
V	33.3
Cr	23.5
Mn	27.4
Fe	27.4
Co	27.7
Ni	30.8
Cu	24.5

3.5.1. Acid Catalysis

Substitution in the presence of acid can be considered

$$Cr(H_2O)_5X^{2+} + H_3O^+ \xrightarrow{k_1} Cr(H_2O)_6^{3+} + HX \qquad (3.15)$$

$$Cr(H_2O)_5X^{2+} + H_2O \xrightarrow{k_0} Cr(H_2O)_6^{3+} + X^- \qquad (3.16)$$

as Reaction 3.15 and the amount of rate acceleration can be seen by comparing k_1 to k_0. Such a comparison is shown in Table 3.14. The mechanism of acid catalysis apparently is protonation of the leaving group

$$[(H_2O)_5Cr-F]^{2+}$$
$$\uparrow$$
$$H^+$$

The protonation weakens the M–X bond and facilitates X^- removal as HX. Since NH_3 does not have an electron pair, NH_3 removal is usually not acid-catalyzed. Aquation of $Ru(NH_3)_6^{2+}$ is acid-catalyzed.[41]

$$Ru(NH_3)_6^{2+} + H_2O \xrightarrow{H^+} Ru(NH_3)_5H_2O^{2+} + NH_4^+ \qquad (3.17)$$

Table 3.14. ACID CATALYZED
AQUATION REACTIONS OF $Cr(H_2O)_5X^{2+}$ [5]

X^-	k_0	k_1
N_3^-	2.6×10^{-8}	9.3×10^{-7}
F^-	6.2×10^{-10}	1.4×10^{-8}
CN^-	1.1×10^{-5}	5.9×10^{-4}
NH_3	no acceleration	

The rate depends directly on the hydrogen ion concentration,

$$\text{rate} = k[\text{H}^+][\text{Ru(NH}_3)_6^{2+}] \tag{3.18}$$

Since other hexaamine complexes do not show an acceleration this rate effect was ascribed to protonation of a t_{2g} pair of electrons on the metal as opposed to ligand protonation as described earlier. Thus, the mechanism

$$\text{Ru(NH}_3)_6^{2+} + \text{H}^+ \rightleftharpoons \text{Ru(NH}_3)_6\text{H}^{3+} \tag{3.19}$$

$$\text{Ru(NH}_3)_6\text{H}^{3+} + \text{H}_2\text{O} \xrightleftharpoons{\text{rapid}} \text{Ru(NH}_3)_5(\text{H}_2\text{O})(\text{H})^{3+} + \text{NH}_3 \tag{3.20}$$

$$\text{Ru(NH}_3)_5(\text{H}_2\text{O})\text{H}^{3+} \rightarrow \text{Ru(NH}_3)_5(\text{H}_2\text{O})^{2+} + \text{H}^+ \tag{3.21}$$

accommodates a direct dependence of the rate on the hydrogen ion concentration. It was suggested that protonation of a filled t_{2g} orbital is much more likely for second- and third-row metals where the d-orbitals extend further into the coordination sphere.

Acid-assisted dechelation (removal of a chelate) is the easiest way to remove a chelating ligand. As discussed earlier, chelate complexes are more stable primarily because of the large effective concentration of the chelating ligand. This effective concentration of the dissociated end of the chelate is eliminated by protonation of the free end. This is illustrated for the following carbonate complex:

where protonation inhibits reformation of the Co–O bond.

3.5.2. Base Catalysis

Presence of a base also facilitates substitution reactions.[5,6] These reactions can be modeled by replacement of a ligand by OH^-.

$$\text{Co(NH}_3)_5\text{X}^{2+} + \text{OH}^- \rightleftharpoons \text{Co(NH}_3)_5\text{OH}^{2+} + \text{X}^- \tag{3.23}$$

The rate law for these reactions is different than for other octahedral substitutions,

$$\text{rate} = k[\text{Co(NH}_3)_5\text{X}^{2+}][\text{OH}^-] \tag{3.24}$$

showing a dependence on the concentration of the entering ligand. A number of mechanisms could lead to this rate law, but the accepted mechanism is the conjugate-base mechanism, outlined as follows.[5,6]

$$\text{Co(NH}_3)_5\text{X}^{2+} + \text{OH}^- \underset{k_{-1}}{\overset{k_1}{\rightleftharpoons}} \text{Co(NH}_3)_4(\text{NH}_2)(\text{X})^+ + \text{H}_2\text{O} \tag{3.25}$$

$$Co(NH_3)_4(NH_2)X^+ \xrightarrow{k_2} Co(NH_3)_4(NH_2)^{2+} + X^- \tag{3.26}$$

$$Co(NH_3)_4(NH_2)^{2+} + H_2O \xrightarrow{\text{fast}} Co(NH_3)_5OH^{2+} \tag{3.27}$$

This scheme reduces to the rate law

$$\text{rate} = \frac{K_1 k_2 [Co(NH_3)_5 X^{2+}][OH^-]}{1 + K_1[OH^-]} \tag{3.28}$$

which reduces to the observed rate law if $K_1[OH^-]$ is much less than 1, where K_1 is the equilibrium constant for Reaction 3.25. The key feature of the conjugate-base mechanism is the labilization of X^- by NH_2^-, presumably from stabilization of the transition state by the NH_2^- group.[6]

$$H_2N = M \underset{NH_3}{\overset{NH_3}{\underset{|}{\overset{|}{<}}}} \overset{NH_3}{\underset{NH_3}{<}}$$

The facile deuteration of the NH_3 ligands in the presence of D_2O is consistent with the first step. The dissociative nature of the reaction can be shown since different nucleophiles react at similar rates, steric acceleration is observed, and the enthalpy of the transition state was independent of the nature of the leaving group.[5]

A necessary requirement for the conjugate-base mechanism is that the deprotonated complex undergoes more rapid substitution than does the protonated complex. This cannot be directly checked for the amine complexes; however, comparison of the reactivity of $M(H_2O)_6^{3+}$ and $M(H_2O)_5(OH)^{2+}$ has been accomplished for aquo complexes.[42]

$$M(H_2O)_6^{3+} + *OH_2 \underset{}{\overset{k_{H_2O}}{\rightleftharpoons}} \tag{3.29}$$

$$M(H_2O)_5(OH)^{2+} + *OH_2 \underset{}{\overset{k_{OH}}{\rightleftharpoons}} \tag{3.30}$$

The hydroxo complexes are more reactive with ratios (k_{OH}/k_{H_2O}) of 75 for M = Cr to 750 for M = Fe.[42]

3.6. Stereochemistry of Octahedral Substitution Reactions

The stereochemical course of substitution reactions of octahedral complexes are not as easily generalized as they are for square-planar substitution reactions.[2,5,24] Dissociation of a ligand from an octahedron leads to a five-coordinate species. There is very little energy difference between the two possibilities for five-coordinate com-

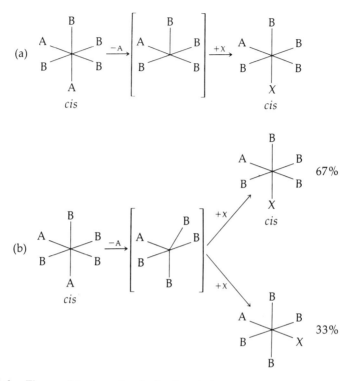

Figure 3.6. The possible stereochemistries for substitution reactions of octahedral Werner-type complexes depending on whether the intermediate is (a) square pyramidal or (b) trigonal bipyramidal.

Table 3.15. STEREOCHEMICAL COURSE OF THE REACTION OF cis-Co(en)$_2(A)(X)$ WITH H$_2$O[6]

A	X	Percentage of cis in Product
OH$^-$	Cl$^-$	100
Br$^-$	Cl$^-$	100
Cl$^-$	Cl$^-$	100
N$_3^-$	Cl$^-$	100
NCS$^-$	Cl$^-$	100
NO$_2^-$	Cl$^-$	100
Cl$^-$	Br$^-$	100

Table 3.16. STEREOCHEMICAL COURSE OF THE REACTION OF *trans*-Co(en)$_2$(A)(X) WITH H$_2$O[6]

A	X	Percentage of *trans* in Product
NO$_2^-$	Cl$^-$	100
NCS$^-$	Cl$^-$	30–50
Cl$^-$	Cl$^-$	65
OH$^-$	Cl$^-$	25
Cl$^-$	Br$^-$	80

plexes, trigonal bipyramidal and square pyramidal. (We will consider this in more detail in Chapter 7.) Both geometries are suggested in different octahedral substitution reactions.[5] A reaction that proceeds through a square pyramidal intermediate would lead to retention of stereochemistry, whereas a reaction proceeding through a trigonal bipyramidal intermediate would lead to a mixture of stereoisomers. These are shown in Figure 3.6 for a *cis* isomer MA_2B_4. In the absence of preferential sites for attack by the incoming ligand, stereochemical retention is an indication of a square pyramidal intermediate, and loss of stereochemistry is evidence for a trigonal bipyramidal intermediate. Data for acid hydrolyses of *cis*-Co(en)$_2$(A)(X) are shown in Table 3.15 and for *trans*-Co(en)$_2$(A)(X) in Table 3.16.[6] The data in Tables 3.15 and 3.16 indicate several features that appear to be general. *Cis* complexes are more likely to proceed with complete stereochemical retention than are *trans* complexes, which often undergo stereochemical loss. Other metal complexes show similar stereochemical behavior with varying amounts of stereochemical retention, depending on the specific complex. *Cis* octahedral substitution reactions often proceed with retention of configuration, and *trans* complexes proceed with some loss, but predictions in the absence of specific data for a given reaction are unreliable.

3.7. Summary

Although it is difficult to generalize for such a large group of complexes as the octahedral coordination complexes, the predominant mechanism of substitution reactions appears to be dissociative interchange. The dissociative mechanism persists even for acid- and base-catalyzed substitution reactions. Substitutional reactivity of octahedral Werner-type complexes depends on: (1) the nature of the leaving group with strongly binding ligands reacting more slowly and with chelate ligands slower than monodentate ligands, (2) the charge on the metal center with more highly charged metals undergoing substitution more slowly, (3) whether the metal is first row, second row, or third row with the reactivity changing first > second > third, and (4) the d-electron configuration of the metal.

3.8. References

1. Several of Werner's papers have been translated and collected in G. B. Kauffman, *Classics in Coordination Chemistry* (Dover Publications, New York, 1968).

2. F. A. Cotton, L. M. Daniels, C. A. Murillo, and J. F. Quesada, *Inorg. Chem.* **32**, 4861 (1993).

3. K. F. Purcell and J. C. Kotz, *Inorganic Chemistry* (W. B. Saunders Company, Philadelphia, 1977).

4. F. A. Cotton and G. Wilkinson, *Advanced Inorganic Chemistry* (Interscience, New York, 1972).

5. R. G. Wilkins, *The Study of Kinetics and Mechanism of Reactions of Transition Metal Complexes* (Allyn and Bacon, Inc., Boston, 1974).

6. F. Basolo and R. G. Pearson, *Mechanisms of Inorganic Reactions* (Wiley and Sons, New York, 1967).

7. M. L. Tobe, *Inorganic Reaction Mechanisms* (Nelson, 1922).

8. E. B. Fleischer, S. Jacobs, and L. Mestichelli, *J. Am. Chem. Soc.* **90**, 227 (1968).

9. R. G. Wilkins, *Acc. Chem. Res.* **3**, 408 (1970).

10. D. Thusius, *J. Am. Chem. Soc.* **93**, 2629 (1971).

11. D. M. Tully-Smith, P. K. Kurimoto, D. A. House, and C. S. Garner, *Inorg. Chem.* **6**, 1524 (1967).

12. D. A. Buckingman, B. M. Foxman, and A. M. Sargenson, *Inorg. Chem.* **9**, 1790 (1970).

13. P. A. Lay, *Inorg. Chem.* **26**, 2144 (1987).

14. A. Haim, *Inorg. Chem.* **9**, 426 (1970).

15. Reference 4, p. 205.

16. J. H. Espenson, *Inorg. Chem.* **8**, 1554 (1969).

17. D. K. Lin and C. S. Garner, *J. Am. Chem. Soc.* **91**, 6637 (1969).

18. Reference 4, p. 222.

19. R. van Eldik, T. Asano, and W. J. le Noble, *Chem. Rev.* **89**, 549 (1989).

20. T. W. Saddle, *Inorg. Chem.* **19**, 3203 (1980).

21. C. H. Langford, *Inorg. Chem.* **18**, 3288 (1979).

22. T. W. Swaddle, *Coord. Chem. Rev.* **14**, 217 (1974).

23. A. Haim and W. K. Wilmarth, *Inorg. Chem.* **1**, 573, 583 (1962).

24. D. Thusius, *Inorg. Chem.* **10**, 1106 (1971).

25. T. Ramasami and A. G. Sykes, *J. C. S. Chem. Comm.* 378 (1978).

26. S.T.D. Lo and D. W. Watte, *Aus. J. Chem.* **28**, 491, 501 (1975).

27. A. D. Hugi, L. Helm, and A. E. Merbach, *Inorg. Chem.* **26**, 1763 (1987).

28. R. Akesson, L.G.M. Petersson, M. Sandström, and U. Wahlgren, *J. Am. Chem. Soc.* **116**, 8705 (1994).

29. C. H. Langford and H. B. Gray, *Ligand Substitution Process* (W. A. Benjamin, New York, 1965).

30. D. W. Carlyle and J. H. Espenson, *Inorg. Chem.* **8**, 575 (1969).

31. H. Brintzinger and G. G. Hammes, *Inorg. Chem.* **5**, 1286 (1966).

32. P. Hemmes and S. Petrucci, *J. Phys. Chem.* **72**, 3986 (1968).

33. R. G. Pearson and J. W. Moore, *Inorg. Chem.* **3**, 1334 (1964).

34. S. P. Ferraris and E. L. King, *J. Am. Chem. Soc.* **92**, 1215 (1970).

35. A. Haim and H. Taube, *Inorg. Chem.* **2**, 1199 (1963).

36. R. B. Jordan, *Reaction Mechanisms of Inorganic and Organometallic Systems* (Oxford University Press, New York, 1991).

37. D. W. Margerum, G. R. Cayley, D. C. Weatherburn, and G. K. Pagenkopf, *American Chemical Society Monograph Series* **174**, 1 (1970).

38. (a) A. G. Desai, H. W. Dodgen, and J. P. Hunt, *J. Am. Chem. Soc.* **92**, 798 (1970). (b) ibid. **91**, 5001 (1969).

39. (a) A. E. Merbach, *Pure Appl. Chem.* **54**, 1479 (1982). (b) ibid. **59**, 161 (1987).

40. R. Akesson, L.G.M. Pettersson, M. Sandström, P.E.M. Siegbahn, and U. Wahlgren, *J. Phys. Chem.* **97**, 3765 (1993).

41. P. C. Ford, J. K. Kuempel, and H. Taube, *Inorg. Chem.* **7**, 1976 (1968).

42. F.-C. Xu, H. R. Krouse, and T. W. Swaddle, *Inorg. Chem.* **24**, 267 (1985) and references therein.

3.9. Problems

3.1. The rates of substitution at cobalt(III)

$$Co(NH_3)_5X^{2+} + Y^- \rightarrow Co(NH_3)_5Y^{2+} + X^-$$

are usually independent of the group Y^-. The primary exception is $Y = OH^-$ where the rates are much faster than for other Y^- and for which there is a first-order dependence on the $[OH^-]$. Explain these observations in terms of mechanistic considerations.

3.2. Order the rates of substitution reactions for the following complexes:

$$Cr(NH_3)_6^{3+}, \ Co(NH_3)_6^{3+}, \ Rh(NH_3)_6^{3+}, \ Ir(NH_3)_6^{3+}, \ Mn(H_2O)_6^{2+},$$
$$Ni(H_2O)_6^{2+}, \ Cu(H_2O)_6^{3+}, \ Cr(H_2O)_6^{3+}$$

3.3. In associative reactions of octahedral complexes a pentagonal bipyramid is a possible intermediate. Given the following data derive the order of reaction rates expected for the different electron configurations in an octahedral to pentagonal bipyramid reaction scheme.

PENTAGONAL BIPYRAMID CRYSTAL
FIELD STABILIZATION ENERGIES

	Strong Field	Weak Field
d^0	0	0
d^1	5.28	5.28
d^2	10.56	10.56
d^3	7.74	7.74
d^4	13.02	4.93
d^5	18.30	0.0
d^6	15.48	5.28
d^7	12.66	10.56
d^8	7.74	7.74
d^9	4.93	4.93
d^{10}	0	0

3.4. There has been active discussion of the significance of the volumes of activation in mechanistic differentiation. Two points of view are presented in Correspondences to Inorganic Chemistry. (1) T. W. Swaddle, *Inorg. Chem.* **19**, 3205 (1980). (2) C. H. Langford, *Inorg. Chem.* **18**, 3288 (1979). Briefly summarize the primary features of these arguments.

3.5. Substitution reactions are not always straightforward. Reactions of $Ru(NH_3)_4$-$(P(OEt)_3)(H_2O)^{2+}$ with different entering ligands show the following dependence. (D. W. Franco and H. Taube, *Inorg. Chem.* **17**, 571 (1978).) Interpret these data.

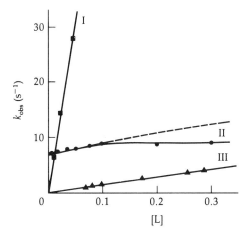

Dependence of k_{obs} on ligand concentration for the reaction of *trans*-$[Ru(NH_3)_4P(OEt)_3$-$(H_2O)]^{2+}$ with: (I) SO_3^{2-}, ■; (II) Me/Pyr$^+$, ●; (III) imN, ▲. [Reprinted with permission from D. W. Franco and H. Taube, *Inorg. Chem.* **17**, 571 (1978). Copyright 1978 American Chemical Society.]

3.6. Base hydrolysis data for $Co(NH_3)_5X^{2+}$ show very little dependence of the activation enthalpy on the leaving group X. [D. A. House and H.K.J. Powell, *Inorg. Chem.* **10**, 1583 (1971).]

X	$\Delta H^{\ddagger}$
Cl$^-$	26.6
Br$^-$	27.0
I$^-$	27.2
NO$_3^-$	28.2

Explain how these data are consistent with a dissociative mechanism.

3.7. Given the following reactions that describe chelate formation:

$$Ni(H_2O)_6^{2+} + L\text{-}L \rightleftarrows [Ni(H_2O)_6, L\text{-}L]\ K_0, \text{rapid}$$

$$[Ni(H_2O)_6, L\text{-}L] \rightleftarrows Ni(H_2O)_5L\text{-}L + H_2O\ K_2, k_2, k_{-2}$$

$$Ni(H_2O)_5L\text{-}L \rightleftarrows Ni(H_2O)_4L\text{-}L + H_2O\ K_3, k_3, k_{-3}$$

derive the rate law for chelate formation and discuss the step that is different for chelate ligands instead of two monodentate ligands. [R. G. Wilkins, *Acc. Chem. Res.* **3**, 408 (1970).]

3.8. Decarboxylation reactions of *trans*-Rh(en)$_2$(X)(OCO$_2$) show a pH dependence as follows at 25°C: [R. van Eldik, D. A. Palmer, H. Kelm, and G. M. Harris, *Inorg. Chem.* **19**, 3679 (1980).]

pH	$k_{obs} s^{-1}$
0.30	0.48
0.60	0.51
1.30	0.51
2.00	0.55
3.25	0.58
3.90	0.58
4.88	0.59
5.68	0.54
6.00	0.44
6.33	0.36
6.45	0.31
6.75	0.22
6.88	0.17

Suggest a mechanism to explain this pH dependence.

3.9. The kinetic *trans* effect for acid hydrolysis of *trans*-Co(cyclam)ACl$^+$ complexes was studied, where A = various anions and Cl$^-$ is the leaving group. Hammett-type substituent constants based on Cl$^-$ = 0.0 were assigned from the observed kinetics; $\sigma(\text{Cl}^-) = 0.0$; $\sigma(\text{NCS}^-) = -3.00$; $\sigma(\text{OH}^-) = 4.04$. The first-order rate constant for hydrolysis of the Co(cyclam)Cl$_2^+$ complex is 3.5×10^{-3} sec^{-1}.

a. Estimate the rate constants for the Co(cyclam)(NCS)Cl$^+$ and Co(cyclam)(OH)Cl$^+$ complexes.

b. Account for the differences in rates in terms of the electronic properties of the substituents.

3.10. In the reaction

$$\text{Cr(H}_2\text{O)}_6^{3+} + \text{SCN}^- \rightarrow \text{Cr(H}_2\text{O)}_5\text{NCS}^{2+} + \text{H}_2\text{O}$$

the rate law is of the form

$$d[\text{CrNCS}^{2+}]/dt = [\text{Cr}^{3+}][\text{SCN}^-][k_1 + k_2/(\text{H}^+)]$$

(waters omitted for convenience)

a. Assuming sufficiently acid conditions so that only the first term in the rate law is important, does the form of the rate law require that the anation process proceed by an associative mechanism? Explain. (For example, if your answer is *no*, show how a stoichiometric mechanism other than associative could lead to the observed rate law.)

b. Account for the existence of the second term in the rate law.

3.11. Explain, using crystal field theory, how the crystal field activation energies would differ for dissociation of a ligand from a low spin Fe(II) as compared with that of a high spin Fe(II) complex.

3.12. Using linear free energy relationships, and citing appropriate literature, present an argument that suggests that substitution of $Co(NH_3)_5X^+$ complexes is dissociative in character.

Organometallic Substitution Reactions

Organometallic chemistry has enjoyed tremendous growth. The use of organo-metallic complexes to model heterogeneous catalyst systems and as homogenous catalysts themselves provides a practical reason for studying fascinating complexes with very different properties than Werner-type complexes. The development of organometallic chemistry has followed that of many other areas of chemistry: first are synthetic studies, next structural studies, and then studies of reactivity.

Organometallic compounds are defined as complexes that contain a M–C bond. The presence of the M–C bond creates a number of very unusual properties for transition metal complexes: volatility (many complexes may be purified by sublimation), solubility in hydrocarbon solvents, and very low oxidation states for the metal (as low as −4). Two features of organometallic chemistry are especially important to our discussion of substitution reactions, ligand bonding and electron counting.

4.1. Ligand Bonding

The nature of the low-oxidation state metal center necessitates ligands with different bonding properties than those commonly observed for Werner-type complexes. The bonding of carbon monoxide, illustrated in Figure 4.1, shows the dual bonding that is important to organometallic complexes. There is donation of the electron density on C to the metal, a σ donation. There is also donation of electron density from the filled d-orbitals on the metal into the CO antibonding orbitals, π-back bonding. Each type of bonding reinforces the other (σ bonding increases the electron density on the metal, which enhances the π-back bonding), creating a synergic interaction.

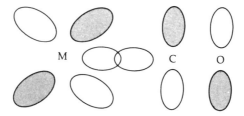

Figure 4.1. Bonding diagram for CO to a transition metal showing both the σ donation and π acceptance by CO.

Back-bonding into the CO antibonding orbital weakens the C–O bond, as shown by a lengthening of the C–O bond and a decrease in the infrared stretching frequencies of carbonyls with respect to carbon monoxide. Synergic or dual binding is also observed for phosphorus bases and ethylene as shown in Figure 4.2. The σ bonding of ethylene involves donation of the bonding electron density of the ethylene to the metal and back-bonding into the antibonding orbital of ethylene. Phosphorus bases [phosphites ($P(OR)_3$) and phosphines (PR_3)], may utilize the d-orbitals to accept π-electron density. There has been much discussion in the literature about the relative amounts of π bonding and σ bonding in a given ligand. A number of techniques including infrared spectroscopy,[1] ^{13}C and ^{31}P NMR,[2,3] molecular mechanics[4] and reaction chemistry[5] have been used to assess the binding capabilities of phosphorus ligands. A summary including electronic and steric factors for 69 li-

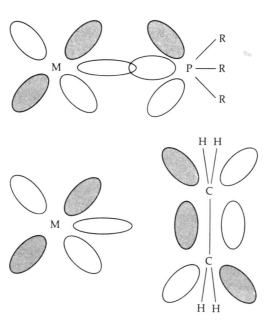

Figure 4.2. Bonding schemes for phosphines and ethylene in organometallic complexes.

gands has been published.[4] The following ordering is generally accepted for the ligands commonly used in organometallic substitution reactions:
for σ-donation

$$PBu_3 > P(OMe)_3 > PPh_3 > NR_3 > P(OPh)_3 > CO$$

and for π-acceptance

$$CO > P(OPh)_3 > P(OMe)_3 > PPh_3 > PBu_3 > NR_3$$

The bonding of phosphorus ligands has been an active area of discussion for a number of years with the relative degrees of σ donation and π acceptance of electron density the focus of the discussion. It has been suggested that phosphines, PR_3, do not accept electron density from a metal.[6] This may be correct for alkyl phosphines (PMe_3, PEt_3, PBu_3, etc.), although NMR data on $W(CO)_5PMe_3$ have been interpreted as requiring π acceptance by PMe_3.[7] The predominant bonding of alkyl phosphines is certainly from the very strong electron donation. For these ligands the most correct statement is that alkyl phosphines have a d-orbital that *may* accept electron density. For aryl phosphines such as PPh_3, however, π acceptance appears to be significant. The most straightforward evidence is from the $Cr–PPh_3$ bond lengths of *trans*-$Cr(CO)_4LPPh_3$ [$L = PBu_3$, $P(OMe)_3$, $P(OPh)_3$, CO] shown in Table 4.1.[8] The shortening of the $Cr–PPh_3$ bond as the donor strength of the *trans* ligand increases is most easily accommodated in terms of π acceptance of PPh_3.

Any discussion of the electronic binding of these ligands is complicated by steric interactions of the rather bulky phosphine and phosphite ligands. The relative size of the ligands is[9]:

$$PPh_3 > PBu_3 > P(OPh)_3 > P(OMe)_3 > CO$$

Each of these ligand parameters is important in discussing reactivity of organometallic compounds. As we proceed through our discussion of reactions of organometallic complexes, we will often refer to the σ-bonding ability, π-accepting ability, or steric size. A graphical method of separating σ, π, and steric effects (quantitative analysis of ligand effects, QALE) has been used.[10]

The majority of organometallic complexes have 18 electrons in the valence shell of the metal. The stable binary metal carbonyls shown in Table 4.2 form a nice series illustrating the importance of the 18-electron rule.[11] Electron counting is a very useful concept for predicting stable complexes. As we consider reactions of

Table 4.1. *trans* EFFECT OF A PHOSPHINE LIGAND ON $Cr–PPh_3$ BOND LENGTHS

Ligand *trans*	$Cr–PPh_3$ Bond Length (Å)
PBu_3	2.345(3)
$P(OMe)_3$	2.362(2)
$P(OPh)_3$	2.393(1)
CO	2.422(1)

Table 4.2. SELECTED BINARY METAL CARBONYLS
OF THE FIRST ROW AND THEIR ELECTRON COUNT

$Cr(CO)_6$		$Mn_2(CO)_{10}$		$Fe(CO)_5$		$Co_2(CO)_8$		$Ni(CO)_4$	
Cr°	$6e^-$	Mn°	$7e^-$	Fe°	$8e^-$	Co°	$9e^-$	Ni°	$10e^-$
6CO	$12e^-$	5CO	$10e^-$	5CO	$10e^-$	4CO	$8e^-$	4CO	$8e^-$
	$18e^-$	Mn–Mn bond	$1e^-$		$18e^-$	Co–Co bond	$1e^-$		$18e^-$
			$18e^-$				$18e^-$		

these complexes, the electron count in the starting complex and the intermediate will be useful in determining the reaction type. It has been proposed that organometallic reactions occur by $16 \rightleftarrows 18$ electron sequences.[11] While there are exceptions to this rule, it is quite valuable in discussing organometallic reactions.

4.2. Metal Carbonyl Substitution Reactions

The majority of studies of organometallic substitution reactions have been accomplished on transition metal carbonyl complexes.[12] These complexes are readily prepared and purified, are stable and have very characteristic infrared spectra that allow quantitative analysis. The reactions can be represented

$$M(CO)_n + L \longrightarrow M(CO)_{n-1}L + CO \qquad (4.1)$$

where M is a transition metal and L is an entering ligand, usually a phosphine or phosphite. Two examples that were among the first to be investigated kinetically were $Ni(CO)_4$ and $Mn(CO)_5Br$.[13,14]

$$Ni(CO)_4 + L \longrightarrow Ni(CO)_3L + CO \qquad (4.2)$$

$$Mn(CO)_5Br + L \longrightarrow cis\text{-}Mn(CO)_4LBr + CO \qquad (4.3)$$

Kinetic data, which are shown in Table 4.3, indicate that the rate law is independent of the concentration of L and first order in metal carbonyl complex. Plots of $ln\,A$ versus time (A = absorbance of infrared stretching mode) were linear for several half-lives confirming the first-order behavior in metal complex. The rate law is illustrated for $Ni(CO)_4$:

$$\text{rate} = k_1[Ni(CO)_4] \qquad (4.4)$$

The accepted mechanism involves rate-determining CO dissociation from the 18-electron complex to form a 16-electron intermediate,[13]

Table 4.3. KINETIC DATA FOR REACTION
4.3 AT 30°C IN $CHCl_3$

L	[L], M	$k(10^5 s^{-1})$
PPh_3	0.133	6.7
PPh_3	0.344	6.8
$AsPh_3$	0.132	6.6
$SbPh_3$	0.135	6.6

$$Ni(CO)_4 \xrightarrow{k_1} Ni(CO)_3 + CO \tag{4.5}$$

$$Ni(CO)_3 + L \xrightarrow{fast} Ni(CO)_3L \tag{4.6}$$

such that the substitution may be considered to be a two step process. The second step (Reaction 4.6) is much more rapid, occurring at a rate of 10^6 s^{-1}M^{-1}, in contrast to the first step (Reaction 4.5), which occurs at a rate of 10^{-4}s^{-1}.[15] The activation parameters were in agreement with a dissociative process with a positive entropy of activation, as shown in Table 4.4.

Substitution of several metal carbonyl complexes shows a small dependence on the nature and concentration of the entering ligand.[18,19]

$$Cr(CO)_6 + L \longrightarrow Cr(CO)_5L + CO \tag{4.7}$$

$$Mo(CO)_5Am + L \longrightarrow Mo(CO)_5L + Am \tag{4.8}$$

Am = amine

The rate law for these substitutions, under pseudo first-order conditions, had two terms, as shown for $Cr(CO)_6$:

$$\text{rate} = k_1[Cr(CO)_6] + k_2[Cr(CO)_6][L] \tag{4.9}$$

The second-order term was always much smaller than the first-order term.[18,19] The original suggestion that this rate law indicated competing ligand attack on the metal and CO dissociation,[18]

Table 4.4. ACTIVATION PARAMETERS FOR THE
REACTION OF $Ni(CO)_4$ WITH NUCLEOPHILES[13,16,17]

L	$\Delta H^{\ddagger}$ (kcal/mole)	$\Delta S^{\ddagger}$ (eu)	$k(10^4 s^{-1})$ at 20°C
PPh_3	24	13	50
$C^{18}O$	24	14	52

$$Cr(CO)_6 \xrightarrow{k_1} Cr(CO)_5 + CO \tag{4.10}$$

$$\downarrow L$$

$$Cr(CO)_6 + L \xrightarrow{k_2} Cr(CO)_5L + CO$$

is better described by a dissociative interchange (I_d) mechanism that has been suggested for the $Mo(CO)_5Am$ system.[19] This mechanism is outlined as shown.[19]

$$M(CO)_n \underset{k_{-f}}{\overset{k_f}{\rightleftarrows}} <M(CO)_{n-1}S, CO> \underset{k_{-10}}{\overset{k_{10}}{\rightleftarrows}} M(CO)_{n-1}S + CO \tag{4.11}$$

$$M(CO)_{n-1}S + L \longrightarrow M(CO)_{n-1}L + S \tag{4.12}$$

$$M(CO)_n + L \underset{k_{-i}}{\overset{k_i}{\rightleftarrows}} <M(CO)_n, L> \xrightarrow{k_2} M(CO)_{n-1} L + CO \tag{4.13}$$

Here, S represents the solvent, and the angular brackets enclose solvent-encased substrate and a species occupying a favorable site for exchange. These equations reduce to the observed rate law. This process seems more reasonable than the concurrent formation of 16- and 20-electron intermediates that would be required in competing dissociative and associative steps. Thus, the body of evidence for the simple metal carbonyls indicates CO dissociation is the mechanism of ligand substitution reactions.

4.2.1. Metal Effects on Reactivity

The effect of the metal center on organometallic reactivity is not as clearly defined as it is for coordination complexes. We will examine (1) the effect of charge, (2) first-row, second-row, and third-row effects, and (3) the effect of d-electron count. As will be discussed, the enhanced reactivity of odd-electron complexes is a major effect of the metal center.

4.2.1.1. Effect of Charge

The effect of charge on M–CO bonding has been frequently examined, with the extent of π-acceptance being greater as the charge on the metal is lowered. The effect of charge on reactivity is not as clear. $V(CO)_6^-$, $Cr(CO)_6$, and $Mn(CO)_6^+$ are relatively inert to CO dissociation. Similarly, $Mn(CO)_5^-$ and $Fe(CO)_5$ are relatively inert, although the fact that $Co(CO)_5^+$ is unknown may indicate that it would have considerable reactivity. The best indication of the expected effect of charge on reactivity is in the greater reactivity of $Ni(CO)_4$ than $Co(CO)_4^-$ (ready substitution versus no reaction in 48 h).[20]

The failure (in many cases) to see the expected large increase in rate of CO loss with increase in charge must indicate that the destabilization of the ground state is negated by a similar destabilization of the transition state. This suggests that the effect of charge on the bonding of CO in the 16-electron transition state for CO dissociation is similar to that for the 18-electron ground state.

4.2.1.2. Row Effects

In general, reactivity is greatest for the second-row metal and least for the third-row metal. This order is different from the order of reactivity observed for classical coordination complexes, for which the reactivity decreases down a row for analogous complexes. The data shown in Table 4.5 show the effect for two homologous series.[18,21] In each case described in Table 4.5 the reaction proceeds by dissociation of a ligand. In general for organometallic substitution reactions the second-row complex reacts most rapidly, but it is more difficult to generalize between first- and third-row complexes. The data in Table 4.5 illustrate this difficulty. For carbonyl complexes the third-row complex usually reacts more slowly in dissociative reactions.

4.2.1.3. Electronic Effects

The effect of changing the metal across a row has not been thoroughly investigated, although the reactivity changes significantly. From qualitative, semi-quantitative and quantitative data the following order of reactivity for the metal carbonyl complexes can be derived.

$$Co(CO)_4, Mn(CO)_5 > V(CO)_6 > Ni(CO)_4 > Ti(CO)_3(P\ P)_2$$
$$> Cr(CO)_6 > Fe(CO)_5$$

This order spans approximately 10 orders of magnitude in rate, indicating a large change in reactivity with the metal center, which is comparable to that observed for Werner-type complexes. For analogous compounds of chromium, iron, and nickel, which have been quantitatively investigated, the order of reactivity is

$$Ni(CO)_2L_2 \gg Cr(CO)_4L_2 \gg Fe(CO)_3L_2$$

Table 4.5. RATE CONSTANTS FOR SUBSTITUTION REACTIONS OF $M(CO)_6$ AND $M(P(OEt)_3)_4$[18,21]

$M(CO)_6$	$k_{obs}(130°C)^a$	$M(P(OEt)_3)_4$	$k_{obs}(40°)^b$
$Cr(CO)_6$	1.4×10^{-4}	$Ni(P(OEt)_3)_4$	3.7×10^{-6}
$Mo(CO)_6$	2.0×10^{-3}	$Pd(P(OEt)_3)_4$	1200
$W(CO)_6$	4.0×10^{-6}	$Pt(P(OEt)_3)_4$	0.3

^{a} Reaction with PPh_3. Values for Mo and W are estimated from other temperatures.
^{b} For Ni the reaction was with cyclohexylisocyanide; for Pd and Pt the values were estimated from other temperatures for exchange reactions.

Table 4.6.   RATE CONSTANTS FOR LIGAND
DISSOCIATION FROM $Cr(CO)_4L_2$, $Fe(CO)_3L_2$,
AND $Ni(CO)_2L_2$ AT 25°C[22]

L	$Cr(CO)_4L_2$	$Fe(CO)_3L_2$	$Ni(CO)_2L_2$
CO	1.9×10^{-12}	—	1.2×10^{-2}
$P(OPh)_3$	4.6×10^{-11}	4.8×10^{-12}	1.0×10^{-7}
PPh_3	3.1×10^{-6}	5.0×10^{-11}	5.6×10^{-4}

with around 10^8 difference in rate between Ni and Fe complexes. Data for different
ligands are shown in Table 4.6.[22] In each case the reaction proceeds by first-order
kinetics and has been assigned to ligand dissociation. No data exist which suggest
that the reactivity order shown in Table 4.6 arises predominantly by ground-state
energy differences.[22] A possible explanation of the variation of the rate of substitution with the metal has been offered in terms of the crystal field activation energies
for these substitution reactions.[22] A primary uncertainty in applying crystal field
activation energies is the geometry of the transition state. Values of the crystal field
activation energy for substitution reactions of the first-row mononuclear carbonyl
complexes are shown in Table 4.7 for several possible geometries. Since Ni(0) is
d^{10} there would be no CFAE. For Cr the predominant evidence is that the intermediate would be square pyramidal. Allowing the geometry of $Fe(CO)_3L$ to be approxi-
mated as a tetrahedron gives an excellent inverse correlation of CFAE with reactivity. The geometry suggested by matrix isolation for $Fe(CO)_4$ is a C_{2v} distorted
tetrahedron.[23] Replacement of one CO with a more sterically demanding ligand
should favor tetrahedral geometry. The observed dependence of rates of ligand
substitution on the metal [Ni(0) > Cr(0) > Fe(0)] can be accounted for by the
crystal field activation model.

For substitution reactions of the Ru and Os complexes the d^8, 16-electron transition state/intermediate should tend toward square planar and be considerably more
reactive than the analogous iron complexes. This is observed; while iron com-

Table 4.7.   CRYSTAL FIELD ACTIVATION ENERGY (IN D_q) FOR A DISSOCIATIVE
MECHANISM IN ORGANOMETALLIC SUBSTITUTIONS

Complex	System	Ground State		Transition State		
		Geometry	CFSE[a]	Geometry	CFSE[a]	CFAE
$Ni(CO)_2L_2$	d^{10}	T_d	0	Trigonal Planar	0	0
$Cr(CO)_4L_2$	d^6	0_h	-24	SqPy	-12	4
$Fe(CO)_3L_2$	d^8	TBP	-14.6	T_d	-4.6	10
				Sq. Pl.	-25	-10

[a]   All values are given in D_q.

plexes, $Fe(CO)_3L_2$, are much slower than are Cr complexes, analogous Ru complexes react more rapidly than do Mo complexes, and Os complexes react more rapidly than do W complexes.[22]

Table 4.8 is composed of dissociative reactions for quite different classes of compounds. These data indicate several features that appear to be general. (1) Organometallic complexes with a d^{10} configuration are relatively labile. (2) Complexes with a d^6 configuration are relatively inert. (3) Dissociative reactivity apparently follows the order $d^{10} > d^4 > d^6 > d^8$ for a homologous series.

Comparison of analogous compounds shows no evidence that the observed dependence of the rate on the metal center arises from ground state properties. Thus, a primary factor in the metal center reactivity is the transition state. This is similar to the conclusions regarding substitutional reactivity of classical coordination complexes, suggesting that similar interpretations may be possible.

4.2.2. Solvent Effects

In contrast to substitution reactions of Werner-type complexes where solvent effects are quite important, solvent effects in metal carbonyl substitution reactions are normally very small, and are usually within a factor of 10. This is understandable since a neutral complex dissociates a neutral ligand (CO) leading to a neutral intermediate; solvation changes should be minimal during the course of the reaction. The rate constants for substitution on $Mn(CO)_5Br$ in various solvents are shown in Table 4.9.[14] The most rapid reactions are observed in hydrocarbon solvents. This was interpreted as an indication that the transition state is less polar than is the ground state.[14] An alternate explanation lies in the cohesive forces (solvent–solvent interactions), which can affect the reaction rate in a manner similar to the way external pressure affects reaction rate. The cohesive energy density (ced) is a measure of the internal pressure of the solvent. For nonpolar reactions the internal

Table 4.8. RATES OF REACTIONS AT 30°C
FOR DIFFERENT ORGANOMETALLIC COMPLEXES
COMPARED WITH THE NUMBER OF d-ELECTRONS[24]

Compound	Rate	d-Electrons
$CpMn(CO)_3$	Very slow at 140°C	6
$CpV(CO)_4$	4.7×10^{-13} s^{-1}	4
$Cr(CO)_6$	1×10^{-12} s^{-1}	6
$Fe(CO)_3(PPh_3)_2$	1×10^{-11} s^{-1}	8
$Cr(CO)_4(PPh_3)_2$	5×10^{-11} s^{-1}	6
$CpRu(CO)_3Br$	3×10^{-10} s^{-1}	6
$CpFe(CO)_2I$	5×10^{-8} s^{-1}	6
$CpMo(CO)_3I$	6.2×10^{-8} s^{-1}	4
$Ni(P(OEt)_3)_4$	1.0×10^{-6} s^{-1}	10
$Ti(CO)_3(P\ P)_2$	Labile at room temperature	4
$Ni(CO)_4$	1×10^{-2} s^{-1}	10

Table 4.9. EFFECT OF SOLVENT
ON THE REACTION OF $Mn(CO)_5Br$
WITH $AsPh_3$ AT 40.0°C[14]

Solvent	k (10^5, s^{-1})
cyclohexane	7.44
toluene	4.54
chloroform	3.29
acetone	1.79
nitrobenzene	1.08

pressure will influence reaction rates in the same direction as external pressure.[25] For a metal carbonyl dissociative process a high internal pressure would thus slow the reaction rate. Table 4.10 includes the internal pressures of a few solvents compared with the reaction rates. The agreement is at least as good as with the dielectric constant.

4.2.3. Nature of the Intermediate

The process under consideration involves loss of CO or other ligand from a metal center with 18 electrons in the valence shell orbitals of the metal. The 16-electron intermediate that is generated is of a lowered coordination number. Competition ratio studies have shown the 16-electron intermediate to be nondiscriminatory toward ligands of widely varying nucleophilicity.[17,19,26] In these studies the concentration of the dissociating ligand is varied, and the change in rate of formation of product can be used to determine the preference the intermediate shows for one ligand over another. A value of 1 for the competition ratio indicates no preference. Studies of this sort have been carried out on $Ni(CO)_3$, with CO and PPh_3 as competing ligands[17]; $Fe(CO)_4$ with CO and olefins competing[27]; $Mo(CO)_5$ with amines competing with phosphines[26]; and $Mo(CO)_4PPh_3$ with amines competing with phosphines and CO.[19] In all of these cases the competition ratios range between 0.5 and 5 (close to 1), which indicates that the unsaturated 16-electron intermediates are discrete and highly reactive, reacting with any nucleophile regardless of its nucleophilicity, with very low energy of activation.

Table 4.10. COMPARISON OF DIELECTRIC CONSTANT
AND COHESIVE ENERGY DENSITY TO RATE CONSTANTS
FOR SOLVENT EFFECTS

Solvent	Dielectric Constant[14]	Rate Constant[14]	ced[25]
cyclohexane	1.99	7.44	2038
carbontetrachloride	2.20	5.45	3034
toluene	2.34	4.54	3278
acetone	19.60	1.79	3853

Unfortunately, there is very little direct evidence regarding the rates of reaction of coordinatively unsaturated metal carbonyl intermediates with nucleophiles. $Cr(CO)_5$, which is generated in a flash photolysis experiment, reacts with CO in solution with a bimolecular rate constant of 3×10^6 $M^{-1}s^{-1}$.[15] This is only about three orders of magnitude lower than the diffusion-controlled encounter rate. Thus, even if all the departure from the diffusion-controlled rate of collisions between $Cr(CO)_5$ and CO were enthalpic in nature, a small activation energy is involved. The evidence, therefore, suggests that the intermediates in metal carbonyl dissociation reactions containing 16 electrons in the valence orbitals of the metal are reactive, and that they combine with nucleophiles in reactions characterized by small energies of activation.

4.2.4. Stereochemistry of CO Dissociation

A complex with nonequivalent CO groups allows the stereochemistry of CO dissociation to be determined. This has prompted several studies of exchange reactions with $Mn(CO)_5Br$.[28,29] Substitution reactions always lead to cis-$Mn(CO)_4LBr$.[14] Substitution reactions, however, do not indicate the stereochemistry of dissociation. An exchange reaction, by the principle of microscopic reversibility, must reflect the site of dissociation. For $Mn(CO)_5Br$ one would anticipate that the carbonyls cis to the bromide would dissociate most readily since those cis to Br are $trans$ to a CO and must share the π electron density, while the CO $trans$ to Br should be strongly bound. The COs cis to Br dissociate much more rapidly than does the CO $trans$ to Br. After an induction period in the ^{13}CO exchange reaction labeled CO does appear in the axial position; this was nicely fit by a fluxional five-coordinate intermediate.[29] Fluxional processes will be further discussed in Chapter 7.

In a further study of the stereochemistry of CO dissociation, ^{13}CO exchange reactions with cis-$Mn(CO)_4LBr$, L = PPh_3, $P(OPh)_3$, Py showed that the carbonyls cis to both L and Br dissociate.[30] Simple bonding arguments predict this result.

4.3. Dissociation of Other Ligands

There have been relatively few studies of dissociation of ligands other than CO from organometallic complexes. Because many of the ligands are large one must analyze the data in terms of steric size, in addition to σ- and π-bonding ability. Indeed, rates of L dissociation from NiL_4 correlate very nicely with ligand "cone angles," as shown in Table 4.11, which suggests a dominant role for steric effects in dissociation of L from NiL_4.[9]

$$NiL_4 + L' \longrightarrow NiL_3L' + L \tag{4.14}$$

This is readily understandable because tetrasubstituted complexes should have very significant steric interactions. Steric interactions are also shown to be important in dissociations from cis-$Mo(CO)_4L_2$ complexes as shown by the data in Table 4.12.[31]

Table 4.11. CONE ANGLES
FOR COMMON LIGANDS[9]

L	cone angle ($°$)
PPh_3	145
$AsPh_3$	142
PBu_3	132
$P(OPh)_3$	128
$P(OMe)_3$	107

$$cis\text{-}Mo(CO)_4L_2 + CO \longrightarrow Mo(CO)_5L + L \qquad (4.15)$$

The ordering of dissociation rates shows an effect on the steric size, but electronic effects must also be involved since phosphites of the same size as phosphines dissociate more slowly. As suggested previously, it is very difficult to separate steric and electronic factors.

The complexes $Cr(CO)_5L$, $trans\text{-}Cr(CO)_4L_2$, and $trans\text{-}Cr(CO)_4LL'$ were investigated for their reaction with carbon monoxide.[32-34]

$$Cr(CO)_5L + CO \longrightarrow Cr(CO)_6 + L \qquad (4.16)$$

The first order rate constants and activation parameters derived from those rate constants are shown in Table 4.13.[32] A comparison of Cr(0)–L and Pt(II)–L relative bond energy data is shown in Table 4.14. The order of bond energies for Cr–L bonds where the ligand is bound to Cr(0) is remarkably similar to the order for $trans\text{-}MePt(PMe_2Ph)_2L^+$ where the ligand is bound to Pt(II). The ordering of Cr–L bond stabilities in $Cr(CO)_5L$ is

$$Cr\text{-}PBu_3 \gg Cr\text{-}P(OPh)_3 > Cr\text{-}PPh_3 \sim Cr\text{-}CO > Cr\text{-}AsPh_3$$
$$> Cr\text{-}Py$$

which indicates that the strength of bonding ability is more significant than whether the bonding is σ or π in nature.[32] The similarity in M–L bond strengths for two

Table 4.12. RATES OF DISSOCIATION
FROM $cis\text{-}Mo(CO)_4L_2$ AT 70°C IN C_2Cl_4[31]

L	Cone Angle	Rate, s^{-1}
PMe_2Ph	122	$<1.0 \times 10^{-6}$
$PMePh_2$	136	1.3×10^{-5}
PPh_3	145	3.2×10^{-3}
$PPhCy_2$	162	6.4×10^{-2}
$P(OPh)_3$	128	$<1.0 \times 10^{-5}$
$P(O\text{-}o\text{-}tolyl)_3$	141	1.6×10^{-4}

Table 4.13. RATES AND ACTIVATION
PARAMETERS FOR L DISSOCIATION
FROM $Cr(CO)_5L$ AT 130°C[32]

L	$k \times 10^6$ s^{-1}	$\Delta H^{\ddagger}$	$\Delta S^{\ddagger}$
$P(OMe)_3$	0.55	—	—
$P(OPh)_3$	16.	32.	0
PPh_3	100	36.	12.
CO	130	40.	23.
$AsPh_3$	12000	36.	22.

metal centers as different as $Cr(CO)_5$ and $MePt(PMe_2Ph)_2^+$ suggests that this is a general order of bond strengths to organometallic centers.

4.4. Ligand Effects

In addition to directing the stereochemistry of CO dissociation, which is readily interpreted as a ground state effect, noncarbonyl ligands activate the complex toward CO dissociation by stabilizing the 16e$^-$ transition state. The data in Table 4.15 illustrate the magnitude of the effect and the order for CO dissociation from $Cr(CO)_5L$.[36] The labilization order

$$CO, H^- < P(OPh)_3 < PPh_3 < I^- < NC_5H_5 < Br^-, CH_3CN$$
$$< Cl^- < CH_3CO^- < NO_3^-$$

was derived from data for octahedral metal carbonyl complexes of group VI and VII. These ligands for a given metal span eight orders of magnitude in dissociation rate for *cis* COs and compare in magnitude with the *trans* effect seen in the substitution reactions of square planar complexes as discussed in Chapter 2.

Stabilization of the 16e$^-$ transition state by an electron donor ligand does not depend on the stereochemistry of the dissociating ligand. A *trans* effect of octa-

Table 4.14. METAL LIGAND BOND ENERGIES.

L	ΔG_{rxn} (kcal/mole)^a	$-\Delta H$(kcal/mole)^b
Py	−7.1	12.2
$AsPh_3$	−3.8	12.8
CO	0.0	—
PPh_3	0.1	19.5
$P(OPh)_3$	1.5	21.4
PBu_3	8.1	24.3 (L = PEt_3)

^{a} Free energy changes for the reaction $Cr(CO)_5L + CO \rightarrow Cr(CO)_6 + L$ at 130°C.[32]
^{b} Enthalpy data for formation reactions of $MePt(PMe_3Ph)_2L^+$.[35]

Table 4.15. FIRST-ORDER RATE
CONSTANTS AT 30°C FOR THE
SUBSTITUTION REACTIONS OF
CHROMIUM CARBONYL COMPLEXES[36]

Compound	$k_1(s^{-1})$
$Cr(CO)_6$	1×10^{-12}
$Cr(CO)_5PR_2R'$	1.5×10^{-10}
$Cr(Co)_5PPh_3$	3.0×10^{-10}
$Cr(CO)_5C(OCH_3)CH_3^-$	4×10^{-7}
$Cr(CO)_5I^-$	$<10^{-5}$
$Cr(CO)_5Br^-$	2×10^{-5}
$Cr(CO)_5Cl^-$	1.5×10^{-4}

hedral metal carbonyl substitution reactions was observed in reactions of *trans*-$Cr(CO)_4LL'$ with CO.[34]

$$\text{*trans*-}Cr(CO)_4LL' + CO \longrightarrow Cr(CO)_5L + L' \qquad (4.17)$$

The reactions progressed by L' dissociation showing good first-order kinetics and appropriate activation parameters. Two series of reactions indicate the effect of the *trans* ligand on the dissociation rate.[34] For L' = PPh_3 the complexes were prepared and reactions were investigated for L = PPh_3, $P(OPh)_3$, PBu_3, $P(OMe)_3$, and CO, and for L' = $P(OPh)_3$, L = $P(OPh)_3$, PPh_3, $P(OMe)_3$, and CO. The results are shown in Table 4.16. Thus, the ordering of *trans*-effect on dissociation of PPh_3 and $P(OPh)_3$ is identical and is shown:

$$PPh_3 > PBu_3 > P(OPh)_3 > P(OMe)_3 > CO$$

This order is very similar to that expected for steric size, but it is difficult to visualize a significant steric interaction from *trans* ligands. Crystal structure determinations showed no correlation between the $Cr-PPh_3$ or $Cr-P(OPh)_3$ bond lengths and rates of PPh_3 and $P(OPh)_3$ dissociation from *trans*-$Cr(CO)_4LL'$.[34] Thus, the rate acceleration by L from $Cr(CO)_4LL'$ does not arise by a ground state destabiliza-

Table 4.16. EFFECT OF *trans* LIGAND
ON DISSOCIATION RATES OF L'
FROM $Cr(CO)_4LL'$ AT 130°C[34]

L	$k(s^{-1})L' = PPh_3$	$k(s^{-1})L' = P(OPh)_3$
PPh_3	3.9	1.3×10^{-2}
PBu_3	2.4×10^{-1}	—
$P(OPh)_3$	1.3×10^{-2}	4.0×10^{-4}
$P(OMe)_3$	8.1×10^{-3}	1.8×10^{-4}
CO	1.0×10^{-4}	1.6×10^{-5}

tion which would be manifested by lengthening of Cr–L' bonds. Rather a transition state stabilization by L is indicated.

The similarity in the orders for both phosphine and phosphite dissociation also strongly suggests that a stabilization of the transition state is involved. Dissociation of either ligand would lead to the same intermediate, $Cr(CO)_4L$. This *trans*-effect order on dissociation is similar to the *cis*-labilization order observed for $Mn(CO)_4LBr$, which is also a transition state effect.[28,29] The ligand in *cis* labilization occupies a basal position in the transition state, while for the *trans* effect it occupies an apical position. The stabilization of the transition state by the ligand apparently does not depend on the site that the ligand occupies but reflects an ability to release electrons to the unsaturated, 16-electron intermediate.[34] A stabilization of the 16-electron intermediate $Cr(CO)_3(o\text{-phenanthroline})$ by electron-releasing o-phenanthroline ligands was noted a number of years ago.[37] This simple concept is also applicable to *cis*-labilization of CO dissociation and to the *trans* effect noted for dissociation from $Cr(CO)_4LL'$. It is also probably generally applicable to organometallic substitution reactions that proceed through an unsaturated intermediate.[8] To completely understand the effect one must consider both ground state and transition state effects. In *cis* labilization the labilization arises by stabilization of the 16-electron intermediate by the presence of a donating ligand. The *cis* stereospecificity results from a stronger M–CO bond *trans* to the donor ligand (a ground state effect). A complete interpretation of substitution reactions of metal carbonyl complexes requires knowledge of site specificity, steric effects in both the ground state and the transition state, and the ground state bond energies; however, the gross effects, especially in a series of complexes, can be accounted for in terms of a stabilization of the electron deficient transition state by electron donating groups.

The extent of substitution depends on the size of the substituting ligand. A series of substitutions on $Mn(CO)_5Br$ have been studied for the effect of ligand size on the number of ligands that can be substituted and the geometry of substitution.[38] These data are shown in Table 4.17. The smaller ligands are able to substitute to a larger extent than are the larger ligands. Steric effects are certainly significant in complexes of this type. A crystal structure determination of $fac\text{-}Cr(CO)_3(PEt_3)_3$ shows the effects on bond length and geometry, with lengthened Cr–P bonds and larger than 90° P–Cr–P angles.[39]

Table 4.17. COMPLEXES PREPARED BY LIGAND SUBSTITUTION ON $Mn(CO)_5Br^a$

Complex	L
trans-$Mn(CO)_3L_2Br$	PPh_3, $P(OPh)_3$, $AsPh_3$
mer-$Mn(CO)_2L_3Br$	PMe_2Ph, PMe_3, $P(OMe)_2Ph$
trans-$Mn(CO)_2L_3Br$	$P(OMe)_3$, $P(OEt)_3$
trans-$Mn(CO)L_4Br$	$P(OMe)_3$

^{a} The geometries are shown in Figure 4.3.

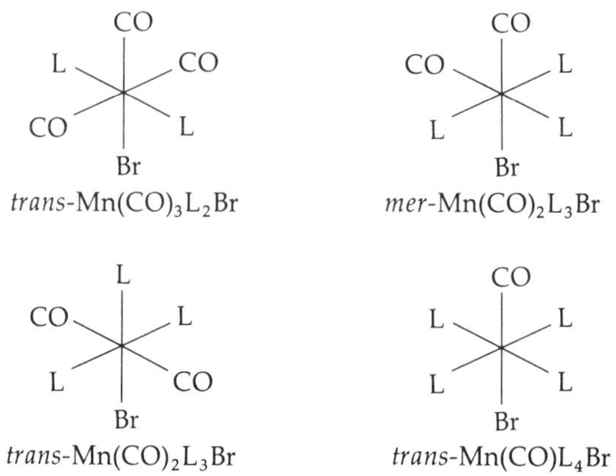

Figure 4.3. Substituted derivatives of manganese pentacarbonyl bromide.

4.5. Complexes with 17 Electrons

While most organometallic complexes contain sixteen or eighteen electrons, complexes containing seventeen electrons have received increasing attention in recent years.[40] Seventeen-electron complexes are formed through oxidation or reduction reactions or through M–M bond homolysis of dimers. The 17-electron complexes are, in general, much more reactive than are 18-electron complexes.[40] This offers the possibility to activate inert complexes through a process termed *electron-transfer catalysis*.

In addition to substitution reactions, which will be discussed in the next paragraph, 17-electron complexes undergo dimerization and electron transfer reactions (see Chapter 8). The dimerization reactions,

$$2Mn(CO)_5 \cdot \longrightarrow Mn_2(CO)_{10} \qquad (4.18)$$

occur at rates approaching diffusion controlled.[40] For example, dimerization of $[Mn(CO)_5]\cdot$ occurs with a rate constant of 9×10^8 $M^{-1}s^{-1}$ and dimerization of $[Re(CO)_5]\cdot$ occurs with a rate constant of 3×10^9 $M^{-1}s^{-1}$.[41]

Seventeen-electron complexes undergo ligand substitution through an associative reaction and a 19-electron intermediate. This can be illustrated by the reactions of $[V(CO)_6]\cdot$:

$$[V(CO)_6] \cdot + L \longrightarrow [V(CO)_6L] \cdot \longrightarrow [V(CO)_5L] \cdot + CO \qquad (4.19)$$

The activation parameters and dependence on L are shown in Table 4.18.[42] These data are fully consistent with an associative reaction. The 17-electron complex $V(CO)_6$ has an associative substitution reaction that is $>10^{10}$ more facile than for the 18-electron $Cr(CO)_6$ complex. The V complexes are among the most inert of the 17-electron complexes. Table 4.19 shows the rate constants for substitution of

Table 4.18. RATE CONSTANTS AND ACTIVATION PARAMETERS FOR SUBSTITUTION OF $V(CO)_6$[42]

L	$k(M^{-1} s^{-1})$	$\Delta H^{\ddagger}$ (kcal mol^{-1})	$\Delta S^{\ddagger}$ (e.u.)
PMe$_3$	132	7	-23
PBu$_3$	50	7	-25
PMePh$_2$	4	9	-26
P(OMe)$_3$	0.7	11	-23
PPh$_3$	0.2	10	-28
AsPh$_3$	0.02	—	—

several complexes.[41] As expected from size considerations, substituting a phosphine ligand for a CO decreases the rate for an associative reaction.

The two dominant characteristics for substitution reactions of 17-electron complexes are very rapid reactions and associative mechanisms. Each of these features is in contrast to reactions of 18-electron complexes. The reactivity has been attributed to formation of a three-electron bond between the entering nucleophile and the 17-electron complex.[43] Electron density analysis supports stabilization of the 19-electron transition state as the primary source for the rapid substitution reactions.[44]

Utilizing the tremendously enhanced reactivity of odd-electron organometallic complexes remains a challenge.

4.6. Substitution on Metal Carbonyl Complexes Containing M–M Bonds

A number of organometallic complexes form M–M bonds to achieve 18 electrons at each metal. Examples are $Mn_2(CO)_{10}$, $Co_2(CO)_8$, $Cp_2Cr_2(CO)_6$, $Fe_3(CO)_{12}$, and $Co_4(CO)_{12}$. Electron-counting for these M–M-bonded systems are shown in Figure 4.4. Substitution reactions on these complexes occur by replacement of a CO by an entering ligand.[45]

Table 4.19. RATE CONSTANTS FOR SUBSTITUTION OF 17-ELECTRON COMPLEXES[41]

Complex	k (M^{-1} s^{-1})^a
[V(CO)$_6$]•	100
[V(CO)$_5$PBu$_3$]•	5×10^{-4}
[Re(CO)$_5$]•	2×10^9
[Mn(CO)$_5$]•	1×10^9
[Mn(CO)$_4$PPh$_3$]•	7×10^2

^{a} The entering ligand is an alkyl phosphine.

$Mn_2(CO)_{10}$		$Co_2(CO)_8$		$Cp_2Cr_2(CO)_6$	
$[Mn(CO)_5]_2$		$[Co(CO)_4]$		$[CpCr(CO)_3]_2$	
Mn	$7e^-$	Co	$9e^-$	Cr^+	$5e^-$
5CO	$10e^-$	4CO	$8e^-$	Cp	$6e^-$
	$\overline{17e^-}$		$\overline{17e^-}$	3CO	$6e^-$
Mn—Mn	$1e^-$	Co—Co	$1e^-$		$\overline{17e^-}$
	$\overline{18e^-}$		$\overline{18e^-}$	Cr—Cr	$1e^-$
					$\overline{18e^-}$

$Fe_3(CO)_{12}$		$Co_4(CO)_{12}$	
$[Fe(CO)_4]_3$		$[Co(CO)_3]_4$	
Fe	$8e^-$	Co	$9e^-$
4CO	$8e^-$	3CO	$6e^-$
	$\overline{16e^-}$		$\overline{15e^-}$
2Fe—Fe	$2e^-$	3Co—Co	$3e^-$
	$\overline{18e^-}$		$\overline{18e^-}$

Figure 4.4. Electron counting of M–M-bonded systems.

$$Mn_2(CO)_{10} \xrightarrow{+L,\ -CO} Mn_2(CO)_9L \xrightarrow{+L,\ -CO} Mn_2(CO)_8L_2 \qquad (4.20)$$

Features that are different than mononuclear complexes are possible substitution at different metals and the effect of the M–M bond on reactivity. In addition, there are complexes that contain metals multiply bonded to each other. Reactions of these are addition reactions that will not be considered here.

4.6.1. Metal Carbonyl Dimers

The reactions that have been most often studied are those of the Group VII dimers, $Mn_2(CO)_{10}$, $Tc_2(CO)_{10}$, $Re_2(CO)_{10}$, and $MnRe(CO)_{10}$.[45–48] These reactions may lead to either the *mono-* or *bis*-substituted complex, depending on the dimer and the reaction conditions as shown in Equation 4.20. The ligands are almost invariably substituted axially, as shown in Figure 4.5. Kinetic studies on these dimers, summarized in Table 4.20, show the rate law

$$\text{rate} = (k_1 + k_2[L])[M_2(CO)_{10}] \qquad (4.21)$$

where the k_2 term is very much smaller than the k_1 term. This rate law is identical to that seen for substitution at $Cr(CO)_6$, and is best ascribed to a dissociative interchange mechanism.[48] Considerable controversy existed over whether the mechanism involves CO dissociation[45,48] or homolytic cleavage[46,47] of the M–M bond. The ultimate experiment to differentiate between metal–metal bond homolysis and other routes has been performed for $Re_2(CO)_{10}$.[50] Two complexes with different rhenium isotopes were prepared [$^{185}Re_2(CO)_{10}$ and $^{187}Re_2(CO)_{10}$] and utilized in substitution and exchange reactions.[50] No mixed isotopic species were observed

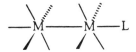

Figure 4.5. Structures of substituted derivatives of the decacarbonyl dimers.

after several half lives under the reaction conditions, ruling out the M–M bond homolysis mechanism in thermal substitution and CO exchange reactions of $Re_2(CO)_{10}$. While the metal isotopes are not available for manganese, similar reactions of $Mn_2(CO)_{10}$ and $Mn_2(^{13}CO)_{10}$ also are fully consistent with CO dissociation and inconsistent with M–M bond cleavage.[51] Thus, definitive evidence has been presented to refute the suggestion that M–M bond homolysis is involved in ligand exchange or substitution reactions of the Group VII dimers.

$$^{185}Re_2(CO)_{10} \qquad\qquad ^{185}Re_2(CO)_8L_2 \qquad\qquad (4.22)$$

$$+ \qquad \xrightarrow[-4CO]{4L} \qquad +$$

$$^{187}Re_2(CO)_{10} \qquad\qquad ^{187}Re_2(CO)_8L_2$$

As noted in Table 4.19 substitution on $MnRe(CO)_{10}$ leads predominantly to the Re isomer, $(CO)_5MnRe(CO)_4L$. One would anticipate that CO dissociation from Mn would occur at a rate around 100 times that from Re by comparison of the reactivities of $Mn_2(CO)_{10}$ with $Re_2(CO)_{10}$ and the reactivities of analogous mononuclear complexes.[48] A mass spectral investigation of $(CO)_5MnRe(^{13}CO)_5$ showed

Table 4.20. SUBSTITUTION REACTIONS ON GROUP VII CARBONYL DIMERS, EACH REACTION WAS FIRST-ORDER IN METAL CARBONYL DIMER[49]

$M_2(CO)_{10}$	L	Product
$Mn_2(CO)_{10}$	PPh_3, $P(OPh)_3$, etc.	$Mn_2(CO)_9L$ and $Mn_2(CO)_8L_2$
$Mn_2(CO)_{10}$	PPh_3	$Mn_2(CO)_9PPh_3$ and $Mn_2(CO)_8(PPh_3)_2$
$Re_2(CO)_{10}$	PPh_3	$Re_2(CO)_8(PPh_3)_2$
$Tc_2(CO)_{10}$	PPh_3	$Tc_2(CO)_9PPh_3$ and $Tc_2(CO)_8(PPh_3)_2$
$MnRe(CO)_{10}$	PPh_3	$(CO)_5MnRe(CO)_4PPh_3$
$MnRe(CO)_{10}$	PPh_3, $P(OPh)_3$, PBu_3	$(CO)_5MnRe(CO)_4L$ and $MnRe(CO)_8L_2$

exclusive dissociation from manganese.[52] To rationalize the substitution on Re it was suggested that CO dissociates from the manganese and that the intermediate/transition state contained a bridging CO such that the unsaturation at one metal center was shared by the other metal. The ligand would then attack at the most favorable site, which would be the Re for steric reasons. Such a scheme is shown in Figure 4.6.

The effect of substitution of one ligand for CO on the rate of substitution at the other metal has only been investigated in one study.[45] The rate of L addition to $Mn_2(CO)_9L$ was studied, and the rates are shown in Table 4.21. These rates, which were ascribed to CO dissociation, are very similar to the effects seen in *cis*-labilization of mononuclear complexes. This can either be a long-range effect, where the phosphine on one metal center affects the rate of dissociation at the second metal center, or the labilization by the phosphine could occur at the carbonyls *cis* to the phosphine at the same metal center. The unsaturation could be transferred from the substituted metal center to the unsubstituted through a bridging CO in a scheme similar to that shown in Figure 4.6 for $MnRe(CO)_{10}$.

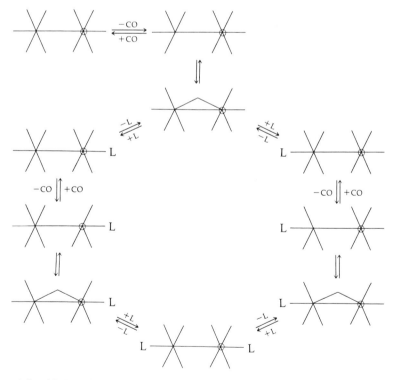

Figure 4.6. $MnRe(CO)_{10}$ mechanism involving CO dissociation. The open circles represent Re atoms. [Reprinted with permission from D. Sonnenberger and J. D. Atwood, *J. Am. Chem. Soc.* **102**, 3484 (1980). Copyright 1980 American Chemical Society.]

Table 4.21. VARIATION WITH L OF
THE RATE CONSTANTS FOR FURTHER
SUBSTITUTION ON $Mn_2(CO)_9L$[45]

L	Relative Rate
PPh_3	44
PBu_3	12
$P(OPh)_3$	1.5
CO	1

Substitution reactions of $Co_2(CO)_8$ show more variation than the Group VII dimers. Reaction with ^{13}CO, $AsPh_3$, and H_2 show a rate law that is independent of the concentration of the incoming nucleophile at temperatures from $-15°$ to $30°C$.[53]

$$Co_2(CO)_8 + AsPh_3 \longrightarrow Co_2(CO)_7AsPh_3 + CO \qquad (4.23)$$

The activation parameters ($\Delta H^{\ddagger} = 22$ kcal/mole and $\Delta S^{\ddagger} = 10$ eu) and lack of dependence on the incoming ligand are consistent with CO dissociation.

$$Co_2(CO)_8 \longrightarrow [Co_2(CO)_7 + CO] \qquad (4.24)$$
$$\downarrow AsPh_3, \text{ rapid}$$
$$Co_2(CO)_7AsPh_3$$

The reaction with $AsPh_3$ leads to the disubstituted complex at a much slower rate, which indicates that the ligand affects reactivity of $Co_2(CO)_8$ differently than it does the Group 7 metal carbonyl dimers. Substitution of $Co_2(CO)_8$ by phosphines is quite different,

$$Co_2(CO)_8 + 2PR_3 \longrightarrow [Co(CO)_3(PR_3)_2]^+[Co(CO)_4^-] + CO \qquad (4.25)$$

leading to ionic products at rates that are more rapid than CO dissociation and which depend on the concentration and the nucleophilicity of the entering ligand. In addition, the apparent order with respect to $Co_2(CO)_8$ varied within the range 1.0–1.5, depending on the reaction conditions. The kinetic data for substitution by PBu_3 at $-15°$ are shown in Figure 4.7 with a 1.40 dependence on $[Co_2(CO)_8]$. The reaction with PBu_3 was studied at -15, -10, 5, and $10°C$, with the order of the reaction in $Co_2(CO)_8$ as 1.40, 1.35, 1.20, and 1.0, respectively. A radical chain mechanism was suggested to account for these observations in the substitution of phosphine bases on $Co_2(CO)_8$.[53]

$$Co_2(CO)_8 + L \underset{k_2}{\overset{k_1}{\rightleftharpoons}} Co_2(CO)_8L \qquad (4.26)$$

$$Co_2(CO)_8L \overset{k_3}{\longrightarrow} \cdot Co(CO)_3L + \cdot Co(CO)_4 + CO \qquad (4.27)$$

$$\cdot Co(CO)_3L + Co_2(CO)_8 \overset{k_4}{\longrightarrow} Co(CO)_3L^+ + \cdot Co_2(CO)_8^- \qquad (4.28)$$

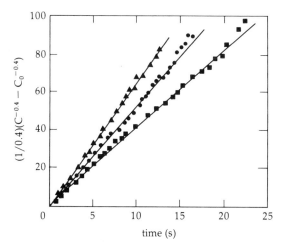

Figure 4.7. Pseudo 1.40-order plots for reaction of $Co_2(CO)_8$ with PBu_3 at three different concentrations of PBu_3.

$$\cdot Co_2(CO)_8^- \xrightarrow{k_5} \cdot Co(CO)_4 + Co(CO)_4^- \qquad (4.29)$$

$$Co(CO)_3L^+ + L \xrightarrow{k_6} Co(CO)_3L_2^+ \qquad (4.30)$$

Thus, two distinct mechanisms exist for $Co_2(CO)_8$—a CO dissociative route for reaction with weak nucleophiles and a radical chain pathway for reaction with strong nucleophiles.

4.6.2. Transition Metal Clusters

Reactions of Lewis bases with metal clusters may yield either mononuclear or polynuclear products. Substitution reactions on $Fe_3(CO)_{12}$ represent the features that may be seen. Reaction with L at 50°C leads to substituted metal clusters,

$$Fe_3(CO)_{12} + nL \xrightarrow{50°C} Fe_3(CO)_{12-n}L_n + nCO \qquad (4.31)$$

$$Fe_3(CO)_{12} + L \xrightarrow{80°C} Fe(CO)_4L + Fe(CO)_3L_2 \qquad (4.32)$$

while reaction at 80°C produces substituted mononuclear fragments.[54,55] The Ru and Os analogues have less tendency to fragment, presumably due to the stronger M–M bonds.[56,57]

There are two possible sites for substitution in most metal clusters, axial or equatorial, as shown for $M_3(CO)_{12}$ and $M_4(CO)_{12}$ in Figure 4.8. Substitution has been observed in both sites, and it often appears to be controlled by steric interactions. The first PPh_3 substituted onto $Ir_4(CO)_{12}$ occupies an axial site.[58] When two

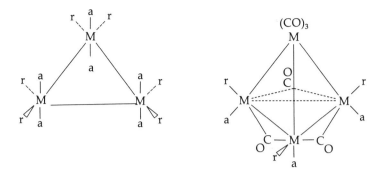

Figure 4.8. Axial (a) and radial (r) positions for $M_3(CO)_{12}$ and $M_4(CO)_{12}$.

PPh_3 ligands are substituted onto $Ir_4(CO)_{12}$, one occupies an axial site and the second an equatorial site of a different basal iridium. Additional substitution on the *bis*-substituted complex leads to substitution at two equatorial and one axial sites. These substituted products are shown in Figure 4.9. At the conditions where substitution reactions occur metal cluster complexes are fluxional (i.e., the three PPh_3 ligands are equivalent and the nine carbonyl ligands are equivalent) (fluxional processes will be discussed in more detail in Chapter 7).

The Group 8 clusters, $M_3(CO)_{12}$, provide models for many cluster substitution reactions. Substitution on $Ru_3(CO)_{12}$ and $Os_3(CO)_{12}$ proceed by CO dissociation to the *tris*-substituted clusters.[56,59]

$$Ru_3(CO)_{12} + 3PPh_3 \longrightarrow Ru_3(CO)_9(PPh_3)_3 + 3CO \qquad (4.33)$$

$$Os_3(CO)_{12} + 3PPh_3 \longrightarrow Os_3(CO)_9(PPh_3)_3 + 3CO \qquad (4.34)$$

Substitution on $Fe_3(CO)_{12}$ gives $Fe_3(CO)_{11}L$, which either further substitutes or fragments depending on L.[59] Figure 4.10 shows formation of the substituted triiron clusters sequentially for L = $P(OPh)_3$. For L = PPh_3 the monosubstituted cluster fragments to the products shown in Reaction 4.31.[59] Table 4.22 shows the relative

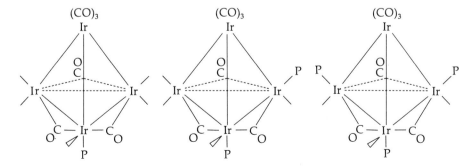

Figure 4.9. Substituted products of $Ir_4(CO)_{12}$. (P is a phosphorus donor ligand).

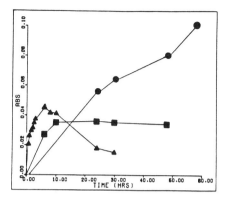

Figure 4.10. Absorptions for $Fe_3(CO)_{11}P(OPh)_3$ (▲), $Fe_3(CO)_{10}(P(OPh)_3)_2$ (■), and $Fe_3(CO)_9(P(OPh)_3)_3$ (●) during the reaction of $P(OPh)_3$ with $Fe_3(CO)_{12}$ at 30°C in hexane. The pattern is typical for consecutive substitution reactions. [Reprinted with permission from A. Shojaie and J. D. Atwood, *Organometallics* **4**, 190 (1985). Copyright 1985 American Chemical Society.]

rates for substitution on $M_3(CO)_{12}$ and on mixed-metal clusters. Substitution on $M_3(CO)_{12}$ does not follow the pattern of reactivity observed for the mononuclear $M(CO)_5$ compounds (see earlier material in this chapter). The cluster reactivity M = Fe > Ru > Os for $M_3(CO)_{12}$ is the same as observed for CO dissociation from $HM_3(\mu\text{-COMe})(CO)_{10}$.[62,63] The discrepancy between the clusters and $Fe(CO)_5$ may be a further indication that crystal field activation plays a role in the reactivity of $Fe(CO)_5$.

The mixed-metal clusters allow assessment of the effect of an adjacent metal center on reactivity. For the sequence $Fe_2M(CO)_{12}$ (M = Fe, Ru, Os) CO dissociation occurs from iron, but M asserts an effect on the rate,[60,61] as shown in Table 4.22, M = Ru > Fe > Os. Activation of an iron center by ruthenium was also observed for $(\mu\text{-H})FeRu_2(\mu\text{-COMe})(CO)_{10}$.[63]

These group 8 clusters show fragmentation to mononuclear complexes, substitution of the cluster and activation of a metal by an adjacent metal center towards CO dissociation.

Table 4.22. RATE CONSTANTS FOR CO DISSOCIATION FROM TRIMETALLIC CLUSTER COMPLEXES AT 30°C[56,59–61]

Cluster	$k(s^{-1})$	$\Delta H^{\ddagger}$ (kcal/mol)	$\Delta S^{\ddagger}$ (eu)
$Fe_3(CO)_{12}$	4.0×10^{-5}	29.5 ± 0.8	19 ± 2
$Ru_3(CO)_{12}$	8.6×10^{-6}	32 ± 2	23 ± 4
$Os_3(CO)_{12}$	1×10^{-9}	38.6 ± 0.5	24 ± 2
$Fe_2Ru(CO)_{12}$	1.8×10^{-4}	26.8 ± 0.4	6.7 ± 0.4
$FeRu_2(CO)_{12}$	1.0×10^{-4}	26 ± 1	9 ± 4
$Fe_2Os(CO)_{12}$	4.2×10^{-6}	27 ± 3	7 ± 7

Substitution on $Ir_4(CO)_{12}$ has been the most thoroughly studied cluster reaction.[64-67]

$$Ir_4(CO)_{12} + nL \longrightarrow Ir_4(CO)_{12-n}L_n + nCO \tag{4.35}$$

The product observed depends on the reaction conditions and the ligand L. The rate law observed has two terms as shown.

$$rate = (k_1 + k_2[L])[Ir_4(CO)_{12}] \tag{4.36}$$

This rate law is of the same form as that seen for substitutions on other metal carbonyl complexes, but the relative value of the two terms are very different than observed in other systems.[65] The ligand-dependent term dominates for ligands that are reasonably good nucleophiles [CNR, PBu_3, PPh_3, $P(OPh)_3$].[65,68] As shown by the graph of [L] versus k_{obs} in Figure 4.11, the value of k_1 is independent of the nature of L, and may be ascribed to a CO dissociative pathway. The relative values of k_2 (Table 4.23) show a strong dependence on the nucleophilicity of the entering ligand, indicating nucleophilic attack on the metal complex. The site of attack is uncertain although it was argued that the accessible metal center (a third-row metal with 100° angles between COs) was most likely.[65]

As the tetrairidium cluster is substituted some changes are noted.[66,67] Substitution on $Ir_4(CO)_{11}L$, L = PPh_3, $P(OPh)_3$, and $AsPh_3$ occurs by a primarily ligand independent mechanism, probably CO dissociation. For more nucleophilic entering

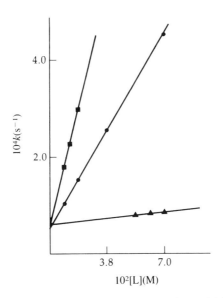

Figure 4.11. Plot of k_{obs} versus [L] for a series of reactions with $Ir_4(CO)_{12}$ at 109°C in chlorobenzene: ●, PPh_3; ▲, $AsPh_3$; ■ $P(OPh)_3$. [Reprinted with permission from D. Sonnenberger and J. D. Atwood, *Inorg. Chem.* **20**, 3243 (1981). Copyright 1981 American Chemical Society.]

Table 4.23. SECOND-ORDER RATE
CONSTANTS FOR SUBSTITUTION
OF $Ir_4(CO)_{12}$ BY L[65]

L	$10^5 k_2$, $M^{-1}s^{-1}$
$AsPh_3$	27
PPh_3	530
$P(OPh)_3$	1300
PBu_3	too fast to measure

ligands (PBu_3, CN-t-Bu) the ligand dependent path still dominates.[65,66,68] The CO dissociation rates from $Ir_4(CO)_{11}L$ show the same trends with changes in L as mononuclear complexes and metal carbonyl dimers. The relative ordering is as follows:

$$PBu_3 > PPh_3 > AsPh_3 > P(OPh)_3 > CO$$

The same trends continue to the *bis*-substituted tetrairidium clusters, $Ir_4(CO)_{10}L_2$, with a continued increase in rates of CO dissociation, as shown in Table 4.24. An acceleration in rate with substitution was also seen in reactions of $Ru_3(CO)_{12}$ with PPh_3.[69]

The dependence on the ligands present on the metal cluster is very similar to that seen for mononuclear complexes both in order and magnitude. This suggests that the donor ability of the ligand is important in stabilizing the unsaturated metal cluster, which is generated by ligand loss from the saturated cluster; it may also indicate dissociation from a substituted metal center, although the metal cluster may be capable of transmitting electronic effects between metal centers.

In metal clusters that are more highly substituted with phosphine ligands steric effects may be significant. The rates of phosphine dissociation from $Ir_4(CO)_8L_4$ show evidence of a steric effect.[70,71]

$$Ir_4(CO)_8L_4 + CO \longrightarrow Ir_4(CO)_9L_3 + L \tag{4.37}$$

$$L = PMe_3, PEt_3, PBu_3$$

Data for substitution at 85°C are shown in Table 4.25. The steric interactions in $Ir_4(CO)_8(PMe_3)_4$ are sufficient to cause an elongated Ir–Ir bond.[71] A similar,

Table 4.24. RELATIVE REACTION
RATES FOR SUBSTITUTED DERIVATIVES
OF $Ir_4(CO)_{12}$[66,67]

L	$Ir_4(CO)_{10}L_2$	$Ir_4(CO)_{11}L$
PBu_3	360	93
PPh_3	700	47
$AsPh_3$	700	34
$P(OPh)_3$	15	5.7
CO	0.7	0.7

Table 4.25. A COMPARISON OF THE
CONE ANGLE OF THE LIGAND WITH THE
RATE CONSTANT FOR DISSOCIATION
FROM $Ir_4(CO)_8L_4$ (85°C)

L	Cone Angle	$k(s^{-1})$
PMe$_3$	118	6.35×10^{-7}
PEt$_3$	132	2.09×10^{-3}
PBu$_3$	132	1.00×10^{-3}

though smaller rate effect on CO dissociation from $Ir_4(CO)_9L_3$, has also been attributed to steric interactions.[70] The rate of CO dissociation was investigated by ^{13}CO exchange; the comparison of rate with ligand is shown in Table 4.26. While the solvents are not the same, it is interesting to note that there appears to be no significant increase in CO dissociation from $Ir_4(CO)_{10}(PBu_3)_2$ to $Ir_4(CO)_9(PBu_3)_3$. The lack of a significant enhancement in rate may indicate that the electronic effect does not continue to the third substitution and that steric interactions are not significantly different between the *bis-* and *tris-*substituted complexes.

Substitution reactions on metal clusters offer mechanistic possibilities unavailable for mononuclear complexes. The ligand may replace a M–M bond, as was shown in the reaction of $Cp(CO)_2Mn(\mu\text{-PPh})Fe_2(CO)_8$ with PPh_3.[72] This reaction is shown in Figure 4.12. Reaction with a nucleophile led to an adduct without one Mn–Fe bond. The reaction was reversible, dissociation of L led to formation of the Mn–Fe bond. Despite this observation there appears to be no kinetic evidence for this as an important step in metal cluster substitution reactions, although only a few clusters have been studied.

Mixed metal clusters (clusters with at least two different metals) have considerable potential for mechanistic investigations of metal cluster reactions. Several reactions have been examined for the tetranuclear mixed-metal cluster $H_2FeRu_3(CO)_{13}$.[63]

$$H_2FeRu_3(CO)_{13} + 4CO \longrightarrow Ru_3(CO)_{12} + Fe(CO)_5 + H_2 \qquad (4.38)$$

Kinetics studies of this reaction showed a rate law,

$$\frac{-d[H_2FeRu_3(CO)_{13}]}{dt} = (k_1 + k_2[CO])[H_2FeRu_3(CO)_{13}] \qquad (4.39)$$

Table 4.26. RATE CONSTANTS FOR CO
DISSOCIATION FROM $Ir_4(CO)_9L_3$ AT 80°C

L	Cone Angle	$k(10^3 s^{-1})$
PMe$_3$	118	0.17
PEt$_3$	132	1.33
PBu$_3$	132	1.27
P(i-Pr)$_3$	160	6.76

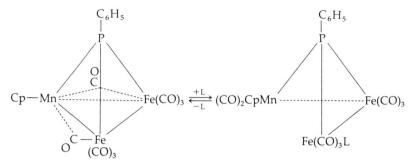

Figure 4.12.   Reaction of a cluster with a ligand (L = CO, PPh$_3$) which leads to reversible M–M bond breaking.

although the first-order term was negligible. The plot of k_{obs} versus [CO] is shown in Figure 4.13. The suggested mechanism involves association of CO with the intact cluster concomitant with cleavage of one M–M bond to give a butterfly cluster as an intermediate.[73] The activation parameters ($\Delta H^{\ddagger}$ = 20.0 kcal/mole; $\Delta S^{\ddagger}$ = −25.4 kcal/mole) are consistent with this mechanistic suggestion. Reaction of H$_2$FeRu$_3$(CO)$_{13}$ with phosphine and phosphite ligands led to substituted derivatives of the tetranuclear cluster in moderate yields.[73] The reaction with PPh$_3$ occurs at lower temperatures than does the reaction with CO, which leads to cluster break up, and with a rate law that is independent of the [PPh$_3$]. A CO dissociative route was suggested for this substitution reaction.[73] Substitution led to two different isomers, both involving replacement of a CO on Ru, in varying quantities depending on the ligand size and basicity.

It is rather surprising that the clusters thus far investigated show the same basic

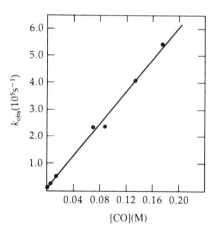

Figure 4.13.   Plots of k_{obs} versus [CO] for reaction of CO with H$_2$FeRu$_3$(CO)$_{13}$ showing the linear dependence of rate on [CO].

mechanistic features as do mononuclear complexes. The primary reaction model seen thus far is CO dissociation, although large metals are susceptible to nucleophilic attack at the metal. As more clusters are prepared and more reactions are studied we may anticipate more divergence in the reactions of metal clusters from those of mononuclear complexes.

4.7. Ligand Substitution Reactions on Alkyl Complexes

Alkyl complexes undergo ligand substitution processes,

$$CH_3Mn(CO)_5 + L \longrightarrow CH_3Mn(CO)_4L + CO \tag{4.40}$$

similar in stoichiometry to other substitution reactions. An important difference is that in many reactions of this type an acyl intermediate is observed.

$$CH_3Mn(CO)_5 + L \longrightarrow CH_3C(O)Mn(CO)_4L \longrightarrow \tag{4.41}$$
$$CH_3Mn(CO)_4L + CO$$

Reactions of this type are termed *CO insertion* or *alkyl migration* since the net reaction in forming the acyl intermediate is cleavage of the CH_3–Mn bond and formation of a methyl carbon bond as shown in Figure 4.14. These reactions are very important in catalytic reactions, such as hydroformylation, methanol carbonylation, and homogeneous CO reduction.

4.7.1. Kinetics and Rate Law

Most of the studies of kinetics of substitutions of alkyl complexes have involved $RMn(CO)_5$. The rate is dependent on the concentration of the metal complex and the

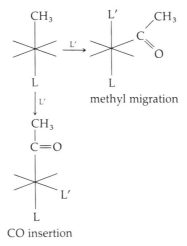

Figure 4.14. Product difference for methyl migration and CO insertion.

entering ligand.[74,75] The rate law that best accommodates the observations is shown in the following.

$$\text{rate} = \frac{k_1 k_2 [L][RMn(CO)_5]}{k_{-1} + k_2 [L]} \tag{4.42}$$

The mechanism most consistent with this rate law involves a rapid preequilibrium,

$$RMn(CO)_5 \underset{k_{-1}}{\overset{k_1}{\rightleftharpoons}} [RC(O)Mn(CO)_4] \tag{4.43}$$

$$RC(O)Mn(CO)_4 \xrightarrow{k_2,\ L} RC(O)Mn(CO)_4L \tag{4.44}$$

followed by reaction of the 16-electron acyl intermediate with L.[75] As discussed earlier, the characteristics of this type of rate law are independent of the rate on ligand concentration at high ligand concentration ($k_2[L] \gg k_{-1}$) and linear dependence on the ligand concentration at low concentration of ligand. A plot of k_{obs} versus [L] is shown in Figure 4.15. The activation parameters observed for these reactions

$$CH_3Mn(CO)_5 + CO \longrightarrow CH_3C(O)Mn(CO)_5 \tag{4.45}$$

$$\Delta H^{\ddagger} = 14.2 \text{ kcal/mole}, \ \Delta S^{\ddagger} = -21.1 \text{ eu}$$

$$CH_3MoCp(CO)_3 + L \longrightarrow CH_3C(O)MoCp(CO)_2L \tag{4.46}$$

$$\Delta H^{\ddagger} = 16.1 \text{ kcal/mole}, \ \Delta S^{\ddagger} = -25 \text{ eu}$$

are consistent with the mechanism suggested in Equations 4.43 and 4.44.[74,75]

The rate of carbonylation of an alkyl group has been shown to depend on the electron-withdrawing ability of the R group.[76] The results of the carbonylation of substituted methylmanganese pentacarbonyl complexes are shown in Table 4.27.

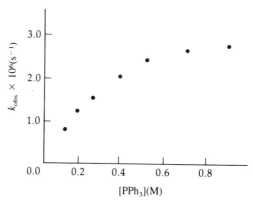

Figure 4.15. Dependence on [PPh₃] of the rate of substitution for the reaction with $CH_3FeCp(CO)_2$.[56]

Table 4.27. RATES OF CARBONYLATION
FOR A SERIES OF SUBSTITUTED METHYL–
MANGANESEPENTACARBONYL COMPLEXES
AT 30° C IN β,β' DIETHOXYDIETHYL ETHER

R	$\sigma^*(RCH_2)^a$	$k(10^5 \, M^{-1} \, s^{-1})$
C_2H_5	−0.115	14000
CH_3	−0.10	12000
cyclo-C_6H_{11}	−0.06	2500
H	0.00	1200
CH_3O	0.52	25
C_6H_5	0.22	12
HOC(O)	1.05	<5

^{a} Taft parameter.

$$RCH_2Mn(CO)_5 + CO \longrightarrow RCH_2C(O)Mn(CO)_5 \qquad (4.47)$$

When R is an electron-withdrawing group the rate is dramatically slowed, most likely by affecting the preequilibrium (Eq. 4.43).

4.7.2. Solvent Effect

The formation of a 16-electron intermediate in a preequilibrium would allow binding of solvent and suggests a large effect on the specific solvent employed. This is observed in reactions of alkyl complexes. Coordinating solvents accelerate the reaction as shown in Table 4.28 for $CH_3Mn(CO)_5$ and $CH_3MoCp(CO)_3$. This dependence on the solvent is consistent with coordination of the solvent to the unsaturated intermediate, although the variation of K_{eq} with solvent indicates that solvation of the alkyl and acyl complexes are different in polar solvents.[74]

$$CH_3Mn(CO)_5 \underset{-S}{\overset{S}{\rightleftharpoons}} CH_3C(O)Mn(CO)_4S \qquad (4.48)$$

Solvated acyl complexes have not been isolated although spectroscopic evidence has been reported for $CH_3Mn(CO)_4(THF)$ and $CH_3C(O)FeCp(CO)$ (DMSO).[74,77] It

Table 4.28. SOLVENT EFFECT ON THE METHYL
MIGRATION REACTION[74–76]

$CH_3Mn(CO)_5$			$CH_3MoCp(CO)_3$	
Solvent	K_{eq}	Relative Rate	Solvent	Relative Rate
mesitylene	220	1	toluene	1
$(n\text{-Bu})_2O$	120	2	THF	10
n-octyl chloride	200	4	nitromethane	70
DMF	3000	50	DMF	8000

seems most likely that the intermediate is solvent coordinated when the substitution is run in reasonably nucleophilic solvents. A general solvation is probable in solvents such as hexane or toluene.

For the molybdenum complex, MeMoCp(CO)$_3$, where solvent effects on the substitution reaction are large (Table 4.28), the role of solvent in the coordination sphere of the metal was demonstrated.[78] The separation of donor and polarity effects was accomplished by using a series of methyl-substituted tetrahydrofurans that have similar dielectric constants, but which vary widely in donor ability. The results are shown in Table 4.29.[78] These data support a CH$_3$C(O)MoCp(CO)$_2$(THF) complex formed as the intermediate.

4.7.3. Effect of Entering Ligand

The entering ligand competes with solvent for the 16-electron intermediate; thus, the reaction shows a dependence on the nucleophilicity of the entering ligand. Data for reaction of CH$_3$MoCp(CO)$_3$ with PBu$_3$, PPh$_3$, P(OBu)$_3$, and P(OPh)$_3$ are shown in Table 4.30.[75]

$$CH_3MoCp(CO)_3 + L \longrightarrow CH_3C(O)MoCp(CO)_2L \qquad (4.49)$$

$$L = PBu_3, PPh_3, P(OBu)_3, P(OPh)_3$$

These data show several features. As discussed in Section 4.7.2., the reactions are much slower in toluene than they are in THF. While the ordering of ligand dependence is the same for each solvent and parallels the nucleophilicity of the ligands,

$$PBu_3 > P(OBu)_3 > PPh_3 > P(OPh)_3$$

the rate differences are larger in toluene. The dependence on the nucleophilicity is not as large as would be expected for direct nucleophilic attack in either solvent, which provides further confirmation of the mechanism suggested.

4.7.4. Stereochemical Considerations

Much attention has been directed toward the stereochemistry of substitution reactions of alkyl complexes. Substitution of CH$_3$Mn(CO)$_5$ initially leads to *cis-*

Table 4.29. RATES OF SUBSTITUTION OF MeMoCp(CO)$_3$ IN SUBSTITUTED TETRAHYDROFURANS[78]

Solvent	$10^4 k_1$, s^{-1a}
THF	7.8
3-MeTHF	6.5
2-MeTHF	1.48
2,5-MeTHF	0.23

[a] The rate constant is for the first order, solvent dependent step.

Table 4.30. LIGAND DEPENDENCE OF THE REACTION OF $CH_3MoCp(CO)_3$ WITH L AT 50.7°C IN THF OR TOLUENE[75]

L	$k_{obs} \times 10^4$ s^{-1} (THF)	$k_{obs} \times 10^5$ s^{-1} (toluene)
PBu_3	4.41	3.13
$P(OBu)_3$	4.17	2.58
PPh_3	3.83	1.10
$P(OPh)_3$	3.17	0.20

$CH_3C(O)Mn(CO)_4L$, although the *cis* complex does eventually isomerize to the *trans*.[79]

$$CH_3Mn(CO)_5 + L \longrightarrow cis\text{-}CH_3C(O)Mn(CO)_4L \xrightarrow{\text{slow}} \qquad (4.50)$$
$$trans\text{-}CH_3C(O)Mn(CO)_4L$$

As discussed previously the stereochemistry of substitution reactions do not necessarily provide information on the stereochemistry of the intermediates. In this case, however, formation of a kinetically stable product [$cis\text{-}CH_3C(O)Mn(CO)_4L$] indicates that the substitution reaction probably does provide stereochemical information. This is confirmed by carbonylation with ^{13}CO, which leads exclusively to $cis\text{-}CH_3C(O)Mn(CO)_4(^{13}CO)$.

$$CH_3Mn(CO)_5 + {}^{13}CO \rightarrow cis\text{-}CH_3C(O)Mn(CO)_4({}^{13}CO) \qquad (4.51)$$

The fact that the labeled CO does not appear in the acyl confirms that the initial preequilibrium step (Eq. 4.43) occurs and that the CO does not directly insert into the CH_3–Mn bond. A more difficult question to answer concerned whether the methyl group was migrating to a CO or whether a CO was inserting into the CH_3–Mn bond. These processes are termed *methyl migration* and *CO insertion*, respectively. The distinction between the two groups lies in which is moving in the formation of the 16-electron acyl intermediate. This question was addressed for the manganese complex by consideration of the decarbonylation of labeled $CH_3C(O)Mn(CO)_5$ complexes.[80]

$$CH_3C(O)Mn(CO)_5 \longrightarrow CH_3Mn(CO)_5 + CO \qquad (4.52)$$

This reaction is the microscopic reverse of the carbonylation such that determining which group moves for the decarbonylation will also prove which group moves in the carbonylation (i.e., whether the reaction is a methyl migration or CO insertion). The results expected for both methyl migration and CO insertion are shown in Figure 4.16. The observed distribution of *cis* to *trans* product was 2 to 1, which is in agreement with methyl migration.[80]

Reaction of optically active alkyl groups show that the carbonylation proceeds with retention of configuration at the carbon.[81]

$$Me_3CC(H)(D)C(H)(D)FeCp(CO)_2 + PPh_3 \longrightarrow \qquad (4.53)$$
$$Me_3CC(H)(D)C(H)(D)C(O)FeCp(CO)PPh_3$$

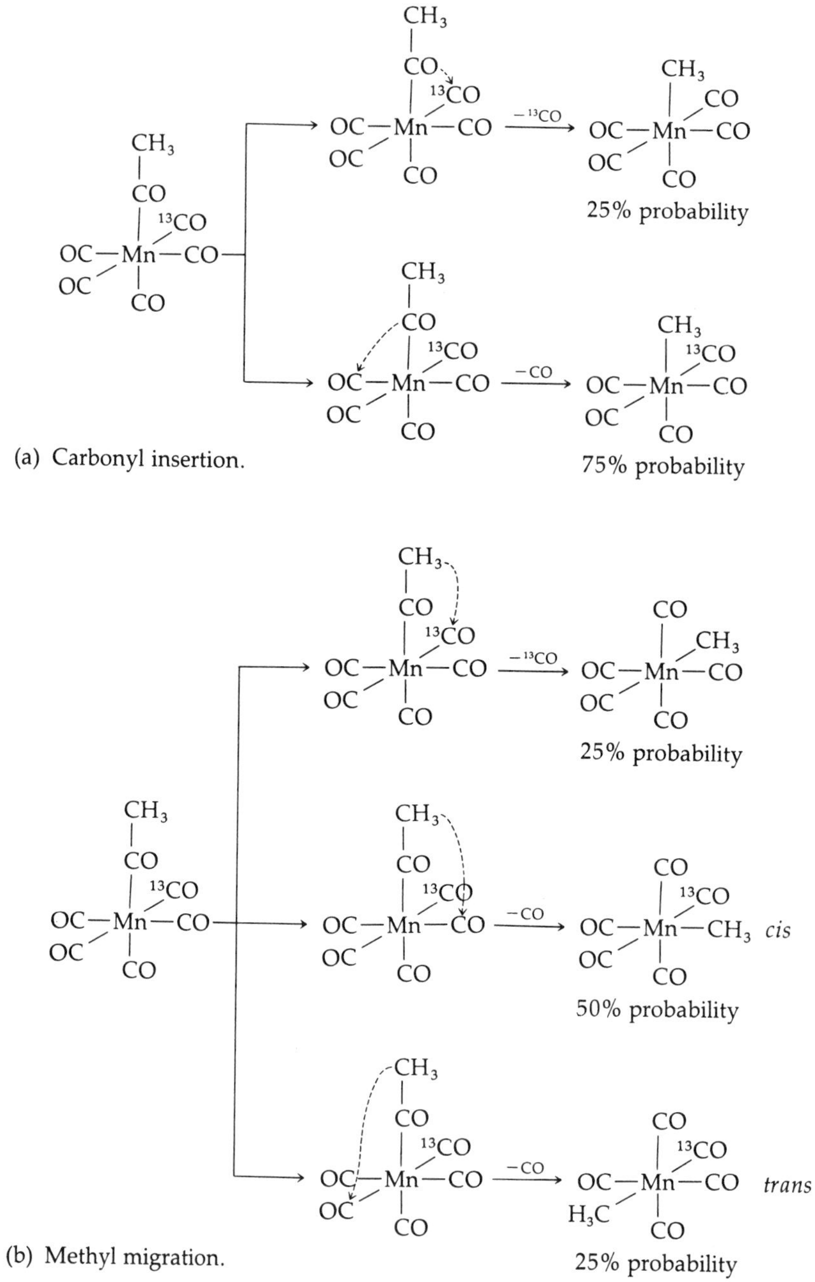

Figure 4.16. Decarbonylation of $CH_3C(O)Mn(CO)_5$ illustrating the differences in (a) carbonyl insertion and (b) methyl migration. [Reprinted with permission from K. Noack and F. Calderazzo, *J. Organomet. Chem.* **10,** 101 (1967). Copyright 1967 Elsevier Sequoia S. A.]

[threo-3,3-dimethylbutyl-1,2-d_2](Cp)dicarbonyliron + PPh$_3$ →
[threo-4,4-dimethylpentanoyl-2,3-d_2]
(Cp)carbonyltriphenylphosphineiron

This suggests that the C–Fe bond remains as the C–C bond forms, as shown in the following scheme.

$$\begin{array}{ccc} \text{-C-Fe} & \rightleftharpoons & \text{-C---Fe} \\ | & & \ddots \ | \\ \text{C} & & \text{C} \\ \text{O} & & \text{O} \end{array} \qquad (4.54)$$

However, it does not aid in determining whether the alkyl group migrates or the carbonyl inserts. It is consistent with the results of an extended Hückel calculation of the reaction coordinate for migration of CH$_3$ in CH$_3$Mn(CO)$_5$[82] that the C–Fe bond remains while the C–C bond forms. It was suggested that the transition state for the methyl migration corresponds to the methyl bonded to both the metal and to CO. This is shown in Figure 4.17.

Calculations on Pd(CH$_3$)(H)(CO)(PH$_3$) showed that the carbonylation reaction takes place by methyl migration.[83] The optically active complex CpFe(CO)(PPh$_3$)-(CH$_3$) was carbonylated stereospecifically; comparison with known absolute configurations showed that the acetyl group is located where the carbonyl was in the initial complex.[84]

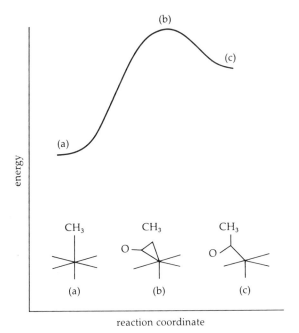

Figure 4.17. Energy profile calculated for methyl migration on CH$_3$Mn(CO)$_5$.

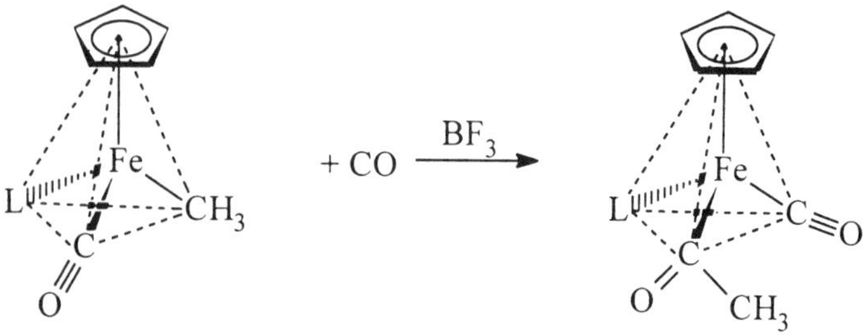

(4.55)

This nicely shows that methyl migration occurs. A methyl migration was also demonstrated for reaction of $Fe(CO)_2(I)(Me)(PMe_3)_2$ with ^{13}CO.[85] Thus, the methyl migration is well established for conversion of an alkyl, carbonyl complex to an acetyl complex.

4.7.5. Nature of the Intermediate

Considerable effort has been expended on trying to identify the 16-electron acyl intermediate, especially for manganese. Photochemical generation of the intermediate by CO dissociation from $CH_3C(O)Mn(CO)_5$ in an inert matrix at low temperature showed the geometry to be trigonal bipyramidal with the acyl in the equatorial plane.[86] Significant amounts of $CH_3Mn(CO)_5$ were also formed, suggesting a very low activation energy for methyl migration from the 16-electron acyl intermediate. Careful consideration of the labeling ratios in $CH_3Mn(CO)_5$ by ^{13}C NMR led to the conclusion that the geometry was square pyramidal with the acyl in the equatorial plane.[87] Figure 4.18 shows these possibilities. The two results may be rationalized by realizing that the ^{13}C NMR study was done in THF/acetone mixtures, where a solvent may occupy the empty coordination site, preventing relaxation of the square pyramid. There was no evidence for an η^2-acyl in either the matrix isolation or the ^{13}C NMR, as has been observed for early transition metal complexes.

A stable "intermediate" was formed by oxidative-addition of various substituted phenylacetyl chlorides to *trans*-chlorobis(triphenylphosphine)dinitrogeniridium.[88]

$$Ir(N_2)(PPh_3)_2Cl + CH_2YC(O)Cl \longrightarrow CH_2YC(O)Ir(PPh_3)_2Cl_2 \qquad (4.56)$$
$$+$$
$$N_2$$

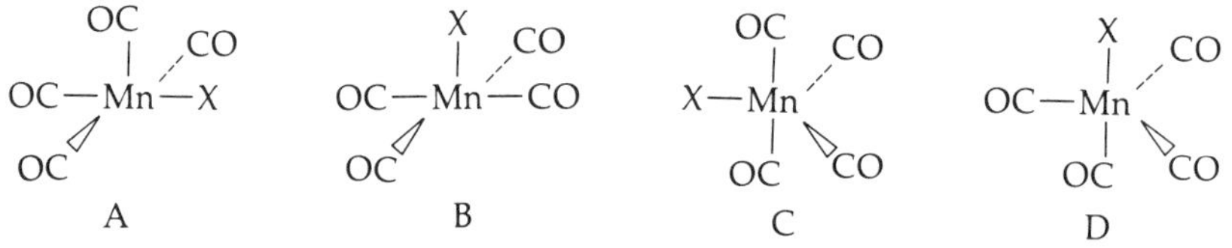

Figure 4.18. Possible intermediates in the reaction of $CH_3Mn(CO)_5$. On the basis of ^{13}C NMR investigations of the reaction intermediate A [X = $C(O)CH_3$] was suggested.

These five-coordinate iridium acyls could be isolated, characterized, and the kinetics of the rearrangement to the six-coordinate benzyl complex investigated.

$$(4.57)$$

$L = PPh_3$; $Y = C_6H_5$, C_6F_5, etc.

As expected the presence of a ligand that could occupy the sixth coordination position inhibits the reaction. As shown in Table 4.31 the presence of electron-releasing substituents on the benzyl group promotes migration and electron-withdrawing substituents retards the migration.[88]

Many of the features of alkyl migration that we have discussed are incorporated into a study of the decarbonylation of ^{13}C-labeled cis-acetylbenzoyltetracarbonyl-rhenate(I) complexes.[89]

$$NMe_4[cis\text{-}(CO)_4Re(C(O)CH_3)(^{13}C(O)Ph] \longrightarrow \qquad (4.58)$$

$$NMe_4(fac\text{-}(CO)_3(^{13}CO)ReC(O)CH_3(Ph)]$$

$$NMe_4[cis\text{-}(CO)_4Re(^{13}C(O)CH_3)(C(O)Ph] \longrightarrow \qquad (4.59)$$

$$NMe_4[cis\text{-}(CO)_4Re(^{13}C(O)Me)(Ph)]$$

$$+$$

$$NMe_4[mer\text{-}(CO)_3(^{13}CO)Re(C(O)Me)(Ph)]$$

Table 4.31. RATE CONSTANTS FOR THE REARRANGEMENT OF PHENYLACETYL, $CH_2YC(O)Ir(PPh_3)_2Cl_2$, COMPLEXES TO BENZYL(CARBONYL) COMPLEXES IN BENZENE AT 30°C[88]

Y	$k \times 10^4$
C_6H_5	7.09
$p\text{-}CH_3OC_6H_4$	9.19
$p\text{-}CH_3C_6H_4$	8.35
$p\text{-}O_2NC_6H_4$	3.90
C_6F_5	0.68[a]

[a] At 40°C.

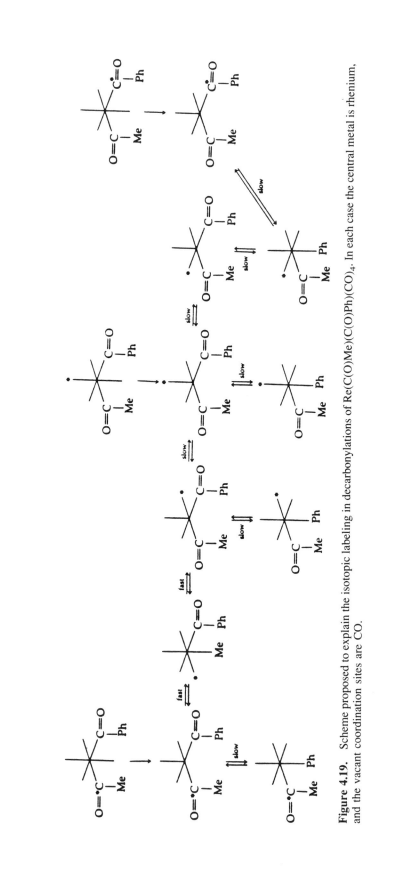

Figure 4.19. Scheme proposed to explain the isotopic labeling in decarbonylations of Re(C(O)Me)(C(O)Ph)(CO)$_4$. In each case the central metal is rhenium, and the vacant coordination sites are CO.

The scheme that was used to explain the labeling of these reactions is shown in Figure 4.19.[89] This detailed scheme is composed of three primary steps: methyl group migration, phenyl group rearrangement, and fluxional behavior of the five-coordinate intermediate. The methyl group migration occurs relatively rapidly, whereas phenyl migration and five-coordinate rearrangement occur at comparable (slower) rates. This scheme includes several aspects that we have discussed. Note that in each case the alkyl group migrates to a position *cis,* that CO dissociation occurs *cis* to both the acetyl and the benzoyl groups, and that equilibration of CO in the five-coordinate intermediate leads to all three possible products.[89] Phenyl migration to different carbonyls of $Cr(CO)_5C(O)Ph^-$ occurred readily at room temperature.[90]

Alkyl migration (CO insertion) is catalyzed by the presence of Lewis acids such as $AlCl_3$.[91] Cyclic products can be formed as shown in Reaction 4.60.

$$CH_3Mn(CO)_5 + AlCl_3 \longrightarrow (CO)_4Mn-C \begin{smallmatrix} {}_{\parallel\parallel}CH_3 \\ {}^{\diagdown}O \end{smallmatrix} \quad\quad (4.60)$$

The effect of several Lewis acids is shown in Table 4.32. As shown by the data in Table 4.32 a Lewis acid can accelerate methyl migration by a factor of 10^8. By coordinating to the oxygen of a CO, the Lewis acid weakens the C–O bond and lowers the activation energy for the migration. Such a mechanism is shown in the following.[91]

$$(CO)_4Mn-C\equiv O + AlCl_3 \rightleftharpoons (CO)_4Mn-C\equiv O-AlCl_3 \quad\quad (4.61)$$

with CH_3 groups on Mn.

$$(CO)_4Mn-C\equiv O\text{-}AlCl_3 \longrightarrow \left[(CO)_4Mn-C \right] \longrightarrow (CO)_4Mn-C \quad\quad (4.62)$$

Table 4.32. KINETIC DATA FOR
REACTIONS SIMILAR TO REACTION 4.61[91]

Lewis Acid	$k_{obs}(s^{-1})$
none	2×10^{-6}
AlClEt$_2$	0.37
AlCl$_2$Et	10
AlCl$_3$	170

4.7.6. Summary

To summarize the results on substitution at alkyl complexes: (1) A preequilibrium exists between the 18-electron alkyl carbonyl complex and a 16-electron acyl that probably has solvent coordinated. (2) The rate of migration depends on the electron-withdrawing ability of the alkyl group with more electron-withdrawing groups reacting more slowly. (3) More nucleophilic entering ligands speed the reaction. (4) The reaction proceeds by methyl migration as the carbanion. (5) The migration may be catalyzed by Lewis acids.

4.8. Hydride Complexes

Hydride complexes are important in many reactions catalyzed by organometallic complexes, yet the state of understanding of substitution reactions of hydride complexes is low. The primary reason is that metal hydrides are quite sensitive to oxygen, which makes quantitative studies difficult.

Another difficulty is that the presence of trace impurities leads to rapid reactions by *radical chain mechanisms*. Most hydride complexes have been observed to undergo facile ligand substitution reactions, probably by radical reactions such as those that have been studied for $HRe(CO)_5$.[92,93] Under rigorously pure conditions, with the exclusion of light, $HRe(CO)_5$ showed no reaction with PBu_3 in 60 days at $25°C$.[92] When such care was not taken the reaction yielded $HRe(CO)_4PBu_3$ and $HRe(CO)_3(PBu_3)_2$ at variable rates. The following sequence of reactions was proposed to explain the radical reactions.

$$\cdot R + HRe(CO)_5 \longrightarrow RH + \cdot Re(CO)_5 \tag{4.63}$$

$$\cdot Re(CO)_5 + L \rightleftharpoons \cdot Re(CO)_4L + CO \tag{4.64}$$

$$\cdot Re(CO)_4L + L \rightleftharpoons \cdot Re(CO)_3L_2 + CO \tag{4.65}$$

$$\cdot Re(CO)_4L + HRe(CO)_5 \longrightarrow \cdot Re(CO)_5 + HRe(CO)_4L \tag{4.66}$$

$$\cdot Re(CO)_3L_2 + HRe(CO)_5 \longrightarrow \cdot Re(CO)_5 + HRe(CO)_3L_2 \tag{4.67}$$

$$\cdot Re(CO)_3L_2 + HRe(CO)_4L \longrightarrow \cdot Re(CO)_4L + HRe(CO)_3L_2 \tag{4.68}$$

$$\cdot Re(CO)_{5-n}L_n + \cdot Re(CO)_{5-m}L_m \longrightarrow Re_2(CO)_{10-m-n}L_{n+m} \tag{4.69}$$

Thus, the substitution reaction is catalyzed by trace impurities (R•) that initiate a self-perpetuating chain mechanism for substitution. Reaction 4.69 is a chain termination step. Substitution reactions of 17-electron complexes (Reactions 4.64 and 4.65) occur readily (see earlier in Chapter 4). A primary piece of evidence for the radical chain mechanism was that photolysis of catalytic amounts of $Re_2(CO)_{10}$ under conditions where M–M bond cleavage is known to occur initiates the radical chain process.[92]

Other than radical processes, metal hydride complexes could undergo substitution by dissociative, associative, or hydride migration processes. Very few studies have been accomplished under conditions where we can have confidence that radical mechanisms are not operative. The reactions of $HMn(CO)_5$ with $L(L = {}^{13}CO$, $AsPh_3$, PPh_3 and PBu_3) have been investigated showing dependence of the rate on the nature and concentration of the entering ligand.[94] The rates for different ligands are shown in Table 4.33. These reactions clearly did not involve a CO dissociative route and were interpreted in terms of a hydride migration mechanism that has often been postulated in reactions of metal hydride complexes.

$$HMn(CO)_5 + L \longrightarrow HC(O)Mn(CO)_4L \qquad (4.70)$$
$$\downarrow -CO$$
$$HMn(CO)_4L$$

The dependence on the nature of the entering ligand is considerably larger than is observed for alkyl migration (Table 4.30). Nucleophilic attack must be much more important for the hydride complexes. Thus, the hydride migration would not be a rapid preequilibrium, and would only be formed as a result of the nucleophilic attack. At this point there is no evidence favoring hydride migration over direct nucleophilic attack on the complex in substitution reactions of $HMn(CO)_5$.

The very rapid reactions of hydride complexes, $H_2Fe(CO)_4$ and $H_2Ru(CO)_4$, which undergo substitutions rapidly at $-70°C$, were interpreted in terms of hydride migration.[95] Since these complexes also undergo facile thermal decomposition it was suggested that a weak M–H bond may be a requirement for both reactions.[95] A weak M–H bond may also facilitate radical mechanisms, and at this point there are no data that allow an unambiguous mechanistic interpretation of metal hydride substitution reactions that are not initiated by radical (17-electron) species. Much more research on reactions of metal hydride complexes is required before the mechanisms of this very important class of complexes are understood.

Table 4.33. VARIATION IN THE RATE OF SUBSTITUTION ON $HMn(CO)_5$ WITH $L = {}^{13}CO$, $AsPh_3$, PPh_3, PBu_3 AT 20°C[94]

L	Rate $(M^{-1}s^{-1})$
${}^{13}CO^a$	1.2×10^{-3}
$AsPh_3{}^b$	3×10^{-4}
PPh_3	1.4×10^{-2}
PBu_3	fast

[a] At 26°C.
[b] At 28°C.

4.9. Nitrosyl, Allyl, and Other Complexes

The nitrosyl ligand, NO, and the allyl ligand, $CH_2 = CH\text{-}CH_2$, may function as either three-electron or one-electron donors to a metal, and it is this feature that dominates their substitution reactions. The different types of binding are shown in Figure 4.20 for NO and C_3H_5. This capability allows associative reactions to occur through 18-electron intermediates.

$$\eta^3\text{-}C_3H_5Mn(CO)_4 + L \longrightarrow \eta^1\text{-}C_3H_5Mn(CO)_4L \tag{4.71}$$

$$\downarrow \text{-CO}$$

$$\eta^3\text{-}C_3H_5Mn(CO)_3L$$

Substitution reactions of nitrosyl complexes, $Mn(CO)_4NO$ and $Co(CO)_3NO$, show a rate law that is consistent with attack of the ligand in an associative reaction.[96,97]

$$\text{rate} = k[\text{Metal nitrosyl}][L] \tag{4.72}$$

The activation parameters as shown in Table 4.34 are also consistent with the associative nature of the substitution reaction. The negative entropies of activation are indicative of an associative process. The dependence on the entering ligand as shown in Table 4.35 shows the nucleophilic nature of the attack.

$$M(CO)_nNO + L \longrightarrow M(CO)_{n-1}(NO)L + CO \tag{4.73}$$

The ordering

$$PBu_3 > PPh_3 > P(OPh)_3 > AsPh_3$$

is in agreement with the nucleophilicities of these ligands. The data are all consistent with the sequence

$$Mn(CO)_4NO + L \longrightarrow [Mn(CO)_4L(NO)] \longrightarrow \tag{4.74}$$

$$Mn(CO)_3L(NO) + CO$$

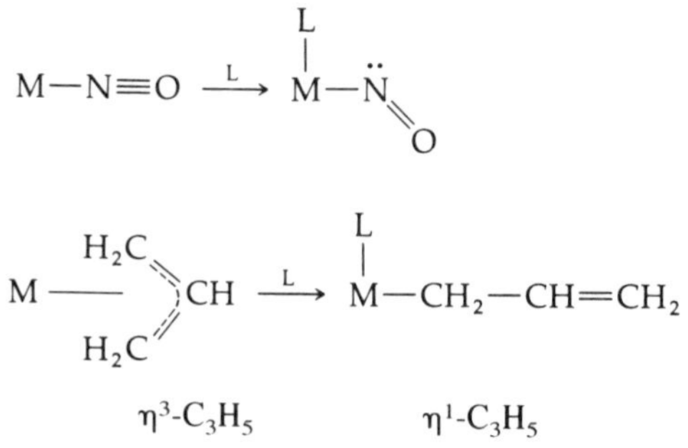

Figure 4.20. Modes of rearrangement for the nitrosyl and allyl groups to accommodate an entering 2-electron nucleophile.

Table 4.34. ACTIVATION PARAMETERS
FOR SUBSTITUTION OF NITROSYL
COMPLEXES[96,97]

Complex	$\Delta H^{\ddagger}$ (kcal/mole)	$\Delta S^{\ddagger}$ (eu)
$Mn(CO)_4NO$	19	-11
$Co(CO)_4NO$	21	-13

where the nitrosyl changes from a three-electron donor in the beginning and final complexes to a one-electron donor (bent configuration) in the intermediate. Each of the complexes in Equation 4.68 contains 18-electrons at the metal.

Further substitution on $Mn(CO)_3L(NO)$ shows a two-term rate law,[96]

$$Mn(CO)_3L(NO) + L \longrightarrow Mn(CO)_2L_2(NO) + CO \tag{4.75}$$

$$\text{rate} = (k_1 + k_2[L])[Mn(CO)_3L(NO)]$$

for L = PPh_3, PBu_3, and $P(OPh)_3$. It was suggested that the presence of the bulky phosphorus ligand inhibits further attack. However, phosphorus ligand exchange on $Mn(CO)_3(P(OPh)_3)NO$ by PBu_3 or PPh_3 occurs by a ligand-dependent pathway at a lower temperature than the further substitution for CO. It would seem most likely that the ligand attacks the complex in an associative step. *Bis*-substituted six-coordinate complexes often undergo relatively rapid

$$Mn(CO)_3(P(OPh)_3)NO + L \rightleftharpoons [Mn(CO)_3(P(OPh)_3)NO(L)] \tag{4.76}$$

phosphine or phosphite dissociation.[32-34] The phosphite would dissociate from the six-coordinate intermediate for phosphorus ligand exchange. For further substitution CO dissociation would require higher temperatures and the ligand attack would be a rapid preequilibrium. The rate law for this process would show two terms as observed. Substitution of other nitrosyl complexes has also shown a two-term rate law, and a similar mechanism may be operative for those complexes.[98,99]

Allyl complexes have the potential to react as the nitrosyl complexes. Reactions of $(\eta^3\text{-}C_3H_5)Mn(CO)_4$ with a ligand lead to $(\eta^1\text{-}C_3H_5)Mn(CO)_4L$, which could be viewed as the intermediate.

Table 4.35. DEPENDENCE OF THE RATE
FOR SUBSTITUTION OF NITROSYL
COMPLEXES ON THE ENTERING
LIGAND FOR EQUATION 4.73

L	M = Co, n = 3	M = Mn, n = 4
$AsPh_3$	.002	0.015
$P(OPh)_3$	0.03	0.08
PPh_3	1	1
PBu_3	90	40

$$(\eta^3\text{-}C_3H_5)Mn(CO)_4 \xrightarrow{L} (\eta^1\text{-}C_3H_5)Mn(CO)_4L \qquad (4.77)$$

$$\downarrow \Delta, \text{-CO}$$

$$(\eta^3\text{-}C_3H_5)Mn(CO)_3L$$

Although the kinetics of this substitution reaction have not been investigated, the mechanism is obvious and similar to that suggested for the analogous nitrosyl complex, $Mn(CO)_4(NO)$. Very few kinetics studies on allyl complexes have been accomplished that allow an unambiguous interpretation of the data. Reaction of (η^3-$C_3H_5)Co(CO)_3$ with PPh_3 under pseudo first-order conditions showed a rate law

$$(\eta^3\text{-}C_3H_5)Co(CO)_3 + PPh_3 \longrightarrow (\eta^3\text{-}C_3H_5)Co(CO)_2PPh_3 + CO \qquad (4.78)$$

$$\text{rate} = k[(\eta^3\text{-}C_3H_5)Co(CO)_3]$$

that is independent of the concentration of PPh_3.[100] This was interpreted as a CO dissociative process, although a rate-determining η^3-$C_3H_5 \rightarrow \eta^1$-$C_3H_5$ rearrangement would also be consistent. It is important to emphasize that for the observed rate law the η^3-allyl $\leftrightarrow$ η^1-allyl conversion cannot be a rapid preequilibrium. The sequence would have to be as follows.

$$(\eta^3\text{-}C_3H_5)Co(CO)_3 \xrightarrow{k} (\eta^1\text{-}C_3H_5)Co(CO)_3 \qquad (4.79)$$

$$(\eta^1\text{-}C_3H_5)Co(CO)_3 + L \xrightarrow{\text{fast}} (\eta^1\text{-}C_3H_5)Co(CO)_3L \qquad (4.80)$$

$$(\eta^1\text{-}C_3H_5)Co(CO)_3L \xrightarrow{\text{fast}} (\eta^3\text{-}C_3H_5)Co(CO)_2L + CO \qquad (4.81)$$

One could consider the first step to be rate-determining dissociation of the olefin. Although this scheme would be consistent with the binding capability of the allyl group it is no more reasonable than the CO dissociative process. Changing the substituent in the two position to either electron-withdrawing (Br, Cl, Ph) or electron-releasing (CH_3) groups increased the rate of reaction, which is not readily explained in terms of either CO dissociation or $\eta^3 \rightarrow \eta^1$ allyl interconversion.[100] Further kinetic studies of substitutions on allyl complexes will be required to understand the details of the reaction.

The capability of a ligand to coordinate with variable numbers of electrons may be important in reactions of organic ligands such as butadienes, cyclopentadienyls, and aromatic complexes. The possible bonding modes are shown in Figure 4.21. Very few kinetic studies have been reported on complexes of this type. Substitution reactions on (η^5-indenyl)tricarbonylmolybdenum halides showed a rate law of the type

$$\text{rate} = (k_1 + k_2[L])[(\eta^5\text{-}C_9H_7)Mo(CO)_3X]$$

and the second-order path was ascribed to an η^5-indenyl $\rightarrow$ η^3-indenyl conversion (cyclopentadienyl $\rightarrow$ allyl).[101] The generality of this mechanism is uncertain. Reaction of $CpRh(CO)_2$ with nucleophiles occurs by a second-order mechanism.[102]

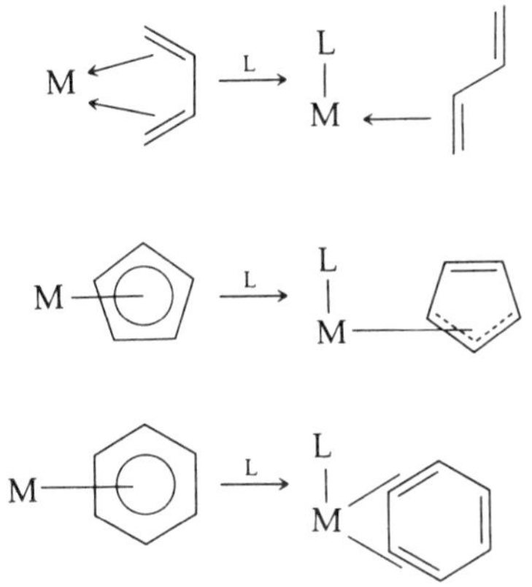

Figure 4.21. Possible modes of rearrangement for diene, cyclopentadiene, and arene complexes to accommodate an entering nucleophile.

$$CpRh(CO)_2 + L \longrightarrow CpRh(CO)L + CO \tag{4.82}$$

Data for different ligands as shown in Table 4.36 indicate the dependence on the nature of the entering nucleophile. The rate law is consistent with direct nucleophilic attack on the complex. This is most readily accomplished by attack leading to an $(\eta^3\text{-}C_5H_5)Rh(CO)_2L$, as follows:

$$CpRh(CO)_2 + L \longrightarrow [(\eta^3\text{-}C_5H_5)Rh(CO)_2L] \longrightarrow \tag{4.83}$$
$$CpRh(CO)L + CO$$

Synthetic evidence for reduced coordination by a cyclopentadienyl has been provided in reaction of PMe_3 with $CpRe(NO)(CO)CH_3$.[103]

Table 4.36. RATE CONSTANTS FOR
REACTION 4.82 AT 40°C SHOWING THE
DEPENDENCE ON THE NATURE OF L[102]

L	$k(M^{-1}s^{-1})$
Py	very slow
AsPh$_3$	very slow
P(OPh)$_3$	7.3×10^{-5}
PPh$_3$	2.9×10^{-4}
P(OBu)$_3$	5.6×10^{-4}
PBu$_3$	4.3×10^{-3}

$$CpRe(NO)(CO)(CH_3) + 2PMe_3 \longrightarrow \underset{Me_3P}{\overset{OC}{\diagdown}}\underset{}{\overset{NO}{\underset{|}{Re}}}\underset{PMe_3}{\overset{CH_3}{\diagup}} \qquad (4.84)$$

In this reaction the η^5-C_5H_5 group rearranges to an η^1-C_5H_5. This is especially interesting since the molecule has two other means to accommodate an incoming nucleophile: methyl migration to an acyl and changing the nitrosyl from a three-electron donor to a one-electron donor. If it is easier for a cyclopentadienyl to rearrange to η^3-C_5H_5 or η^1-C_5H_5 than for methyl migration or nitrosyl rearrangement to occur, then this may be a very common mode of reaction.

Displacement of arene ligands from organometallic complexes also occurs by a ligand-dependent path.[104]

$$(arene)Mo(CO)_3 + 3L \longrightarrow fac\text{-}Mo(CO)_3L_3 + arene \qquad (4.85)$$

The reaction is first order in arene complex and first order in entering nucleophile, is dependent on the nucleophilicity of the ligand L and leads specifically to the facial isomer.[104] The most reasonable mechanism, as shown in the following, involves reduced coordination by the arene from a six-electron donor, to a four-electron donor, to a two-electron donor before loss of arene from the complex.

$$(\eta^6\text{-}arene)Mo(CO)_3 + L \xrightarrow{\;k_1\;} (\eta^4\text{-}arene)Mo(CO)_3L \qquad (4.86)$$

$$(\eta^4\text{-}arene)Mo(CO)_3L + L \xrightarrow{\;k_2\;} (\eta^2\text{-}arene)Mo(CO)_3L_2 \qquad (4.87)$$

$$(\eta^2\text{-}arene)Mo(CO)_3L_2 + L \xrightarrow{\;k_3\;} fac\text{-}Mo(CO)_3L_3 + arene \qquad (4.88)$$

The rate-determining step would be k_1, which accounts for the rate dependencies. The facial stereochemistry requires that the arene remain coordinated through the final addition of L. Arene ligand exchange may also proceed through similar steps where the entering ligand is another arene.[105]

A detailed study of arene displacement from $(\eta^6\text{-}arene)Cr(CO)_3$ showed a process whose rate depended on $[(\eta^6\text{-}arene)Cr(CO)_3]$ and on $[L]$.[106]

$$(\eta^6\text{-}arene)Cr(CO)_3 + 3L \xrightarrow{\;k_3\;} fac\text{-} \text{ or } mer\text{-}Cr(CO)_3L_3 \qquad (4.89)$$

$L = PBu_3$ or PPh_3; arene = benzene, triphenylene, thiophene, etc.

The arene dependence of the reaction rate: benzene < triphenylene < pyrene < phenanthrene < naphthalene < anthracene, spans nearly eight orders of magnitude.[106] This order correlates with the loss of resonance energy on forming the η^4-arene and supports the initial step as rate determining.

Many types of ligands that have variable bonding capabilities undergo substitution reactions by nucleophilic attack on the metal complex. In each case the intermediate contains 18 electrons.

4.10. Summary

Ligand substitutions on eighteen-electron organometallic complexes occur by ligand dissociation (either D or I_d), unless there are special features of the ligands that allow associative steps. Ligands that have this capability are CH_3 and H, which can undergo migrations, NO and allyls, which can function as either three-electron donors or one-electron donors, and possibly other ligands, which can readily assume variable coordination modes such as η^5-$C_5H_5 \rightarrow \eta^3$-$C_5H_5$ conversions.

4.11. References

1. F. A. Cotton and C. S. Kraihanzel, *J. Am. Chem. Soc.* **84**, 4432 (1962).

2. S. O. Grim, D. A. Wheatland, and W. McFarlane, *J. Am. Chem. Soc.* **89**, 5573 (1967).

3. G. M. Bodner, *Inorg. Chem.* **14**, 2694 (1975).

4. T. L. Brown, *Inorg. Chem.* **31**, 1286 (1992).

5. M. J. Wovkulich and J. D. Atwood, *J. Organomet. Chem.* **184**, 77 (1980).

6. J. P. Collman and L. S. Hegedus, *Principles and Applications of Organotransition Metal Chemistry* (University Science Books, Hill Valley, California, 1980).

7. S. P. Wang, M. J. Richmond, and M. Schwartz, *J. Am. Chem. Soc.* **114**, 7595 (1992).

8. J. D. Atwood, M. J. Wovkulich, and D. C. Sonnenberger, *Acc. Chem. Res.* **16**, 350 (1983).

9. C. A. Tolman, *J. Am. Chem. Soc.* **92**, 2956 (1970).

10. (a) Md. M. Rahman, H. Y. Liu, A. Prock, and W. P. Giering, *Organometallics* **6**, 650 (1987). (b) M. N. Golovin, Md. M. Rahman, J. E. Belmonte, and W. P. Giering, *Organometallics* **4**, 1981 (1985).

11. C. A. Tolman, *Chem. Soc. Rev.* **1**, 337 (1972).

12. W. A. Herrmann, *J. Organomet. Chem.* **383**, 21 (1990).

13. J. P. Day, F. Basolo, R. G. Pearson, L. F. Kangas, and P. M. Henry, *J. Am. Chem. Soc.* **90**, 1925 (1968).

14. R. J. Angelici and F. Basolo, *J. Am. Chem. Soc.* **84**, 2495 (1962).

15. J. M. Kelley, H. Herman, and E. K. Von Gustorf, *J. C. S. Chem. Comm.* 105 (1973).

16. F. Basolo and A. Wojcicki, *J. Am. Chem. Soc.* **83**, 520 (1961).

17. J. P. Day, F. Basolo, and R. G. Pearson, *J. Am. Chem. Soc.* **90**, 6927 (1968).

18. J. R. Graham and R. J. Angelici, *Inorg. Chem.* **6**, 2082 (1967).

19. W. D. Covey and T. L. Brown, *Inorg. Chem.* **12**, 2820 (1973).

20. F. Ungvary and A. Wojcicki, *J. Am. Chem. Soc.* **109**, 6848 (1987).

21. M. Meier, F. Basolo, and R. G. Pearson, *Inorg. Chem.* **8**, 795 (1969).

22. S. P. Modi and J. D. Atwood, *Inorg. Chem.* **22**, 26 (1983).

23. M. Poliakoff and J. J. Turner, *J. Chem. Soc., Dalton Trans.* 2276 (1974).

24. J. D. Atwood, *J. Organomet. Chem.* **383**, 59.

25. M. R. J. Dack, *J. Chem. Ed.* **51**, 231 (1974).

26. C. L. Hyde and D. J. Darensbourg, *Inorg. Chem.* **12**, 1286 (1973).

27. G. Cardaci and V. Narciso, *J. C. S. Chem. Comm.* 2289 (1972).

28. A. Wojcicki and F. Basolo, *J. Am. Chem. Soc.* **83**, 525 (1961).

29. J. D. Atwood and T. L. Brown, *J. Am. Chem. Soc.* **97**, 3380 (1975).

30. J. D. Atwood and T. L. Brown, *J. Am. Chem. Soc.* **98**, 3155 (1976).

31. D. J. Darensbourg and A. H. Graves, *Inorg. Chem.* **18**, 1257 (1979).

32. M. J. Wovkulich and J. D. Atwood, *J. Organomet. Chem.* **184**, 77 (1980).

33. M. J. Wovkulich, S. J. Feinberg, and J. D. Atwood, *Inorg. Chem.* **19**, 2608 (1980).

34. M. J. Wovkulich and J. D. Atwood, *Organometallics* **1**, 1316 (1982).

35. L. E. Manzer and C. A. Tolman, *J. Am. Chem. Soc.* **97**, 1955 (1975).

36. J. D. Atwood and T. L. Brown, *J. Am. Chem. Soc.* **98**, 3160 (1976).

37. R. J. Angelici, S. E. Jacobson, and C. M. Ingemanson, *Inorg. Chem.* **7**, 2466 (1968).

38. R. H. Reimann and E. Singleton, *J. Chem. Soc. Dalton*, 841 (1973).

39. A. Holladay, M. R. Churchill, A. Wong, and J. D. Atwood, *Inorg. Chem.* **19**, 2195 (1980).

40. *Organometallic Radical Processes*, ed. W. C. Trogler (Elsevier Science Publishers, 1990) and references therein.

41. T. L. Brown, in *Organometallic Radical Processes*, ed. W. C. Trogler (Elsevier, Amsterdam, 1990), p. 67.

42. W. C. Trogler, *Int. J. Chem. Kinet.* **19**, 1024 (1987).

43. W. C. Trogler, in *Organometallic Radical Processes*, ed. W. C. Trogler (Elsevier, Amsterdam, 1990), p. 306.

44. Z. Lin and M. B. Hall, *J. Am. Chem. Soc.* **114**, 6574 (1992).

45. H. Wawersik and F. Basolo, *Inorg. Chim. Acta* **3**, 113 (1969).

46. L.I.B. Haines, D. Hopgood, and A. J. Poë, *J. Chem. Soc. A*, 421 (1968).

47. L.I.B. Haines and A. J. Poë, *J. Chem. Soc. A*, 2826 (1969).

48. D. Sonnenberger and J. D. Atwood, *J. Am. Chem. Soc.* **102**, 3484 (1980).

49. J. D. Atwood, *Inorg. Chem.* **20**, 4031 (1981).

50. A. M. Stolzenberg and E. L. Muetterties, *J. Am. Chem. Soc.* **105**, 822 (1983).

51. N. J. Coville, A. M. Stolzenberg, and E. L. Muetterties, *J. Am. Chem. Soc.* **105**, 2499 (1983).

52. N. J. Coville and P. Johnston, *J. Organomet. Chem.* **363**, 343 (1989).

53. M. Absi-Halabi, J. D. Atwood, N. P. Forbes, and T. L. Brown, *J. Am. Chem. Soc.* **102**, 6248 (1980).

54. S. M. Grant and A. R. Manning, *Inorg. Chim. Acta* **31**, 41 (1978).

55. A. F. Clifford and A. K. Mukererjce, *Inorg. Chem.* **2**, 151 (1963).

56. J. P. Candlin and A. C. Shortland, *J. Organomet. Chem.* **16**, 289 (1969).

57. S. M. Grant and A. R. Manning, *Inorg. Chim. Acta* **31**, 41 (1978).

58. G. F. Stuntz and J. R. Shapley, *J. Am. Chem. Soc.* **99**, 607 (1977).

59. A. Shojaie and J. D. Atwood, *Organometallics* **4**, 187 (1985).

60. R. Shojaie and J. D. Atwood, *Inorg. Chem.* **26**, 2199 (1987).

61. R. Shojaie and J. D. Atwood, *Inorg. Chem.* **27**, 2558 (1988).

62. (a) L. M. Bavaro and J. B. Keister, *J. Organomet. Chem.* **287**, 357 (1985). (b) L. M. Bavaro, P. Montangero, and J. B. Keister, *J. Am. Chem. Soc.* **105**, 4977 (1983).

63. D. S. Parfitt, J. D. Jordan, and J. B. Keister, *Organometallics* **11**, 4009 (1992).

64. K. J. Karel and J. R. Norton, *J. Am. Chem. Soc.* **96**, 6812 (1974).

65. D. Sonnenberger and J. D. Atwood, *Inorg. Chem.* **20**, 3243 (1981).

66. D. Sonnenberger and J. D. Atwood, *J. Am. Chem. Soc.* **104**, 2113 (1982).

67. D. Sonnenberger and J. D. Atwood, *Organometallics* **1**, 694 (1982).

68. G. F. Stuntz and J. R. Shapley, *J. Organomet. Chem.* **213**, 389 (1981).

69. S. K. Malik and A. Poë, *Inorg. Chem.* **17**, 1484 (1978).

70. D. J. Darensbourg and B. J. Baldwin-Zuschke, *J. Am. Chem. Soc.* **104**, 3906 (1981).

71. D. J. Darensbourg and B. J. Baldwin-Zuschke, *Inorg. Chem.* **20**, 3846 (1981).

72. G. Huttner, J. Schneider, H. D. Muller, G. Mohr, J. van Seyer, and L. Wohlfahrt, *Angew. Chem. Int. Ed. Engl.* **18**, 76 (1979).

73. (a) J. R. Fox, W. L. Gladfelter, and G. L. Geoffroy, *Inorg. Chem.* **19**, 2574 (1980). (b) J. R. Fox, W. L. Gladfelter, T. G. Wood, J. A. Smegal, T. K. Foreman, G. L. Geoffroy, I. Tavanaiepour, V. W. Day, and C. S. Day, *Inorg. Chem.* **20**, 3214 (1981).

74. F. Calderazzo and F. A. Cotton, *Inorg. Chem.* **1**, 30 (1962).

75. I. S. Butler, F. Basolo, and R. G. Pearson, *Inorg. Chem.* **6**, 2074 (1967).

76. J. N. Cawse, R. A. Fiato, and R. L. Pruett, *J. Organomet. Chem.* **172**, 405 (1979).

77. K. Nicholas, S. Raghu, and M. Rosenblum, *J. Organomet. Chem.* **78**, 133 (1974).

78. M. J. Wax and R. G. Bergman, *J. Am. Chem. Soc.* **103**, 7028 (1981).

79. C. S. Kraihanzel and P. K. Maples, *Inorg. Chem.* **7**, 1806 (1968).

80. K. Noack and F. Calderazzo, *J. Organomet. Chem.* **10**, 101 (1967).

81. G. M. Whitesides and D. J. Boschetto, *J. Am. Chem. Soc.* **91**, 4313 (1969).

82. H. Berke and R. Hoffmann, *J. Am. Chem. Soc.* **100**, 7224 (1978).

83. N. Koga and K. Morokuma, *J. Am. Chem. Soc.* **107**, 7230 (1985).

84. H. Brunner, B. Hammer, I. Bernal, and M. Druax, *Organometallics* **2**, 1595 (1983).

85. S. C. Wright and M. C. Baird, *J. Am. Chem. Soc.* **107**, 6899 (1985).

86. T. M. McHugh and A. J. Rest, *J. Chem. Soc., Dalton Trans.* 2323 (1980).

87. T. C. Flood, J. E. Jensen, and J. A. Statler, *J. Am. Chem. Soc.* **103**, 4411 (1981).

88. M. Kubota, D. M. Blake, and S. A. Smith, *Inorg. Chem.* **10**, 1430 (1971).

89. C. P. Casey and L. M. Baltusis, *J. Am. Chem. Soc.* **104**, 6347 (1982).

90. I. Lee and N. J. Cooper, *J. Am. Chem. Soc.* **115**, 4389 (1993).

91. T. G. Richmond, F. Basolo, and D. F. Shriver, *Inorg. Chem.* **21**, 1272 (1982).

92. B. H. Byers and T. L. Brown, *J. Am. Chem. Soc.* **99**, 2527 (1977).

93. N. W. Hoffman and T. L. Brown, *Inorg. Chem.* **17**, 613 (1978).

94. B. H. Byers and T. L. Brown, *J. Organomet. Chem.* **127**, 181 (1977).

95. R. G. Pearson, H. W. Walker, H. Mauermann, and P. C. Ford, *Inorg. Chem.* **20**, 2741 (1981).

96. H. Wawersik and F. Basolo, *J. Am. Chem. Soc.* **89**, 4626 (1967).

97. E. M. Thorsteinson and F. Basolo, *J. Am. Chem. Soc.* **88**, 3929 (1966).

98. G. Innorta, G. Reichenbach, and A. Foffani, *J. Organomet. Chem.* **22**, 731 (1970).

99. G. Cardaci and S. M. Murgia, *Inorg. Chim. Acta* **6**, 222 (1972).

100. R. F. Heck, *J. Am. Chem. Soc.* **85**, 651 (1963).

101. A. J. Hart-Davis, C. White, and R. J. Mawby, *Inorg. Chim. Acta* **4**, 441 (1970).

102. H. G. Schuster-Woldan and F. Basolo, *J. Am. Chem. Soc.* **88**, 1657 (1966).

103. C. P. Casey and W. D. Jones, *J. Am. Chem. Soc.* **102**, 6156 (1980).

104. F. Zingales, A. Chiesa, and F. Basolo, *J. Am. Chem. Soc.* **88**, 2707 (1966).

105. W. Strohemeier and R. Müller, *Z. Phys. Chem.* **40**, 85 (1964).

106. S. Zhang, J. K. Shen, F. Basolo, T. D. Ju, R. F. Lang, G. Kiss, and C. D. Hoff, *Organometallics* **13**, 3692 (1994).

4.12. Problems

4.1. Show that the following are 18-electron complexes:

$W(CO)_6$, $HFe(CO)_4^-$, $CH_3Co(CO)_3PPh_3$, $Ru_3(CO)_{12}$, $Mn(CO)_4NO$, $(\eta^6\text{-}C_6H_6)Cr(CO)_3$, $(\eta^5\text{-}C_5H_5)Fe(CO)_2(\eta^1\text{-}C_5H_5)$.

4.2. Show the electron count for the starting material, intermediates, and products of the following organometallic substitution reactions.

a. $Mn(CO)_5Br + PPh_3 \rightarrow Mn(CO)_4(PPh_3)Br + Co$

b. $CH_3MoCp(CO)_3 + PPh_3 \rightarrow CH_3MoCp(CO)_2PPh_3 + CO$

c. $HRe(CO_5 \xrightarrow{PBu_3} HRe(CO_4)PBu_3 + HRe(CO)_3(PBu_3)_2$

d. $Co(CO)_3NO + PPh_3 \rightarrow Co(CO)_2(PPh_3)NO$

4.3. Discuss in terms of ligand bonding the infrared data for *trans*-$Cr(CO)_4L_2$ complexes as shown:

L	$V(cm^{-1})$
$P(OPh)_3$	1935
$P(OMe)_3$	1916
PPh_3	1884
PBu_3	1873

4.4. Why are the solvent effects for organometallic substitution reactions that proceed by ligand dissociation (Table 4.7) so small in comparison to the solvent effects for ligand dissociation in Werner-type inorganic complexes?

4.5. What information does the competition ratio give regarding energy profiles?

4.6. Rates of decarbonylation of $RC(O)Mn(CO)_5$, which proceed by rate-determining CO dissociation, are shown as a function of R. [J. N. Cawse, R. A. Fiato, and R. L. Pruett, *J. Organomet. Chem.* **172**, 405 (1979)].

R	$10^5 k(s^{-1})$
CF_3	2.8
CH_3OCH_2	58
$C_6H_5CH_2$	160
CH_3	250

Explain these observations.

4.7. Explain the order of ligand effects observed for the further substitution of

$$Mn_2(CO)_9L: PPh_3 > PBu_3 > P(OPh)_3 > CO$$

4.8. Suggest a reason other than steric effects for why phosphines cannot be substituted as extensively as phosphites on metal carbonyl complexes. [Consider the infrared frequencies of fac-$Cr(CO)_3(PEt_3)_3$ 1920(s), 1820(s) cm^{-1}].

4.9. The general order of ligand bond strengths to organometallic centers is given.

$$M\text{-}PBu_3 > M\text{-}P(OMe)_3 > M\text{-}P(OPh)_3 > M\text{-}PPh_3 > M\text{-}NC_5H_5$$

Consider this order and contrast it to σ donation, π acceptance, and steric size. Offer an explanation for the observed order of bond strength.

4.10. Explain in some detail the data presented in Table 4.10.

4.11. Explain the observation that $Mn_2(CO)_9PPh_3$ and $Re_2(CO)_9PPh_3$ are not observed but $MnRe(CO)_9PPh_3$ is observed as intermediates in the substitution reactions (D. Sonnenberger and J. D. Atwood, *J. Am. Chem. Soc.* **102**, 3484 (1980)).

4.12. The order of reactivity for trinuclear metal clusters, $M_3(CO)_{12}$ (M = Fe, Ru, Os) is Fe > Ru > Os, which is different than it is for mononuclear complexes. At this point the reason for this reversal is unknown. Suggest at least one possible reason.

4.13. Show that the rate law for $CH_3Mn(CO)_5$ is that given as derived from Reactions 4.43 and 4.44.

4.14. Explain the observations which led to the suggestion of methyl migration for $Mn(CO)_5CH_3$ and $CpFe(CO)(L)CH_3$.

4.15. Draw an energy profile for the reaction of $CH_3Mn(CO)_5$ with CO.

4.16. Predict the effect that a ligand (PR_3) will have on an alkyl migration (how will replacing a CO with PR_3 effect the energy profile from Prob. 4.15?)

4.17. How does the presence of Lewis-acid affect the energy profile in Problem 4.15?

4.18. Most catalytic reactions of transition metal hydride complexes are suggested to occur by ligand dissociation. Can you rationalize these postulates in terms of the observed stoichiometric reactions of transition metal hydrides?

4.19. $Mn(CO)_4NO$ and $Co(CO)_3NO$ react by associative reactions. Can you justify their relative reactivity by consideration of crystal field activation energies (refer to Table 3.9)?

4.20. Reactions of a number of complexes of bidentate ligands [i.e., $(Ph_2AsCH_2CH_2AsPh_2)$-$Cr(CO)_4$] occur by ligand-dependent pathways. Suggest a scheme that conforms to the $18e^- \rightarrow 16e^- \rightarrow 18e^-$ transition and show how the ligand dependency arises. [G. R. Dobson, *Acct. Chem. Res.* **9**, 300 (1976)].

4.21. Based on the scheme from Problem 4.20 and your knowledge of metal ligand bond strengths, suggest the initial (mild condition) product of the reaction of $Cr(CO)_4(P^\wedge N)$ where $P^\wedge N$ is a bidentate ligand with a phosphorus and a nitrogen donor atom. [W. J. Knebel and R. J. Angelici, *Inorg. Chem.* **13**, 627 (1974)].

4.22. Room temperature reaction of $Cr(CO)_5^{2-}$ with $Ph^{13}C(O)Cl$ and subsequent methylation with Me_3O^+ led to the carbene $(CO)_5CrC(OMe)Ph$ with ^{13}CO scrambled throughout the carbonyls. Suggest a mechanism. [I. Lee and N. J. Cooper, *J. Am. Chem. Soc.* **115**, 4389 (1993)].

4.23. Given the following data on $\Delta H^\ddagger$ for arene displacement and ΔH_{rxn} for arene exchange on $(\eta^6\text{-arene})Cr(CO)_3$ provide a detailed reaction coordinate diagram for arene displacement by a phosphine ligand.

Arene	$\Delta H^\ddagger$ (kcal/mole)	ΔH_{rxn} (kcal/mole)
benzene	23	0.0
triphenylene	14	2.4
pyrene	13	6.9
naphthalene	10	4.9

S. Zhang, J. K. Shen, F. Basolo, T. D. Ju, R. F. Lang, G. Kiss, and C. D. Hoff, *Organometallics* **13**, 3692 (1994).

4.24. What rate law would one observe for each of the following reactions?

    a. $Ni(CO)_4 + PPh_3 \rightarrow$
    b. $CH_3Mn(CO)_5 + PPh_3 \rightarrow$
    c. $CpRh(CO)_2 + PPh_3 \rightarrow$
    d. $Fe_3(CO)_{12} + {}^{13}CO \rightarrow$

4.25. A linear correlation has been found between the ^{103}Rh chemical shift of $Cp^*Rh[C(O)p\text{-}C_6H_4X]I(PPh_3)$ and the rate constant for aryl migration to form the product.

$$Cp^*Rh(CO)(I)(p\text{-}C_6H_4X) \xrightarrow[+PPh_3]{k} Cp^*Rh[C(O)p\text{-}C_6H_4X](I)(PPh_3)$$

[V. Tedesco and W. von Philipsborn, *Organometallics* **14**, 3600 (1995).] What are the implications of this correlation for the mechanism of aryl migration?

4.26. $Cp^*Co(CO)$ displays markedly different behavior in Kr(l) than does the rhodium analogue. The differences include much more rapid reaction with the starting dicarbonyl, a reaction that is unaffected by the presence of Xe(l) or cyclohexane. Suggest a reason for the different reactivity of $Cp^*Co(CO)$ and $Cp^*Rh(CO)$. [A. A. Bengali, R. G. Bergman, and C. B. Moore, *J. Am. Chem. Soc.* **117**, 3879, (1995)].

4.27. Photolysis of $Mn(C(O)Me)(CO)_5$ [IR: 2113(w), 2051(m), 2012(s) and 1661(w)] in cyclohexane resulted in a product [IR: 1991(s), 1952(s) and 1607(w) cm^{-1}] that is different than the product of photolysis in THF [IR: 1981(br) and 1931(Br)]. Suggest a reason for these differences. [W. T. Boese and P. C. Ford, *J. Am. Chem. Soc.* **117**, 8381 (1995).]

5

Oxidative Addition and Reductive Elimination

Oxidative addition and reductive elimination are reactions often used in synthesis. They are involved in most schemes proposed for homogeneously catalyzed reactions. Oxidative addition reactions lead to an expansion of the coordination sphere of a metal with a concomitant increase of the formal oxidation state. Reductive elimination reactions are the reverse of oxidative addition reactions, and they lead to a decrease in the coordination sphere of the metal and to a decrease in the formal oxidation state of the metal. Although in many cases the oxidative addition and reductive elimination reactions are the microscopic reverse of one another and mechanistic information for one may pertain to the other, studies have typically focused on the reactions separately.

5.1. Oxidative Addition

Oxidative addition has been observed at a number of metal centers, often with different mechanisms.[1-5] The reaction can generally be represented as in Reaction 5.1, where M represents a metal complex and XY is a compound such as hydrogen, halogen, alkyl halide, or metal halide. The mechanism depends on both the metal complex and the adding group.[3] Several examples are shown here and in Figure 5.1.[3]

$$M + XY \rightarrow M(X)(Y) \tag{5.1}$$

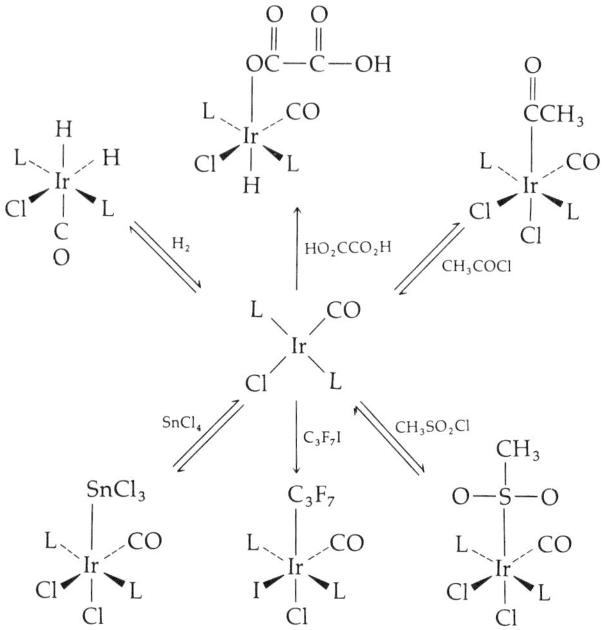

Figure 5.1. Examples of oxidative addition reactions. [J. P. Collman, *Acct. Chem. Res.* **1**, 136 (1968).]

$$trans\text{-}Ir(CO)(PPh_2Me)_2Cl + HBr \longrightarrow$$

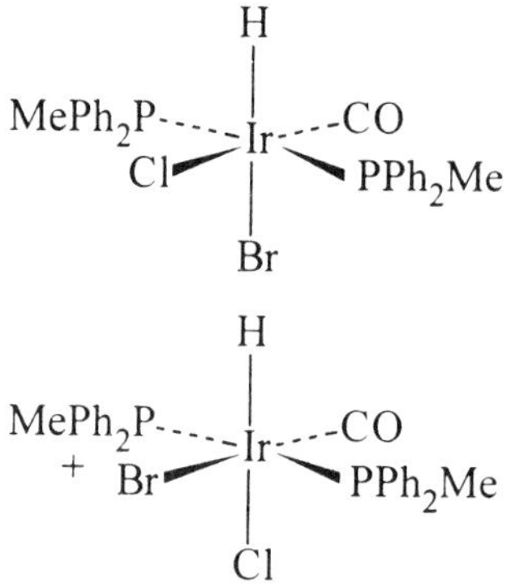

$$(5.2)$$

$$Os(CO)_5 + H_2 \rightarrow cis\text{-}H_2Os(CO)_4 + CO \qquad (5.3)$$

$$2Co(CN)_5^{3-} + CH_3I \rightarrow Co(CN)_5I^{3-} + Co(CN)_5CH_3^{3-} \qquad (5.4)$$

$$Pt(PPh_3)_2 + Ph_3SnCl \rightarrow Pt(PPh_3)_2(SnPh_3)Cl \qquad (5.5)$$

In Reaction 5.2 the square planar iridium(I) complex is transformed into an octahedral Ir(III) complex by oxidative addition of HBr. Oxidative addition of H_2 to $Os(CO)_5$ in Reaction 5.3 causes the trigonal bipyramidal Os(0) complex to be converted to an octahedral Os(II) complex. Reaction 5.4 is an example of addition to two metal centers, where each five-coordinate Co(II) complex is changed to an octahedral Co(III) complex.[2] In Reaction 5.5 a two-coordinate Pt(0) complex is converted to a square-planar Pt(II) complex. These examples indicate different types of oxidative addition reactions. Most of the detailed studies have been accomplished on square-planar complexes such as those shown in Reaction 5.2 and Figure 5.1.

5.1.1. Stereochemistry

Oxidative addition is stereospecific with only one of the possible isomers prepared in most cases.[3–6] Addition of hydrogen to square-planar complexes leads to *cis* distribution of the hydrogen atoms in the octahedral product. Addition to $Ir(CO)(PPh_3)_2Cl$ is shown in Figure 5.1 Hydrogen halides and silanes also add *cis* to square-planar complexes. The stereochemistry of H_2 addition to iridium(I) has been considered experimentally and theoretically.[7–9] Crabtree noted that with H_2 adding *cis* to *trans*-$Ir(CO)L_2X$ two possibilities exist,

$$X\text{----}Ir\text{----}L \quad \xrightarrow{H_2} \quad X\text{----}Ir\text{----}L \quad \text{or} \quad X\text{----}Ir\text{----}L \qquad (5.6)$$

parallel   perpendicular

and used the terms *parallel* and *perpendicular* to describe the possibilities.[7] Table 5.1 shows the products formed for different X and L groups.[7] Theoretical studies by Sargent and Hall show that X is crucial to the stereochemistry with weak donors (Cl⁻) favoring addition in the X–Ir–CO plane (parallel), while strong donors (Me, Ph) favor addition in the L–Ir–L plane (perpendicular).[9] Similar studies of oxidative addition of HI and $HSiEt_3$ to $Ir(Br)(CO)(dppe)$ show a different stereoselection for the two molecules.[10]

Table 5.1. STEREOCHEMISTRY OF OXIDATIVE ADDITION TO *trans*-$Ir(CO)L_2X$[7]

X	L	Product
Cl	PMe_3	parallel
Ph	PMe_3	perpendicular (kinetic); parallel (thermodynamic)
Cl	PPh_3	parallel
Br	PPh_3	parallel
Me	PPh_3	perpendicular[8]

$$\text{(5.7)}$$

The silane addition (nucleophilic) occurs in the P–Ir–CO plane while HI addition (electrophilic) occurs in the P–Ir–Br plane.[10] Isomerization occurred after the initial oxidative addition in each case.

Oxidative addition of other molecules to square-planar complexes occurs *trans* as shown in Figure 5.1 and Reaction 5.2.[6] In a few cases mixtures of isomers are observed in the final product mixture. The reaction of HBr with Ir(PPh$_2$Me)(CO)Cl is one example (Reaction 5.2) where both *cis* and *trans* isomers are formed. The reaction of HCl with Ir(PPh$_2$Me)$_2$(CO)Br produced the same mixture of products.[3]

$$\text{trans-Ir(PPh}_2\text{Me)}_2\text{(CO)Cl + HBr} \longrightarrow \quad \text{(5.8)}$$

$$\text{trans-Ir(PPh}_2\text{Me)}_2\text{(CO)Br + HCl} \longrightarrow \quad \text{(5.9)}$$

The same product ratio from two different routes suggests that an equilibrium has been attained between the two octahedral isomers leading to an equilibrium distribution. Thus, the addition may be stereospecific although a mixture is observed.[3] A solvent dependence of the stereochemical course of oxidative addition of hydrogen halides to *trans*-IrL$_2$(CO)X (L = phosphine) has been observed.[11] In methanol, water, or acetonitrile mixtures of isomers are obtained (as in Reactions 5.6 and 5.7); in benzene or chloroform, the addition leads exclusively to the *cis* isomer.

$$trans\text{-}Ir(CO)L_2X + HCl \longrightarrow \quad \begin{array}{c} L \\ | \\ H\text{----}Ir\text{----}CO \\ Cl\text{---}\!\!\diagdown\!\!| \quad \diagdown X \\ | \\ L \end{array} \qquad (5.10)$$

In polar solvents free halide may be present to exchange with the square planar complex $trans\text{-}Ir(CO)L_2X$ or with the six-coordinate complex $IrL_2(CO)X_2(H)$.[11]

Oxidative additions to five-coordinate, 18-electron complexes have some differences to the preceding square planar examples. The addition is accompanied by ligand loss such that the product of the addition has 18 electrons. The addition occurs with *cis* stereochemistry, even for addition of polar molecules such as CH_3I and HCl.[3]

Enthalpy changes for oxidative addition to Ir(I) have been evaluated by adiabatic titration calorimetric methods.[12]

$$trans\text{-}IrCl(CO)(PMe_3)_2 + RI \rightarrow Ir(R)(I)(Cl)(CO)(PMe_3)_2 \qquad (5.11)$$

$$R = I, H, CH_3, C_2H_5, n\text{-}C_3H_7, \text{ etc.}$$

Values are shown in Table 5.2. These enthalpy changes have a number of implications for bonding and reactions, but for the purposes of our discussion the stronger bond of R to Ir leads to a larger enthalpy for the reaction.[12] In the reaction with Ir each addition has a favorable enthalpy.

5.1.2. Kinetic Studies

Most kinetic studies have been accomplished on Ir(I) complexes, $trans\text{-}IrX(CO)L_2$.[13] Oxidative addition reactions exhibit second-order kinetics, dependent on the concentration of the metal complex and on the adding molecule.

Table 5.2. ENTHALPY CHANGES (kJ/mole) FOR OXIDATIVE-ADDITION OF RI TO $trans\text{-}IrCl(CO)(PMe_3)_2$ IN 1,2 DICHLOROETHANE[12]

RI	ΔH
I_2	-185 ± 7
HI	-160 ± 3
MeI	-117 ± 7
EtI	-110 ± 3
PrI	-103 ± 4
i-PrI	-88 ± 10
$C_6H_5CH_2I$	-95 ± 7
$CH_3C(O)I$	-125 ± 4
$C_6H_5C(O)I$	-121 ± 3

$$\text{rate} = k[\text{Ir}][\text{YZ}] \tag{5.12}$$

Data collected for $IrX(CO)(PPh_3)_2$ show the dependence on X and YZ, as shown in Table 5.3.[13] The rate of oxidative addition depends on the X group to a small extent and to the adding molecule quite significantly.[13] The data in Table 5.3 indicate basic differences for oxidative addition of H_2 and CH_3I to $IrX(CO)(PPh_3)_2$. For H_2 the reactivity changes with X, $Cl < Br < I$, while for CH_3I the order of reactivity is reversed.[13] This difference is shown in both rates and activation parameters. For H_2 addition the enthalpy and entropy of activation are larger than they are for CH_3I. In both cases the negative entropies of activation are consistent with the second-order nature of the reaction. The dependence of the oxidative addition on the phosphine ligand has been examined. The results are shown in Tables 5.4 and 5.5.[14] The rate of addition of CH_3I is shown to be quite dependent on the ligand environment at the metal center, while the rate of H_2 addition is relatively insensitive to the ligand environment. For CH_3I addition the rate is enhanced by electron-donating ligands. The electron donation would increase the nucleophilicity of the Ir. Thus, the importance of the nucleophilicity of the metal center is indicated for oxidative addition of CH_3I. Steric effects may also be important, but they are difficult to separate from electronic effects. Plotting $\log k$ versus the Hammett constant for the para substituents of the phosphine gives a straight line showing a direct relation between the basicity of the metal and the rate of reaction. This plot is shown in Figure 5.2.[14] The effect of both size and basicity of the ligand on oxidative additions at $IrL_2Cl(CO)$, L $= PMe_2Ph$, $PMe_2(o\text{-MeOC}_6H_4)$, $PMe_2(p\text{-MeOC}_6H_4)$, $P(t\text{-Bu})Me_2$, $P(t\text{-Bu})Et_2$, $P(t\text{-Bu})_2Me$ showed that increasing the basicity of the ligand increases the rate of reaction.[15] Increasing the size of the ligands greatly reduces the tendency for oxidative addition.[16]

The effect of added iodide on the addition of CH_3I to an iridium complex, $[Ir(cod)(o\text{-phen})]Cl$, was investigated.[17]

$$Ir(cod)(o\text{-phen})^+ + CH_3I \xrightarrow{I^-} Ir(cod)(o\text{-phen})(CH_3)I \tag{5.13}$$

Catalytic amounts of I^- increase the rate of reaction. This is shown in Figure 5.3 for different concentrations of iodide. These data could be fit to the rate law

Table 5.3. RATES AND ACTIVATION PARAMETERS FOR REACTION OF $IrX(CO)(PPh_3)_2$ WITH H_2 OR CH_3I AT 30°C IN BENZENE[13]

X	YZ	$k(M^{-1}s^{-1})$	$\Delta H^{\ddagger}$ (kcal/mole)	$\Delta S^{\ddagger}$ (eu)
Cl	H_2	0.93	10.8	−23
Br	H_2	14.3	12.0	−14
I	H_2	$>10^2$	—	—
Cl	CH_3I	3.9×10^{-3}	5.6	−51
Br	CH_3I	1.8×10^{-3}	7.6	−46
I	CH_3I	1.05×10^{-3}	8.8	−43

Table 5.4. REACTION OF IrCl(CO)L$_2$ WITH CH$_3$I AT 25°C[14]

L	$k(M^{-1}s^{-1})$	$\Delta H^{\ddagger}$ (kcal/mole)	$\Delta S^{\ddagger}$ (eu)
PPh$_3$	$3.3 \pm 0.1 \times 10^{-3}$	7.0	-47
PEtPh$_2$	$1.2 \pm 0.1 \times 10^{-2}$	9.8	-34
PEt$_2$Ph	$1.4 \pm 0.1 \times 10^{-2}$	9.9	-34
P(p-C$_6$H$_4$CH$_3$)$_3$	$3.3 \pm 0.2 \times 10^{-2}$	13.8	-20
P(p-C$_6$H$_5$F)$_3$	$1.5 \pm 0.2 \times 10^{-4}$	17.0	-20
P(p-C$_6$H$_5$Cl)$_3$	$3.7 \pm 0.1 \times 10^{-5}$	14.9	-28

$$\text{rate} = \left(\frac{k_1 + k_2 K_e[I^-]}{1 + K_e[I^-]} \right) [CH_3I][Ir(cod)(o\text{-phen})^+] \tag{5.14}$$

where the constants are defined in Figure 5.4.[17]

Reactions of the rhodium complex, $trans$-Rh(PR$_3$)$_2$(CO)X, with alkyl halides have also been investigated.[18] The reaction does not stop at the alkyl, but instead rearranges to the acyl complex.

$$trans\text{-Rh(PR}_3)_2\text{(CO)Cl} \underset{-CH_3I}{\overset{CH_3I}{\rightleftharpoons}} \text{Rh(PR}_3)_2\text{(CO)(Cl)(CH}_3\text{)(I)} \tag{5.15}$$

$$\downarrow$$

$$\text{Rh(PR}_3)_2\text{(C(O)Me)(Cl)(I)}$$

The kinetics of both steps were evaluated, but our primary interest is in the first step.[18] Values for the rates and equilibrium constants are shown in Table 5.6. The reversibility of the Rh reaction indicates that the reaction is not as energetically favorable as for the Ir analogues. The rate of addition to Rh(I) is more affected by the nature of the phosphine than it is for Ir, as seen by comparison of the data in Tables 5.5 and 5.6. For these Rh complexes the addition reaction was autocatalytic, the formation of the acetyl product speeded the reaction.[18] Oxidative addition of C$_2$H$_5$I or C$_4$H$_9$I did not occur. An alternate proposal for the preparation of the acyl product has been offered.[19]

Table 5.5. SECOND-ORDER RATE CONSTANTS FOR REACTION OF $trans$-IrCl(CO)(P(p-C$_6$H$_4$Y)$_3$)$_2$ WITH H$_2$ AT 30°C[14]

Y	$k(10M^{-1}s^{-1})$	$\Delta H^{\ddagger}$ (kcal/mole)	$\Delta S^{\ddagger}$ (eu)
OCH$_3$	6.6	6.0	-39
CH$_3$	5.3	4.3	-45
H	9.3	10.8	-23
F	2.5	11.6	-22
Cl	1.6	9.8	-28

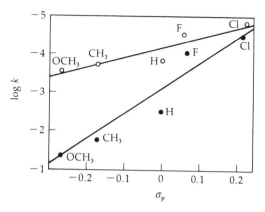

Figure 5.2. Plots of the rate constant for oxidative-addition of CH_3I (solid circle) and $C_6H_5CH_2Cl$ (open circles) versus the basicity of the metal as measured by the Hammett constants for the substituents on the phosphines of *trans*-IrCl(CO)[P(p-C_6H_4Y)$_3$]$_2$. [Reprinted with permission from R. Ugo, A. Pasini, A. Fusi, and S. Cenini, *J. Am. Chem. Soc.* **94**, 7364 (1972). Copyright 1972 American Chemical Society.]

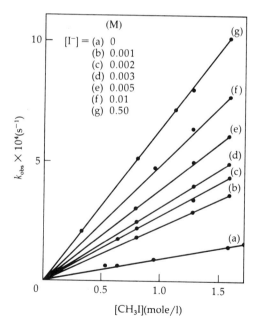

Figure 5.3. The effect of added I^- on the rate of CH_3I addition to (cod)Ir (o-phenanthroline)$^+$. [Reprinted with permission from D.J.A. de Waal, T.I.A. Gerber, and W. J. Louw, *Inorg. Chem.* **21**, 1260 (1982). Copyright 1982 American Chemical Society.]

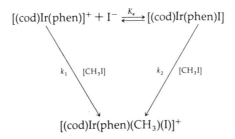

Figure 5.4. A scheme that explains the dependence of the rate of oxidative-addition on added I^-.

$$RhCl(CO)(PR_3)_2 \rightleftarrows PR_3 + RhCl(CO)PR_3 \tag{5.16}$$

$$PR_3 + CH_3I \rightleftarrows PR_3CH_3^+I^- \tag{5.17}$$

$$RhCl(CO)(PR_3) + I^- \rightleftarrows [RhCl(I)(CO)PR_3]^- \tag{5.18}$$

$$RhCl(I)(CO)PR_3^- + CH_3I \rightarrow [RhCl(I)_2(CH_3)(CO)PR_3]^- \tag{5.19}$$

$$RhCl(I)_2(CH_3)(CO)PR_3^- \rightarrow [RhCl(I)_2(C(O)CH_3)PR_3]^- \tag{5.20}$$

$$RhCl(I)_2(C(O)CH_3)PR_3^- + PR_3 \rightarrow RhCl(I)(C(O)CH_3)(PR_3)_2 + I^- \tag{5.21}$$

The primary evidence for the mechanism was an improvement in the fit of the complicated kinetic data.[19]

5.1.3. Mechanisms

No one mechanism holds for all oxidative addition reactions. Three different mechanisms (concerted addition, nucleophilic attack, and free radical) have been observed, depending on the reactants and the conditions for the reaction. The data pertinent to each mechanism will be considered separately.

5.1.3.1. Concerted Addition

Concerted addition is the suggested mechanism for addition of H_2, an addition that invariably occurs *cis*. Concerted addition of H_2 can be viewed as donation of the

Table 5.6. RATE AND EQUILIBRIUM CONSTANTS FOR THE ADDITION OF CH_3I TO *trans*-$Rh(PR_3)_2(CO)Cl$ AT 25°C

R	$k_1(\times 10^4 s^{-1})$	K_{eq}
Ph	5.5	0.93
p-MeOC$_6$H$_4$	44.3	6.1
p-FC$_6$H$_4$	1.2	0.82

bonding electron density of the hydrogen molecule to the metal, a simultaneous formation of two M–H bonds while weakening the H–H bond. This is illustrated in Figure 5.5. Molecular hydrogen complexes provide interesting models, and perhaps intermediates, for the oxidative addition of H_2.[20] The concerted mechanism is consistent with the observed kinetics, activation parameters, and stereochemistry of addition. Both the small deuterium isotope effect ($k_H/k_D = 1.22$) and calculations indicate that only a little H_2 bond weakening in the transition state exists.[21,22] Only the most reactive complexes form stable adducts by oxidative addition of hydrogen. The addition of H_2 is usually rapid and reversible. It is favored by donor ligands, is greater for iridium complexes than for analogous rhodium complexes, and has a reactivity order $I^- > Br^- > Cl^-$. Addition of silanes and hydrogen halides in nonpolar solvents also occur by concerted mechanisms.

5.1.3.2. Nucleophilic Attack

Evidence for nucleophilic attack by the metal arises from studies of the oxidative addition of alkyl halides, although not all oxidative additions of alkyl halides proceed by nucleophilic attack. Definitive evidence comes from the inversion of configuration at the carbon.[5] Inversion has been demonstrated for addition of alkyl halides to a number of complexes, several examples are shown in Figure 5.6. The utilization of several reactions in which the stereochemistry was known for each reaction, except the oxidative addition, led to the definition of stereochemistry at the carbon. The oxidative addition of a benzyl halide, $PhCH_2X$, to PdL_3 or PdL_4 was followed by carbonylation to the acyl (which proceeds by retention of configuration, as discussed in Chapter 4) and decomposition by CH_3OH to the ester (which occurs with retention of configuration).[5] Since the absolute configuration and the purity of the alkyl halides and the ester products could be determined from their optical rotations, the stereochemistry of the oxidative addition reaction could be determined. The reactions are shown in Figure 5.7 and Table 5.7. These oxidative additions proceed with predominant inversion of configuration, in some cases 100% net inversion.[5] This is consistent with S_N2 (nucleophilic) attack at the carbon by the metal, with inversion just as observed in organic reactions. The reactivity toward RX is also consistent with nucleophilic attack.

$$R = CH_3 > CH_3CH_2 > secondary > cyclohexyl > adamantyl$$

$$X = I > tosylate \sim Br > Cl$$

Figure 5.5.   Concerted addition of H_2 to a square planar metal M.

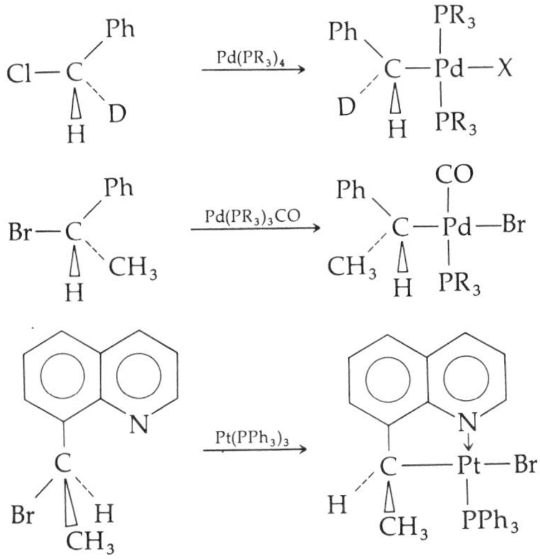

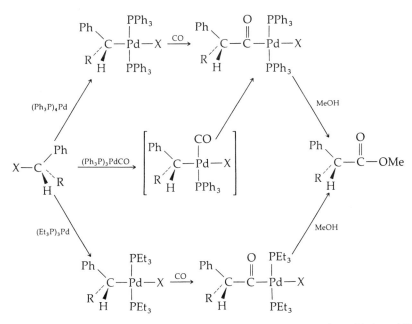

Figure 5.6. Examples of oxidative-addition that proceed with inversion of configuration at carbon.

Figure 5.7. Scheme used to determine the stereochemistry at carbon for oxidative-addition to Pd complexes. [Reprinted with permission from J. K. Stille and K.S.Y. Lau, *Acct. Chem. Res.* **10**, 434 (1977). Copyright 1977 American Chemical Society.]

Table 5.7. STEREOCHEMISTRY FOR THE OXIDATIVE ADDITION OF BENZYL HALIDES (Ph(R)HCX) TO Pd(0) COMPLEXES[5]

Pd(0) Complex	R	X	Net Inversion (%)	Relative Rate
Pd(PEt$_3$)$_3$	D	Br	30	10^3
	D	Cl	72	5 × 10^2
Pd(PPh$_3$)$_4$	D	Cl	74	3
Pd(PPh$_3$)$_4$[a]	D	Cl	100	
Pd(CO)(PPh$_3$)$_3$	D	Cl	100	2
Pd(PPh$_3$)$_4$	CH$_3$	Br	90	1.5
Pd(PPh$_3$)$_3$(CO)	CH$_3$	Br	90	1
Pd(PPh$_3$)$_4$	CF$_3$	Cl	<10	very slow

[a] Under CO

This order was determined for reactions of the rhodium complex shown in Figure 5.8.[23,24] The transition state for nucleophilic attack is shown in Figure 5.9. The reactions presumably go through an ionic complex, [RM]$^+$X$^-$, which can be isolated in some cases before the final product formation.

Reaction of Pt(bipy)(Me)$_2$ with CD$_3$I allowed isolation of the cationic complex in CH$_3$CN.[25]

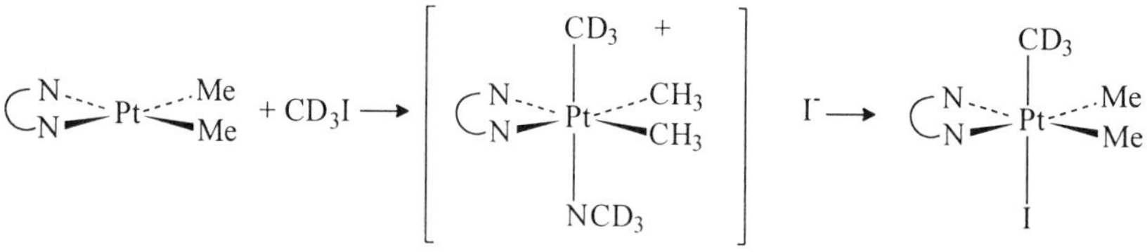

$$(5.22)$$

The role of added halide is demonstrated by oxidative addition of PhI to Pd(PPh$_3$)$_2$ generated in situ.[26]

$$\text{Pd(PPh}_3)_2 + \text{Cl}^- \longrightarrow \text{Pd(Cl)(PPh}_3)_2^- \xrightarrow[\text{rapid}]{\text{PhI}} \text{Pd(Cl)(I)(Ph)(PPh}_3)_2^- \qquad (5.23)$$

$$\downarrow \text{slow, -Cl}^-$$

$$\text{Pd(I)(Ph)(PPh}_3)_2$$

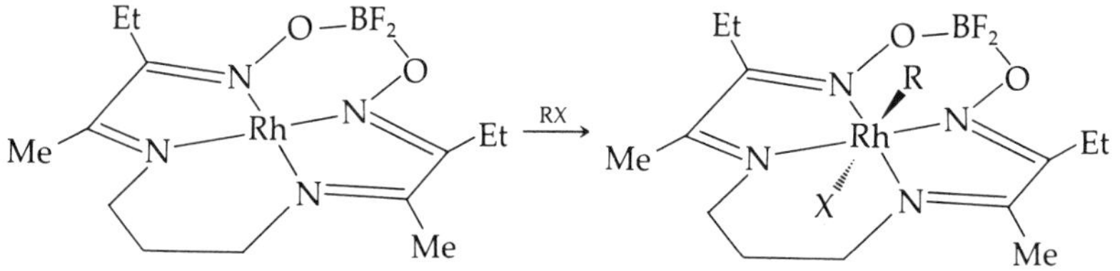

Figure 5.8. Reaction used to determine the dependence of rate on the nucleophilicity of R. [Reprinted with permission from J. K. Stille and K.S.Y. Lau, *Acct. Chem. Res.* **10**, 438 (1977). Copyright 1977 American Chemical Society.]

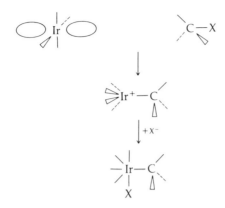

Figure 5.9. Nucleophilic attack on carbon by an Ir complex.

The halide serves to increase nucleophilicity of the palladium. The product is often of *trans* stereochemistry, but it may be *cis*.

5.1.3.3. Radical Mechanisms

Alkyl halides that do not readily undergo nucleophilic attack often proceed by radical mechanisms.[27–31] The characteristics associated with a radical mechanism are loss of stereochemistry, nonreproducible rates, inhibition by radical inhibitors, acceleration by O_2, and acceleration by light. Radical reactions have been observed in a number of cases. Reactions of $IrCl(CO)(PMe_3)_2$ showed radical characteristics, and the following scheme was suggested,[28,29]

$$R\bullet + IrCl(CO)(PMe_3)_2 \rightarrow IrCl(CO)(PMe_3)_2(R) \tag{5.24}$$

$$IrCl(CO)(PMe_3)_2(R) + RX \rightarrow X-IrCl(CO)(PMe_3)_2(R) + R\bullet \tag{5.25}$$

involving a radical chain process. Reactions of methyl, benzyl, allyl halides, and α-haloethers showed no indications of radical behavior, while other saturated alkyl halides, vinyl and aryl halides, and α-haloesters show characteristics consistent with a radical chain pathway in reaction with $IrCl(CO)(PMe_3)_2$.[28–30] The phosphine ligand is very important in radical reactions, as shown in Table 5.8.[30] A nonchain

Table 5.8. EFFECT OF PHOSPHINE LIGANDS ON REACTIONS OF ALKYL HALIDES (RX) WITH $IrCl(CO)L_2$[30]

L	Radical Reactions[a]	Nucleophilic Attack[b]
PMe_3	100 (<30 min)	5.0×10^{-2}
PMe_2Ph	100 (5h)	1.5×10^{-2}
$PMePh_2$	10 (1 week)	3.5×10^{-3}

[a] RX = $PhCHFCH_2Br$, % of reaction.
[b] RX = MeI; $M^{-1}s^{-1}$.

radical mechanism was proposed for addition of alkyl halides to $Pt(PPh_3)_2$ where the first step was abstraction of the halide from RX.[31,32]

$$Pt(PPh_3)_2 + RX \rightleftharpoons Pt(PPh_3)_2X + R\bullet \qquad (5.26)$$

$$R\bullet + Pt(PPh_3)_2X \rightarrow Pt(PPh_3)_2(R)X \qquad (5.27)$$

The mechanism outlined in Equations 5.26 and 5.27 was based on spin-trapping experiments and the observation of $Pt(PPh_3)_2X_2$ in solvents where hydrogen abstraction by $R\bullet$ can occur.[32,33] A mechanism to accommodate both nucleophilic attack and radical paths was proposed for oxidative addition to NiL_4 (L = phosphine ligand).[27] This oxidative addition to Ni(0) led to the Ni(II) complex, $RNiL_2X$, and to the paramagnetic Ni(I) complex, $XNiL_3$, in varying yields. The following scheme was suggested.[27]

$$NiL_4 \rightleftharpoons NiL_3 + L \qquad (5.28)$$

$$NiL_3 + RX \rightarrow [Ni^IL_3, RX\bullet] \qquad (5.29)$$

$$[Ni^IL_3, RX\bullet] \longrightarrow \begin{cases} RNiL_2X + L \\ \\ NiL_3 + X^- + R\bullet \qquad (5.30) \end{cases}$$

R = aryl

L = PEt_3, PPh_3

A key feature is the electron transfer from the nickel complex to the alkyl halide, forming a radical pair.

5.1.3.4. Other Metal Complexes

Thus far our examples have encompassed only square-planar M(I) complexes (M = Rh and Ir) and M(0) complexes of Ni, Pd, and Pt. In this section we will examine other examples that will reemphasize the mechanistic features. Oxidative additions of a number of organic halides to Cp_2ZrL_2 (L = PMe_2Ph or $PMePh_2$) were shown to proceed by a radical chain process.[34,35]

$$Cp_2ZrL_2 + RX \rightarrow Cp_2ZrLRX + L \qquad (5.31)$$

The mechanistic interpretation was based on characterization of the products and on an ESR investigation during the course of the reaction, which allowed the intermediate, $Cp_2Zr(III)X\text{-}(PPh_2Me)$, a metal-centered radical, to be identified.[34,35] Oxidative addition to the complexes, CpML(CO) (M = Co, Rh, Ir) were investigated. Depending on the metal, different products can be observed.[36]

$$CpCo(CO)PPh_3 + CH_3I \rightarrow CpCo(I)(C(O)CH_3)PPh_3 \qquad (5.32)$$

$$CpRh(CO)PPh_3 + CH_3I \rightarrow CpRh(I)(C(O)CH_3)PPh_3 \qquad (5.33)$$

$$CpIr(CO)PPh_3 + CH_3I \rightarrow [CpIr(CO)(CH_3)PPh_3]^+I^- \qquad (5.34)$$

Each reaction is oxidative addition to an 18-electron complex and the product in each case contains 18 electrons. The reaction is first order in CH_3I and metal complex, $M(Cp)(CO)L$, and is dependent on the nature of L, as shown in Figure 5.10.[36] These reactions proceed by nucleophilic attack to the intermediate, $[CpM(CO)(CH_3)L]^+I^-$, which is the stable product for M = Ir. For M = Co and Rh methyl migration to the CO opens a coordination site for the iodide.

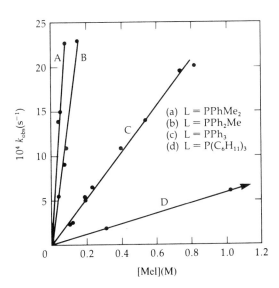

The phosphine ligand dependence as shown in Figure 5.10 represents the steric hinderance by bulky ligands to the expansion of the coordination sphere. Oxidative addition to CpM(CO)L was followed by acyl formation (Reaction 5.35); oxidative addition may also be followed by a decarbonylation as shown for the Rh complex, $Rh(PMe_2Ph)_3Cl$.[37]

Figure 5.10. The dependence of the rate of CH_3I addition on the size of the phosphine ligand on CpCo(CO)L. [Reprinted with permission from A. J. Hart-Davis and W.A.G. Graham, *Inorg. Chem.* **9**, 2658 (1970). Copyright 1970 American Chemical Society.]

$$\text{Rh(PMe}_2\text{Ph)}_3\text{Cl} + \text{RC(O)Cl} \xrightarrow{-P}$$ (5.36)

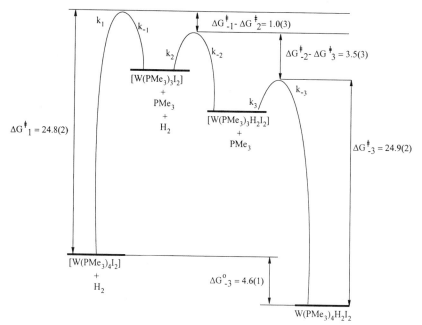

$$P = PMe_2Ph$$

Oxidative addition of H_2 to a six-coordinate W(II) complex, $W(I)_2(PMe_3)_4$ was studied by kinetic and equilibrium experiments that showed a PMe_3 dissociation mechanism.[38]

$$W(I)_2(PMe_3)_4 \underset{+PMe_3}{\overset{-PMe_3}{\rightleftharpoons}} W(I)_2(PMe_3)_3 \underset{-H_2}{\overset{+H_2}{\rightleftharpoons}}$$ (5.37)

$$W(I)_2(H)_2(PMe_3)_3 \underset{-PMe_3}{\overset{+PMe_3}{\rightleftharpoons}} W(I)_2(H)_2(PMe_3)_4$$

The free-energy diagram is shown in Figure 5.11.

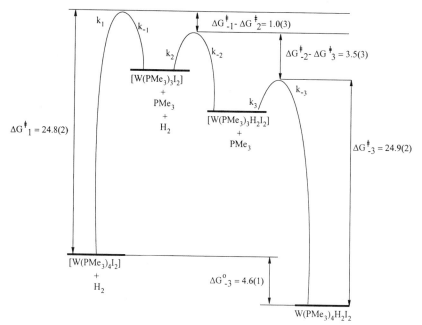

Figure 5.11. Energy surface for oxidative addition of H_2 to $W(PMe_3)_4I_2$ at 60°C. Standard states of the components are 1 M, and energy values are in kilocalories/mole. [Reprinted with permission from D. Rabinovich and G. Parkin, *J. Am. Chem. Soc.* **115**, 354 (1993). Copyright 1993 American Chemical Society.]

5.1.4. Oxidative Addition of C–H Bonds

Activation of hydrocarbons remains an important problem and has focused much attention on activation of C–H bonds.[39] The most comprehensive studies have been on [Cp*ML], M = Rh, Ir; L = CO, PMe$_3$ which are generated in situ, usually by photolytic activation of an 18-electron complex.[40–42]

$$Cp^*Ir(CO)_2 \xrightarrow[-CO]{h\nu} Cp^*Ir(CO) \xrightarrow{RH} Cp^*Ir(CO)(R)(H) \qquad (5.38)$$

$$Cp^*Ir(PMe_3)(H)_2 \xrightarrow[-H_2]{h\nu} Cp^*Ir(PMe_3) \xrightarrow{RH} \qquad (5.39)$$
$$Cp^*Ir(PMe_3)(R)(H)$$

$$Cp^*Rh(PMe_3)(H)_2 \xrightarrow[-H_2]{h\nu} Cp^*Rh(PMe_3) \xrightarrow{RH} \qquad (5.40)$$
$$Cp^*Rh(PMe_3)(R)(H)$$

Kinetic studies were accomplished in Xe(l) and Kr(l) for Cp*Rh(CO)$_2$ by time-resolved IR spectroscopy.[43]

$$Cp^*Rh(CO)_2 \xrightarrow{h\nu} [Cp^*Rh(CO)] \xrightarrow{C_6H_{12}} Cp^*Rh(CO)(C_6H_{11})(H) \quad (5.41)$$

At $-100°C$ in Kr(l) photolysis was shown to result in formation of Cp*Rh(CO)(Kr). In the presence of cyclohexane exponential decay was observed in a process first order in Cp*Rh(CO)(Kr).

$$\text{rate} = k_{obs}[Cp^*Rh(CO)(Kr)] \qquad (5.42)$$

$$k_{obs} = \frac{k_i[C_6H_{12}]}{[C_6H_{12}] + \dfrac{[Kr]}{K_{eq}}} \qquad (5.43)$$

The rate constant (k_{obs}) showed limiting behavior with added C$_6$H$_{12}$, and a two-step mechanism was suggested.

$$Cp^*Rh(CO)(Kr) + C_6H_{12} \rightleftharpoons Cp^*Rh(CO)(C_6H_{12}) + Kr \quad K_{eq} \quad (5.44)$$

$$Cp^*Rh(CO)(C_6H_{12}) \xrightarrow{k_i} Cp^*Rh(CO)(C_6H_{11})(H) \qquad (5.45)$$

An energy of activation of 4.8 ± 0.2 kcal/mole was determined from the temperature dependence.

Competitive studies have been very useful in providing relative rates and activation energies. For example, reaction of Cp*Rh(PMe$_3$) with a 1:1 mixture of benzene and propane showed a preference for oxidative addition of the benzene C–H bond.[41]

$$Cp*Rh(PMe_3) + C_6H_6 + C_3H_8 \longrightarrow Cp*Rh(PMe_3)(C_6H_5)(H) + \quad (5.46)$$
$$\mathbf{4}$$

$$Cp*Rh(PMe_3)(C_3H_7)(H)$$
$$\mathbf{1}$$

The C–H activation was kinetically nonselective, usually within a factor of 10. The energy profile for conversion from the phenyl, hydride complex to the propyl, hydride complex is shown in Figure 5.12. Preparation of Cp*Rh(PMe$_3$)-(C$_6$D$_5$)(H) showed H/D exchange, which was interpreted as indication of an η^2-C$_6$H$_6$ intermediate. The M–C bond strength dominates the thermodynamics of these C–H bond activation systems, not the C–H bond strength.

$$M - C_6H_5 > M - CH = CH_2 > M - CH_3 > M - CH_2R > M$$
$$- CHR_2 > M - CR_3 > M - CH_2C_6H_5$$

Several conclusions on oxidative addition of C–H bonds can be drawn: (1) The reaction is fairly commonly observed, but is reversible and frequently only observed at low temperatures. (2) The metal complex needs to be a coordinatively unsaturated, second- or third-row metal with strong M–C and M–H bonds. The complex should not be sterically crowded or have the possibility of cyclometallating. A filled orbital on the complex is needed to interact with the C–H σ^* orbital. (3) The thermodynamics are controlled by M–C bond strengths. (4) The mechanism involves prior coordination of the alkane. Saturated hydrocarbons appear to coordi-

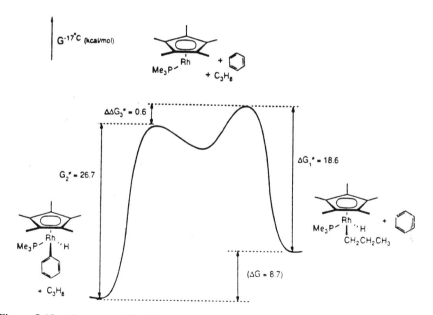

Figure 5.12. Energy profile for competitive reaction of benzene and propane with Cp*Rh(PMe$_3$). W. D. Jones and F. J. Feher, *Acc. Chem. Res.* **22**, 91 (1989).

nate through complexes similar to the molecular hydrogen complexes, while arenes may coordinate through an η^2-arene complex.

5.1.5. β-Elimination

A specific type of C–H activation is β-hydride elimination. The reaction can be represented in the general case as

$$M-CH_2-CH_2R \longrightarrow M-H + CH_2=CHR \qquad (5.47)$$

where the alkyl that possesses a β-hydrogen is converted to a metal hydride and an alkene. The reaction of $(n\text{-Bu})Cu[P(n\text{-Bu})_3]$ showed that the β-hydrogen is transferred to the metal.[44–46]

$$CH_3CH_2CD_2CH_2CuPR_3 \longrightarrow \text{1-butene-}d_1 + d_3\text{-butane} + Cu^0 \quad (5.48)$$
$$+ PR_3$$

$$CH_3CH_2CH_2CD_2CuPR_3 \longrightarrow \text{1-butene-}d_2 + \text{butane -}d_2 + Cu^0 \quad (5.49)$$
$$R = Bu \qquad\qquad\qquad\qquad\qquad\qquad\qquad\qquad + PR_3$$

β-elimination of deuterium from Reaction 5.48 yields $DCuPR_3$, which produces d-$_3$-butane in a binuclear reductive elimination with the alkyl complex. β-elimination of hydrogen from Reaction 5.49 yields the hydrido species and the d_2-butane and 1-butene.[44] The mechanism of β-elimination has been most carefully investigated for the thermal decomposition of di-n-butylbis(triphenylphosphine)platinum(II).[45]

$$(n\text{-butyl})_2Pt(PPh_3)_2 \longrightarrow \text{butane} + \text{1-butene} + [Pt(PPh_3)_2] \qquad (5.50)$$

The data indicate a first-order intramolecular process that is slowed by the presence of excess PPh_3. The following mechanism has been suggested[45]:

$$Bu_2Pt(PPh_3)_2 \rightleftharpoons Bu_2PtPPh_3 + PPh_3 \qquad (5.51)$$

$$Bu_2PtPPh_3 \rightleftharpoons BuPt(H)(\text{1-butene})(PPh_3) \qquad (5.52)$$

$$BuPt(H)(\text{1-butene})PPh_3 \longrightarrow \text{butane} + \text{1-butene} + [Pt(PPh_3)] \qquad (5.53)$$

Scrambling of the deuterium label indicated that the reactions could occur several times and probably also lead to the 2-butene isomer as a transient species. Since the rate-determining step for this reaction involved PPh_3 loss, no deuterium isotope effect was seen.[45] A kinetic isotope effect was evaluated for β-elimination from (n-octyl)Ir(CO)(PPh$_3$)$_2$.[46]

The value of 2.3 ± 0.2 is consistent with Ir inserting into a β–C–H bond. Kinetics of β-elimination from $trans$-Pt(Cl)(Et)L$_2$ ($L = PEt_3$) have also been evaluated.[47]

$$trans\text{-Pt(Cl)(Et)L}_2 \xrightarrow{158^\circ C} trans\text{-Pt(Cl)(H)L}_2 + C_2H_4 \qquad (5.54)$$

$$\text{rate} = k[trans\text{-Pt(Cl)(Et)L}_2] \qquad (5.55)$$

$$E_a = 34 \pm 2 \text{ kcal/mole}$$

The rate was not affected by excess PEt_3 or Cl. A rate-determining dissociation of C_2H_4 was suggested.[47] Evidence for this was that $trans$-Pt(Cl)(CD_2CH_3)L_2 had deuterium scrambled and that the kinetic and thermodynamic isotope effects were similar.

A theoretical examination of β-elimination indicated that an agostic interaction with the C–H donating to the metal was important for oxidative addition.[48] The optimized geometry for the transition state indicated a very late transition state with Pd–H of 1.65 Å and a C–H of 1.64 Å.

β-elimination is involved in alkene isomerization reactions that accompany many homogeneous hydrogenation reactions, decompositions of alkyl complexes, decomposition of alkoxy complexes, and the Wacker process for conversion of ethylene to acetaldehyde (see Chapter 6).

5.2. Reductive Elimination

Reductive elimination has not been studied as thoroughly as oxidative addition. While many complexes undergo oxidative addition leading to stable complexes, only a few reductive elimination reactions proceed from stable starting materials to stable products.[49,50] Thus, there are less systems where reductive elimination can be studied. Reductive elimination is equally important as oxidative addition in catalytic cycles, although the reductive elimination reactions appear to be very rapid in most catalytic cycles. An alkyl hydride has been isolated from a catalytic cycle in only one case.[51]

$$(5.56)$$

The product was characterized at low temperature by spectroscopic measurements. At room temperature the reductive elimination occurred too rapidly for the alkyl hydride to be observed.[51] This is often the case, as shown for elimination of CH_4 from cis-$PtH(CH_3)(PPh_3)_2$, which could be prepared at low temperature but which reductively eliminated at $-25°C$.[52]

$$3(cis\text{-}PtH(CH_3)(PPh_3)_2) \rightarrow 3CH_4 + 2Pt(PPh_3)_3 + Pt \qquad (5.57)$$

The use of other alkyl complexes showed the qualitative order of reactivity.[52]

$$C_6H_5 > C_2H_5 > CH_3 > CH_2CH = CH_2$$

Reductive elimination of fluorotoluene from the fluorophenyl, methyl palladium complexes has been observed.[53]

$$trans\text{-}Pd(CH_3)(C_6H_4F)(PEt_3)_2 \rightarrow Pd + 2PEt_3 + C_6H_4F(CH_3) \qquad (5.58)$$

This reaction was investigated for both $meta$- and $para$-fluorobenzenes and appeared to involve a cis–$trans$ isomerization prior to the reductive-elimination.[53]

Several examples of reductive elimination reactions are provided by the complexes $Ir(CO)(R)(R')L_2X$ (R and R' = H, alkyl, or alkoxy; L = PPh_3 or $P(p\text{-}tolyl)_3$; X = halide or H), which reductively eliminate a number of different products. The reactions are summarized in Table 5.9. The reactions described in Table 5.9 show that reductive elimination from Ir(III) results in formation of different types of bonds. These reductive elimination reactions occur under very similar conditions, indicating that the nature of R and R' does not significantly affect the rate.[54] However, the coupling of two sp^3 carbon centers does not occur. Theoretical studies have suggested that the directionality of the sp^3 hybrid inhibits bond formation.[54]

Despite the fact that a coordination site is not formally required for reductive elimination reactions, ligand dissociation from the metal center apparently facilitates reductive elimination. In the Ir examples described in Table 5.9, halide dissociation is important; however, phosphine ligand dissociation is important in other examples.

Reductive elimination can be induced by the addition of a nucleophile, sometimes a very weak nucleophile. The following examples show a mononuclear and a cluster example where the elimination is reversible.[55,56]

Table 5.9. REDUCTIVE ELIMINATION
REACTIONS FROM $Ir(CO)(R)(R')L_2X$[53]

R	R'	X	Product (R-R')
Me	H	Cl	CH_4
Ph	H	Cl	C_6H_6
OMe	H	Cl	MeOH
Me	C(O)Me	Cl	MeC(O)Me
OMe	C(O)Me	Cl	MeC(O)OMe

$$trans\text{-}PtH_2(PCy_3)_2 \underset{H_2}{\overset{CO}{\rightleftharpoons}} Pt(CO)_2(PCy_3)_2 \qquad (5.59)$$

$$H_3Ru_3(CO)_9CX + CO \rightleftharpoons HRu_3(CO)_{10}CX + H_2 \qquad (5.60)$$

In both examples the hydride complex is thermally stable toward reductive-elimination, but the presence of CO induces the elimination. The ability of such a weak nucleophile as CO to induce reductive elimination may indicate that these reductive eliminations proceed through a preequilibrium where the metal complex and RH are held by a σ-bond (see Reaction 5.44). This would account for the elimination induced by CO and would also be consistent with the C–H activation discussed earlier.

While reductive elimination is very rapid in catalytic cycles and has been observed stoichiometrically as in the examples above, a number of complexes have been prepared that do not undergo reductive elimination.[57–60] A few examples are as follows:

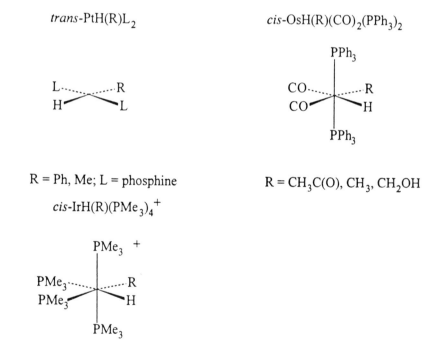

It has not yet been determined why one complex readily reductively eliminates while another does not.

Several polyalkyl complexes that will not eliminate alkanes will eliminate ketones in the presence of CO. The nickel complexes, NiR_2L_2, have been investigated extensively.[61]

$$NiR_2L_2 + CO \rightarrow RC(O)R, RC(O)C(O)R \text{ and } RCH(O) \qquad (5.61)$$
$$+ Ni(CO)_2L_2$$

The exact products depend on the temperature, the nature of L, and the R group.[61] Coordinated acetone is also formed by carbonylation of $(\eta^5\text{-}C_5Me_5)TaMe_4$.[62]

$$(\eta^5\text{-}C_5Me_5)TaMe_4 + CO \rightarrow (\eta^5\text{-}C_5Me_5)TaMe_2(C(O)Me_2) \qquad (5.62)$$

Further reaction of the coordinated acetone was observed, but it is apparently not related to the question of why $CH_3C(O)CH_3$ can form the C–C bond when CH_3CH_3 cannot.

5.2.1. Mechanistic Investigations

Reductive elimination is the microscopic reverse of oxidative addition and similarities in mechanism should be expected. The following sections highlight examples where sufficient kinetic and mechanistic information exist to present a mechanism.

5.2.1.1. Reductive Elimination of H_2

Reductive elimination of H_2 from a platinum dihydride, $Pt(H)_2(PMe_3)_2$, shows a rate law that is first order with no marked effect of added PMe_3.[63]

$$2Pt(H)_2(PMe_3)_2 \rightarrow 2H_2 + Pt(PMe_3)_4 + Pt(0) \qquad (5.63)$$

An inverse isotope effect ($k_H/k_D = 0.45$) indicated rate-determining cleavage of the Pt–H bond.[63] Crossover experiments confirmed that the reaction was not bimetallic. Polar solvents significantly slowed the reductive elimination of H_2 (a factor of 10 decrease from THF to DMSO), which indicates a charge separation in formation of a dihydride such that polar solvents stabilize dihydride formation. For this reductive elimination a late transition state with substantial Pt–H bond breaking and H–H bond formation was indicated.[63]

Activation parameters were determined for reductive elimination of H_2 from $Ir(CO)(dppe)(Et)(H)_2$, $\Delta H^\ddagger = 16 \pm 1$ kcal/mole and $\Delta S^\ddagger = -24 \pm 4$ eu.[64] The negative entropy of activation is unusual for a dissociative process, but it reflects the binding together of two hydrogen atoms.

5.2.1.2. Reductive Elimination of C–H Bonds

Reductive elimination of alkanes from hydrido-alkyl complexes has not been often investigated because synthetic procedures are generally not available for complexes of appropriate stereochemistry. The reaction of $Rh(PMe_3)_3Cl$ with propylene oxide led to one example.[65]

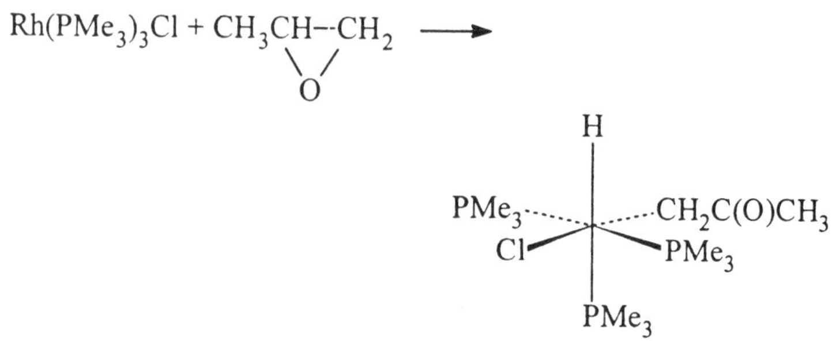

$$Rh(PMe_3)_3Cl + CH_3CH\!-\!CH_2 \longrightarrow \tag{5.64}$$
$$\underset{O}{\diagdown\diagup}$$

$$
\begin{array}{c}
H \\
PMe_3\cdots\!\!\mid\!\!\cdots CH_2C(O)CH_3 \\
Cl\!\!\diagup\!\!\mid\!\!\diagdown PMe_3 \\
PMe_3
\end{array}
$$

The stereochemistry was assigned from the ^{1}H and ^{31}P NMR spectra. Reductive elimination of acetone showed first-order kinetics with $\Delta H^{\ddagger} = 25.0$ kcal/mole and $\Delta S^{\ddagger} = 5.3$ eu.[65]

$$cis\text{-}RhH(CH_2C(O)CH_3)(PMe_3)_3Cl \rightarrow RhCl(PMe_3)_3 + \tag{5.65}$$
$$CH_3C(O)CH_3$$

The presence of PMe$_3$ inhibited the reaction. Mixtures of cis-RhH(CH$_2$C(O)CH$_3$)-(PMe$_3$)$_3$Cl and cis-RhD(CD$_2$C(O)CD$_3$)(PMe$_3$)$_3$Cl led to only CH$_3$C(O)CH$_3$ and CD$_3$C(O)CD$_3$, showing that the elimination is intramolecular.[65] The following scheme was suggested.[65]

$$cis\text{-}RhH(CH_2C(O)CH_3)(PMe_3)_3Cl \rightleftharpoons \tag{5.66}$$
$$RhH(CH_2C(O)CH_3)(PMe_3)_2Cl + PMe_3$$

$$RhH(CH_2C(O)CH_3)(PMe_3)_2Cl \rightarrow Rh(PMe_3)_2Cl + CH_3C(O)CH_3 \tag{5.67}$$

$$Rh(PMe_3)_2Cl + PMe_3 \xrightleftharpoons{\text{fast}} Rh(PMe_3)_3Cl \tag{5.68}$$

The requirement for dissociation of PMe$_3$ is similar to the eliminations observed from Pd.

The complexes Cp$_2$W(H)(CH$_3$) and Cp$^*_2$W(H)(R), R = Me, CH$_2$C$_6$H$_5$ have also provided examples of alkane elimination. At 100°C Cp$^*_2$W(R)(H) undergoes reductive elimination by a first-order process.[66]

$$Cp^*_2W(H)(R) \xrightarrow{100°C} Cp^*(\eta^5, \eta^1\text{-}C_5Me_4CH_2)W(H) + RH \tag{5.69}$$

The rates for elimination of methane and toluene were 1.20×10^{-4}s^{-1} and 1.13×10^{-4}s^{-1}. Activation parameters were determined, $\Delta H^{\ddagger} = 29.3 \pm 0.8$ kcal/mole and $\Delta S^{\ddagger} = 1.5 \pm 2.0$ eu for the benzyl complex.[66] With an extensive discussion of isotope effects, Parkin and Bercaw conclude that the inverse isotope effect, $k_H/k_D = 0.70$ indicates the presence of a σ complex, Cp$^*_2$W(η^2-CH$_4$) (see Fig. 5.13). Intramolecular scrambling of hydrogen was observed between the hydride and the methyl.

$$Cp^*_2W(CH_3)(D) \longrightarrow Cp^*_2W(CH_2D)(H) \tag{5.70}$$

A suggested scheme for these reactions is shown in Figure 5.14.[66] Similar observations were made on Cp$_2$W(H)(CH$_3$).[67] The Cp$^*_2$W(H)(R) complexes provide anoth-

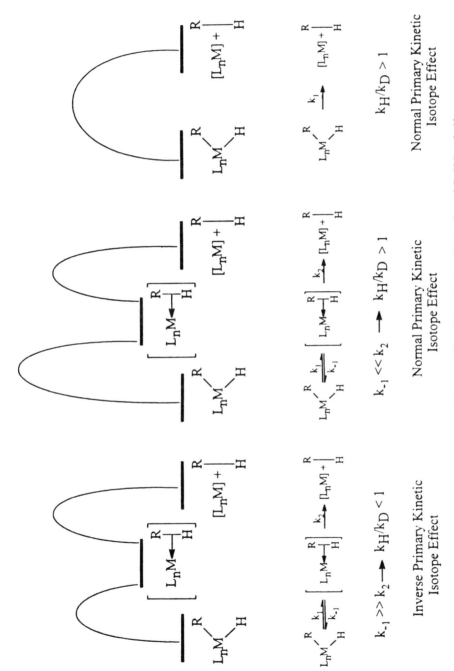

Figure 5.13. An explanation of isotope effects for reductive elimination of C—H bonds.[66]

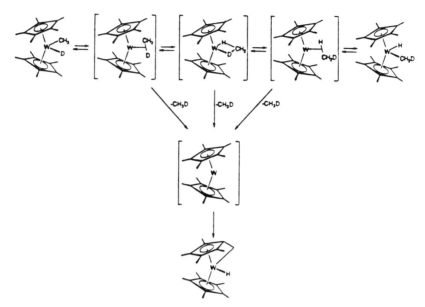

Figure 5.14. H–D scrambling in $Cp_2*W(CH_3)D$. G. Parkin and J. E. Bercaw, *Organometallics* **8**, 1172 (1989).

er example showing preferences for reductive elimination. While reductive elimination forming C–H bonds occurs at 100°C, C–C bond formation from $Cp_2^*W(Me)_2$ is not observed until temperatures were in excess of 200°C. Similarly, H_2 and H_2O were not eliminated from $Cp_2^*W(H)_2$ and $Cp_2^*W(H)(OH)$, respectively.[66]

Although $Cp_2^*W(R)(R')$ shows a strong preference for reductive elimination of C–H bonds over dihydrogen,[66] $k_{H_2} = (9.1 \pm 0.3) \times 10^{-5}s^{-1}$ and $k_{C_2H_6} = (1.77 \pm 0.02) \times 10^{-5}s^{-1}$, a satisfactory explanation of these preferences in reductive elimination has not been presented.

Reductive elimination of benzene from rhodium complexes shows some differences to other C–H reductive elimination systems. Formation of an $\eta^2\text{-}C_6H_6$ complex that is relatively stable results in the reductive elimination being a preequilibrium in one case with rate-determining dissociation of benzene,[68]

$$Cp^*RhL(H)(Ph) \rightleftharpoons Cp^*RhL(\eta^2\text{-}C_6H_6) \xrightarrow{-C_6H_6} [Cp^*RhL] \quad (5.71)$$

$$L = PMe_3, PMe_2Ph, PMePh_2, PPh_3$$

and associative kinetics in a closely related system.[69] The ligand dependence for Reaction 5.71 shows that the rate increases as the size and the donor ability of L increase as expected for rate-determining dissociation of C_6H_6.

5.2.1.3. C–C Bond Formation

Reductive elimination with formation of C–C bonds was studied from the series of dimethyl–palladium compounds shown in Figure 5.15.[70] The rate of reductive elimination was evaluated by the disappearance of the Pd-CH$_3$ NMR resonance. The data shown in Figure 5.16 indicate that these reactions were first order in dimethyl palladium. Only *cis* complexes could directly undergo reductive elimination; elimination of ethane from *trans* complexes was preceded by isomerization.[70] A polar solvent aided the reductive elimination by stabilizing the *cis* complex, by aiding the isomerization of a *trans* complex, and by stabilizing the reduced coordination species that is formed. The rates of reductive elimination parallel the rates of

A   B   C

dissociation of the phosphine. The quantitative data are shown in Table 5.10.[70] A coordination site is not necessary for reductive elimination, but the phosphines that are good σ donors and enhance oxidative addition evidently inhibit the reductive elimination. The complex where the methyl groups are constrained *trans* (3b in Fig. 5.15) does not undergo reductive elimination at 100°C. These organopalladium complexes undergo reductive elimination with retention of configuration at the carbon suggesting a concerted process for the C–C bond-forming step.[71]

Selectivity for C–C bond formation was also demonstrated for Pd(IV) systems.[72,73]

$$PdBrMe_2(CH_2Ph)L_2 \xrightarrow{-C_2H_6} PdBr(CH_2Ph)L_2 \qquad (5.72)$$

L_2 = bipy, o-phen

Table 5.10. RATES OF ETHANE
ELIMINATION FROM *cis* PALLADIUM
COMPLEXES AT 60°C[69]

Complex	$k(s^{-1})$
A	1.04×10^{-3}
B	8.33×10^{-5}
C^a	4.78×10^{-7}

^{a} At 90°C.

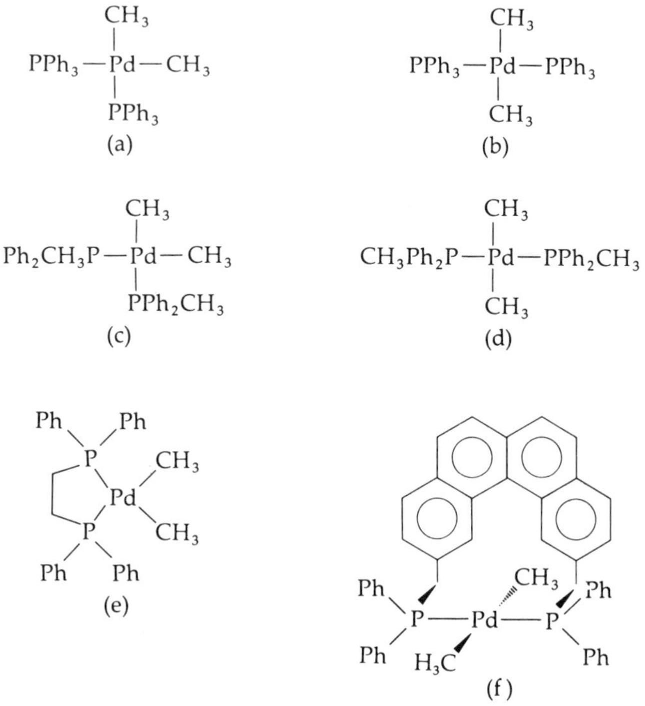

Figure 5.15. Dimethyl Pd complexes that were investigated for reductive elimination. [Reprinted with permission from A. Gillie and J. K. Stille, *J. Am. Chem. Soc.* **102**, 4933 (1980). Copyright 1980 American Chemical Society.]

$$PdIMe_2Ph(bipy) \longrightarrow PdI(Ph)(bipy) + PdI(Me)(bipy) \qquad (5.73)$$

$$+ \qquad\qquad +$$

$$C_2H_6 \qquad\qquad C_7H_8$$

$$4 \qquad : \qquad 1$$

$$PdX(Me)(CH_2Ph)(Ph)(bipy) \xrightarrow{-C_7H_8} PdX(CH_2Ph)(bipy) \qquad (5.74)$$

$$X = Br, I$$

These reactions show the ease of C–C bond formation as follows:

$$CH_3\text{-}CH_3 > CH_3\text{-}C_6H_5 > CH_3\text{-}CH_2Ph, \; C_6H_5\text{-}CH_2Ph$$

These are in marked contrast to reductive elimination from Ir(III) where CH_3-CH_2Ph formed much more readily than CH_3-CH_3.[74]

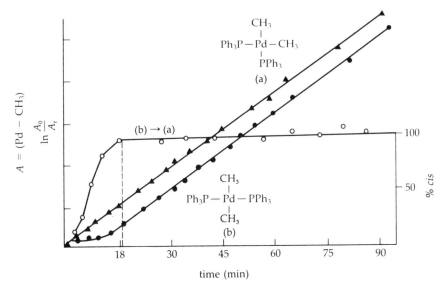

Figure 5.16. Reductive-elimination of ethane from dimethyl Pd complexes. [Reprinted with permission from A. Gillie and J. K. Stille, *J. Am. Chem. Soc.* **102**, 4933 (1980). Copyright 1980 American Chemical Society.]

Reductive elimination from $PtI(dppe)Me_3$ showed competitive C_2H_6 and CH_3I formation.[75] Both products were formed from reactions of the five-coordinate complex formed by I^- dissociation. Ethane elimination was thermodynamically favored, but methyl iodide elimination was kinetically favored.

In a potentially very important result Pedersen and Tilset have shown reductive elimination of C_2H_6 from a 17-electron dimethyl complex is 3×10^9 more facile than from the analogous 18-electron complex.[76]

$$Cp^*Rh(Me)_2(PPh_3) \xrightarrow{-C_2H_6} Cp^*RhPPh_3(S) \qquad (5.75)$$

$$Cp^*Rh(Me)_2(PPh_3)^+ \xrightarrow{-C_2H_6} Cp^*Rh(PPh_3)(S)_2^{2+} \qquad (5.76)$$

If such activation toward reductive elimination by odd electron complexes is general, then electron transfer catalysis may provide a route to eliminations of C–C bonds that currently do not occur.

The reductive elimination of methylarenes from nickel complexes has been investigated showing evidence for two mechanisms: a concerted process and a radical chain process.[77]

$$Ar\text{-}Ni(CH_3)(PEt_3)_2 \longrightarrow ArMe + Ni(PEt_3)_2 \qquad (5.77)$$

The reaction was first order in Ni complex and showed an inverse dependence on the concentration of PEt_3.[77] These data, which are shown in Figure 5.17, suggest a

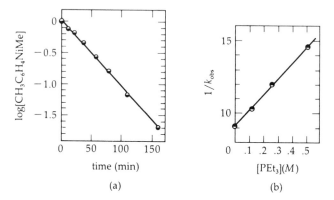

Figure 5.17. Rate data for the disappearance of *o*-tolyl methyl Ni(II) illustrating (a) the first-order nature of the reaction, and (b) the inverse dependence of rate on phosphine concentration. [Reprinted with permission from G. Smith and J. K. Kochi, *J. Organomet. Chem.* **198**, 199 (1980). Copoyright 1980 Elsevier Sequoia S. A.]

concerted elimination from $Ni(Ar)(CH_3)PEt_3$. In the presence of an arylhalide the kinetic behavior showed an induction period (shown in Fig. 5.18) and aryl group scrambling suggestive of a radical chain mechanism.[77] Inhibition of the reaction was seen by radical inhibitors such as duroquinone. Thus, different mechanisms may be seen for reductive elimination depending on the conditions. This is very similar to oxidative addition.[77]

5.2.2. Binuclear Reductive Elimination

In a number of cases binuclear reductive elimination has been observed, where two metal centers participate in the reaction. The decomposition of hydrido carbonyl complexes is an example of this type of reductive elimination.[78,79]

$$2HCo(CO)_4 \longrightarrow Co_2(CO)_8 + H_2 \tag{5.78}$$

$$2HMn(CO)_5 \longrightarrow Mn_2(CO)_{10} + H_2 \tag{5.79}$$

In other cases binuclear reductive elimination of alkanes has been observed.[80,81]

$$Cp_2MH_2 + CH_3Mn(CO)_5 \longrightarrow H_2 + CH_4 \tag{5.80}$$
$$+ Cp(CO)M(\mu\text{-}(\eta^5, \eta^1 - C_5H_4)Mn(CO)_4$$

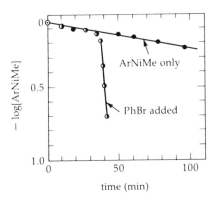

Figure 5.18. Graph illustrating the induction period required upon addition of C_6H_5Br. [Reprinted with permission from G. Smith and J. K. Kochi, *J. Organomet. Chem.* **198**, 199 (1980). Copyright 1980 Elsevier Sequoia S. A.]

$$CpMo(CO)_3H + CpMo(CO)_3R \longrightarrow RCHO + Cp_2Mo_2(CO)_4 \qquad (5.81)$$
$$+ Cp_2Mo_2(CO)_6$$

The mechanism for the latter reaction was suggested as[81]

$$RMoCp(CO)_3 \rightleftharpoons RC(O)MoCp(CO)_2 \xrightarrow{\;HMoCp(CO)_3\;} RC(O)MoCp(CO)_2$$

$$[CpMo(CO)_3]_2 + [CpMo(CO)_2]_2 + RC(O)H$$

$$\text{(5.82)}$$

The sequence is an alkyl migration that opens a coordination site for the bridging hydride, reductive elimination of aldehyde and the combination of metal carbonyl fragments to give products that have a metal single and triple bond, $Cp_2Mo_2(CO)_6$ and $Cp_2Mo_2(CO)_4$. This reaction bears some similarity to hydroformylation, which we will discuss in the next chapter. The reaction of *cis*-RMn(CO)$_4$L and *cis*-HMn(CO)$_4$L where R = benzoyl derivatives and L = CO or P(p-CH$_3$OC$_6$H$_4$)$_3$ were investigated.[82] Four distinct pathways were identified for this binuclear reductive elimination. These are outlined in the following.

$$p\text{-}CH_3O\text{-}C_6H_4CH_2Mn(CO)_5 \rightleftharpoons p\text{-}CH_3O\text{-}C_6H_4CH_2Mn(CO)_4 + CO$$

$$\downarrow HMn(CO)_5 + CO$$

$$p\text{-}CH_3OC_6H_4CH_3 + Mn_2(CO)_{10}$$

$$(5.83)$$

$$p\text{-}CH_3O\text{-}C_6H_4CH_2Mn(CO)_5 + S \rightleftharpoons p\text{-}CH_3OC_6H_4CH_2C(O)Mn(CO)_4S$$

$$\downarrow HMn(CO)_5$$

$$p\text{-}CH_3OC_6H_4CH_2C(O)H + Mn_2(CO)_9S$$

S = polar solvent (5.84)

$$p\text{-}CH_3O\text{-}C_6H_4CH_2Mn(CO)_4P \rightleftharpoons \qquad\qquad (5.85)$$
$$p\text{-}CH_3OC_6H_4CH_2\cdot + \cdot Mn(CO)_4P$$

$$p\text{-}CH_3OC_6H_4CH_2\cdot + HMn(CO)_4P \longrightarrow p\text{-}CH_3OC_6H_4CH_3 \qquad (5.86)$$
$$+ \cdot Mn(CO)_4P$$

$$2\cdot Mn(CO)_4P \longrightarrow Mn_2(CO)_8P_2 \qquad\qquad\qquad (5.87)$$
$$P = P(p\text{-}CH_3OC_6H_4)_3$$

$$p\text{-}CH_3OC_6H_4CH_2Mn(CO)_4P + CO \longrightarrow p\text{-}CH_3OC_6H_4CH_2C(O)Mn(CO)_4P$$

$$\downarrow HMn(CO)_5$$

$$p\text{-}CH_3OC_6H_4CH_2CHO + Mn_2(CO)_9P$$

$$(5.88)$$

The relatively minor changes involved in the completely different mechanisms for binuclear reductive eliminations from these manganese systems suggest that great care must be exercised in interpreting and generalizing from a specific set of experiments. Two additional studies have shown evidence for the alkyl migration scheme (Reaction 5.84). Reaction of cis-$Mn(CO)_4(H)(P(OPh)_3)$ with $Mn(CO)_4(CH_3)$-$(P(OPh)_3)$ results in reductive elimination under conditions where methyl migration occurs but the hydride is unreactive.[83] Reaction of $HMn(CO)_5$ with $MeMn(CO)_5$ gives $Mn_2(CO)_9(CH_3CHO)$ through methyl migration and bimolecular reductive elimination.[84]

The binuclear reductive-elimination reactions considered in the preceding were obviously binuclear since the two groups were on different metal centers. A fascinating set of experiments on cis-$Os(CO)_4RR'$ have shown that formation of RR' also occurs by binuclear elimination.[85] For $R = R' = CH_3$ the elimination does not occur; above 160°C decomposition to CH_4, $(CH_3)_2C(O)$ and unidentified osmium complexes is observed. For $R = R' = H$ the elimination to H_2 and $H_2Os_2(CO)_8$ that occurs at 126°C was indicated as a binuclear reaction with the scheme as shown in the following.[85]

$$Os(CO)_4H_2 \longrightarrow Os(CO)_3H_2 + CO \tag{5.89}$$

$$Os(CO)_3H_2 + Os(CO)_4H_2 \xrightarrow{\text{fast}} H_2Os_2(CO)_7 + H_2 \tag{5.90}$$

$$H_2Os_2(CO)_7 + CO \xrightarrow{\text{fast}} H_2Os_2(CO)_8 \tag{5.91}$$

Elimination of methane from $Os(OC)_4(H)CH_3$ is more facile, occurring at 49°C.

$$2Os(CO)_4(H)CH_3 \longrightarrow CH_4 + HOs(CO)_4Os(CO)_4CH_3 \tag{5.92}$$

Labeling experiments confirmed the binuclear nature of the reaction.[85]

$$Os(CO)_4(H)(CD_3) + Os(CO)_4(D)CH_3 \longrightarrow CH_4 \text{ and } CD_4 \tag{5.93}$$
$$+ \text{ binuclear Os species}$$

Since no scrambling of label in methane occurs under the reaction conditions the products CH_4 and CD_4 could only arise from binuclear elimination. An initial methyl migration was suggested to open a coordination site.[85]

$$Os(CO)_4(H)R \longrightarrow Os(CO)_3(H)C(O)R \tag{5.94}$$

$$Os(CO)_3(H) C(O)R + Os(CO)_4(H)R \longrightarrow HOs(CO)_4Os(CO)_4R \tag{5.95}$$
$$R = CH_3 \qquad\qquad\qquad + RH + CO$$

In the presence of an entering nucleophile such as PEt_3 the elimination of CH_4 is mononuclear.[85]

$$Os(CO)_4(H)CD_3 + Os(CO)_4D(CH_3) \xrightarrow{PEt_3} Os(CO)_4PEt_3 \tag{5.96}$$
$$+ CD_3H + CH_3D$$

It has been suggested that for binuclear elimination to occur one of the eliminating groups must be a hydride and the other complex must be able to open a coordination site for a bridging hydride.

We have seen that reductive elimination, like oxidative addition, can occur through several mechanisms. The specific circumstances (solvent, reactants, additives, etc.) determine which mechanism may function for elimination. Substantially more information is required for a complete understanding of reductive elimination.

5.3. Oxidative Addition and Reductive Elimination

In many cases reductive elimination and oxidative addition occur in concert, as shown in the following[86,87]:

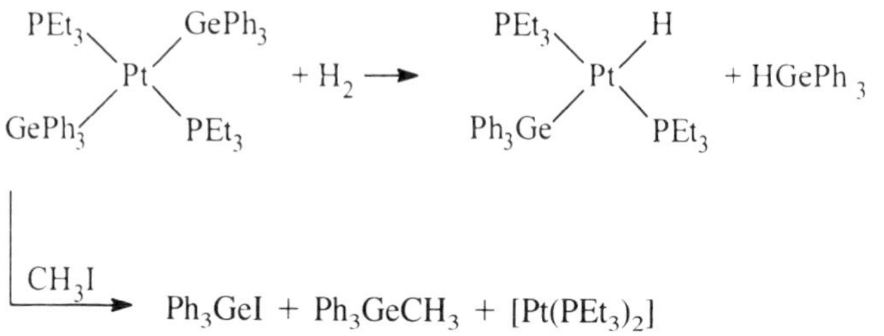

(5.97)

(5.98)

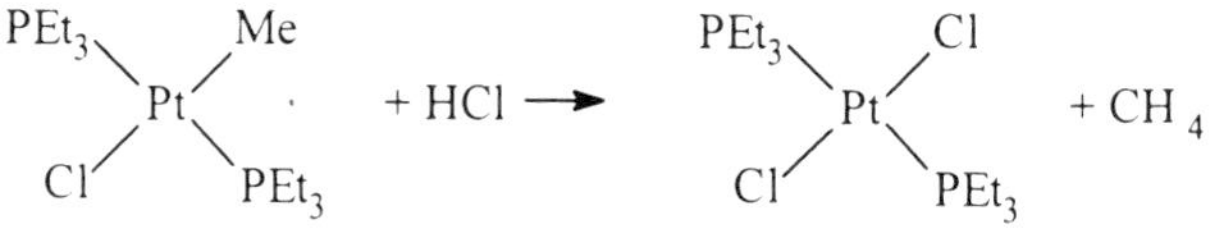

(5.99)

The kinetics of the latter Reaction (5.99) were investigated and the following mechanism was suggested:

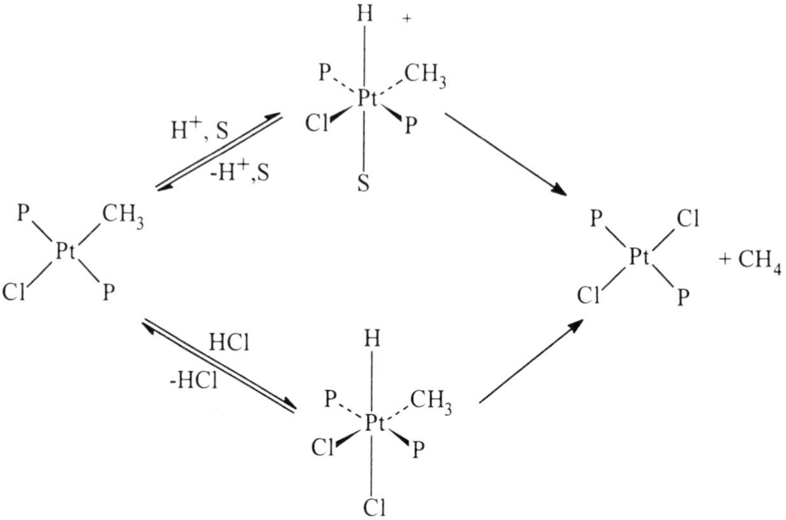

P = PEt$_3$
S = Solvent

(5.100)

The oxidative addition of H$^+$ (which is considered as a hydride when attached to the metal) is followed by reductive elimination of methane.

*Ortho*metallation is another example of an oxidative addition which is followed by reductive elimination. *Ortho*metallation may be considered as the oxidative addition of an aryl C–H bond on a coordinated ligand to a 16-electron metal center.

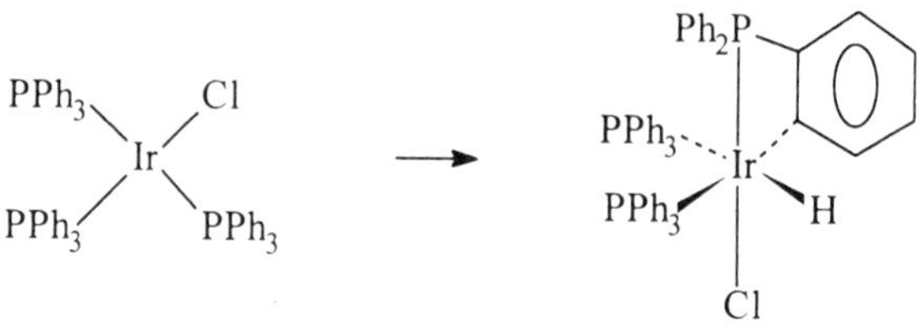

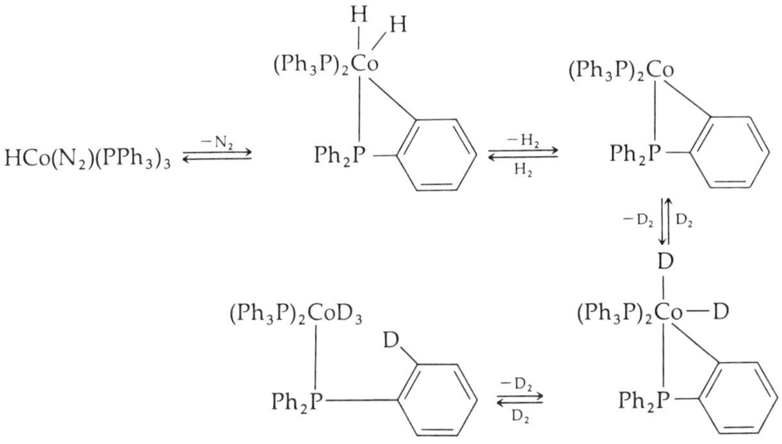

Figure 5.19. The oxidative addition and reductive eliminations via orthometallation that leads to incorporation of 19 deuteriums per mole of $HCo(N_2)(PPh_3)_3$.[55]

The reversibility of *ortho*metallation was shown in reactions of $HCo(N_2)(PPh_3)_3$ with deuterium where 19 hydrogens per mole of Co were exchanged.[88] It was shown that the incorporation into the aryl groups occurred exclusively in the *ortho* position and the scheme shown in Figure 5.19 was indicated.[88]

5.4. Conclusion

Oxidative addition and reductive elimination may proceed through different mechanisms. For reactions involving alkyl complexes, retention of stereochemistry is indicative of nucleophilic attack while loss of stereochemistry indicates radical mechanisms. It would be difficult to predict the mechanism in the absence of data for a given reaction. Oxidative addition may lead to either *cis* or *trans* products, depending on the adding molecule and the metal complex; H_2 addition always occurs *cis*. Reductive elimination from one metal center evidently requires a *cis* geometry but may also occur from two metal centers in a binuclear process. Oxidative addition and reductive elimination are very important in homogeneous catalysis, which will be considered in Chapter 6.

5.5. References

1. L. Vaska, *Acc. Chem. Res.* **1**, 335 (1968).

2. J. Halpern, *Acc. Chem. Res.* **3**, 386 (1970).

3. J. P. Collman and W. R. Roper, *Adv. Organomet. Chem.* **7**, 53 (1968).

4. J. P. Collman, *Acc. Chem. Res.* **1**, 136 (1968).

5. J. K. Stille and K.S.Y. Lau, *Acc. Chem. Res.* **10**, 434 (1977).

6. J. P. Collman and C. T. Sears, Jr., *Inorg. Chem.* **7**, 27 (1968).

7. M. J. Burke, M. P. McGrath, R. Wheeler, and R. H. Crabtree, *J. Am. Chem. Soc.* **110**, 5034 (1988).

8. J. S. Thompson, K. A. Bernard, B. J. Rappoli and J. D. Atwood, *Organometallics* **9**, 2727 (1990).

9. A. L. Sargent and M. B. Hall, *Inorg. Chem.* **31**, 317 (1992).

10. C. E. Johnson and R. Eisenberg, *J. Am. Chem. Soc.* **107**, 6531 (1985).

11. D. M. Blake and M. Kubota, *Inorg. Chem.* **9**, 989 (1970).

12. G. Yoneda and D. M. Blake, *Inorg. Chem.* **20**, 67 (1981).

13. P. B. Chock and J. Halpern, *J. Am. Chem. Soc.* **88**, 3511 (1966).

14. R. Ugo, A. Pasini, A. Fusi, and S. Cenini, *J. Am. Chem. Soc.* **94**, 7364 (1972).

15. E. M. Miller and B. L. Shaw, *J. Chem. Soc. Dalton*, 480 (1974).

16. B. L. Shaw and R. E. Stainbank, *J. Chem. Soc. Dalton*, 223 (1972).

17. D.J.A. DeWoal, T.I.A. Gerber, and W. J. Louw, *Inorg. Chem.* **21**, 1260 (1982).

18. J. C. Douek and G. Wilkinson, *J. Chem. Soc.* (A), 2604 (1969).

19. S. Graks, F. R. Hartley, and J. R. Chipperfield, *Inorg. Chem.* **20**, 3238 (1981).

20. G. J. Kubas, R. R. Ryan, B. I. Swanson, P. J. Vergamini, and H. J. Wasserman, *J. Am. Chem. Soc.* **106**, 450 (1984).

21. J. Chatt and J. M. Davidson, *J. Chem. Soc.*, 843 (1965).

22. K. Kitaura, S. Ohara, and K. Morokuma, *J. Am. Chem. Soc.* **103**, 2891 (1981).

23. J. P. Collman and M. R. MacLaury, *J. Am. Chem. Soc.* **96**, 3019 (1974).

24. J. P. Collman, D. W. Murphy, and G. Dolcetti, *J. Am. Chem. Soc.* **95**, 2687 (1973).

25. M. Crespo and R. R. Puddephatt, *Organometallics* **6**, 2548 (1987).

26. C. Amatore, A. Jutand, and A. Suarez, *J. Am. Chem. Soc.* **115**, 9531 (1993).

27. T. T. Tsou and J. K. Kochi, *J. Am. Chem. Soc.* **101**, 6319 (1979).

28. J. A. Labinger and J. A. Osborn, *Inorg. Chem.* **19**, 3230 (1980).

29. J. A. Osborn, in *Organotransition Metal Chemistry*, eds. Y. Ishii and M. Tsutsui (Plenum Press, New York, 1975), pp. 65–80.

30. J. A. Labinger, J. A. Osborn, and N. J. Coville, *Inorg. Chem.* **19**, 3236 (1980).

31. T. L. Hall, M. F. Lappert, and P. W. Lednor, *J. Chem. Soc.* Dalton, 1448 (1980).

32. A. V. Kramer, J. A. Labinger, J. S. Bradley, and J. A. Osborn, *J. Am. Chem. Soc.* **96**, 7145 (1974).

33. A. V. Kramer and J. A. Osborn, *J. Am. Chem. Soc.* **96**, 7832 (1974).

34. G. M. Williams, K. I. Gell, and J. Schwartz, *J. Am. Chem. Soc.* **102**, 3660 (1980).

35. G. M. Williams and J. Schwartz, *J. Am. Chem. Soc.* **104**, 1122 (1982).

36. A. J. Hart-Davis and W.A.G. Graham, *Inorg. Chem.* **9**, 2658 (1970).

37. M. A. Bennett, J. C. Jeffery, and G. B. Robertson, *Inorg. Chem.* **20**, 323 (1981).

38. D. Rabinovich and G. Parkin, *J. Am. Chem. Soc.* **115**, 353 (1993).

39. J. P. Collman, L. S. Hegedus, J. R. Norton, and R. G. Finke, *Principles and Applications of*

Organotransition Metal Chemistry (University Science Books, Mill Valley, CA), pp. 295–305 and references therein (1987).

40. J. K. Hoyano and W.A.G. Graham, *J. Am. Chem. Soc.* **104**, 3723 (1982).

41. W. D. Jones and F. J. Feher, *Acc. Chem. Res.* **22**, 91 (1989) and references therein.

42. A. H. Janowicz and R. G. Bergman, *J. Am. Chem. Soc.* **104**, 352 (1982).

43. B. H. Weiller, E. P. Wasserman, R. G. Bergman, C. B. Moore, and G. C. Pimantel, *J. Am. Chem. Soc.* **111**, 8288 (1989).

44. G. M. Whitesides, E. R. Stedronsky, C. P. Casey, and S. J. Filippo, Jr., *J. Am. Chem. Soc.* **92**, 1426 (1970).

45. G. M. Whitesides, J. F. Gaasch, and E. R. Stedronsky, *J. Am. Chem. Soc.* **94**, 5258 (1972).

46. J. Evans, J. Schwartz, and P. W. Urquhart, *J. Organomet. Chem.* **81**, C37 (1974).

47. R. L. Brainard and G. M. Whitesides, *Organometallics* **4**, 1550 (1985).

48. N. Koga, S. Obara, K. Kitaura, and K. Morokuma, *J. Am. Chem. Soc.* **107**, 7109 (1985).

49. A. C. Balazs, K. H. Johnson, and G. M. Whitesides, *Inorg. Chem.* **21**, 2162 (1982).

50. K. Tatsumi, R. Hoffmann, A. Yamamoto, and J. K. Stille, *Bull. Chem. Soc., Japan,* **54**, 1857 (1981).

51. A.S.C. Chan and J. Halpern, *J. Am. Chem. Soc.* **102**, 838 (1980).

52. L. Abis, A. Sen, and J. Halpern, *J. Am. Chem. Soc.* **100**, 2915 (1978).

53. (a) J. S. Thompson and J. D. Atwood, *Organometallics* **10**, 3525 (1991); (b) J. S. Thompson, S. L. Randall, and J. D. Atwood, *Organometallics* **10**, 3906 (1991); (c) J. S. Thompson, K. A. Bernard, B. J. Rappoli, and J. D. Atwood, *Organometallics* **9**, 2727 (1990); (d) K. A. Bernard and J. D. Atwood, *Organometallics* **8**, 795 (1989); (e) B. J. Rappoli, J. M. McFarland, J. S. Thompson, and J. D. Atwood, *J. Coord. Chem.* **21**, 147 (1990).

54. (a) J. J. Low and W. A. Goddard, III, *J. Am. Chem. Soc.* **108**, 6115 (1986); (b) J. J. Low and W. A. Goddard, III, *Organometallics* **5**, 609 (1986).

55. G. W. Parshall, *J. Am. Chem. Soc.* **96**, 2360 (1974).

56. G. K. Anderson, H. C. Clark, and J. A. Davies, *Organometallics* **1**, 550 (1982).

57. J. B. Keister, private communication.

58. D. P. Arnold and M. A. Bennett, *J. Organomet. Chem.* **199**, C17 (1980).

59. C.E.L. Headford and W. R. Roper, *J. Organomet. Chem.* **198**, C7 (1980).

60. D. L. Thorn, *Organometallics,* **1**, 197 (1982).

61. T. Yamamoto, T. Kohara, and A. Yamamoto, *Bull. Chem. Soc.* **54**, 2161 (1981).

62. C. D. Wood and R. R. Schrock, *J. Am. Chem. Soc.* **101**, 5421 (1979).

63. D. L. Packett and W. C. Trogler, *Inorg. Chem.* **27**, 1768 (1988).

64. P. P. Deutsch and R. Eisenberg, *J. Am. Chem. Soc.* **112**, 714 (1990).

65. D. Milstein, *J. Am. Chem. Soc.* **104**, 5226 (1982).

66. G. Parkin and J. E. Bercaw, *Organometallics* **8**, 1172 (1989).

67. R. M. Bullock, C.E.L. Headford, K. M. Hennessy, S. E. Kegley, and J. R. Norton, *J. Am. Chem. Soc.* **111**, 3897 (1989).

68. W. D. Jones and V. L. Kuykendall, *Inorg. Chem.* **30**, 2615 (1991).

69. W. D. Jones and E. T. Hessell, *J. Am. Chem. Soc.* **114**, 6087 (1992).

70. A. Gillie and J. K. Stille, *J. Am. Chem. Soc.* **102**, 4933 (1980).

71. M. K. Loar and J. K. Stille, *J. Am. Chem. Soc.* **103**, 4174 (1981).

72. K.-T. Aye, A. J. Canty, M. Crespo, R. J. Puddephatt, J. D. Scott, and A. A. Watson, *Organometallics* **8**, 1518 (1989).

73. B. A. Markies, A. J. Canty, J. Boersma, and G. van Koten, *Organometallics* **13**, 2053 (1994).

74. J. S. Thompson and J. D. Atwood, *Organometallics* **10**, 3525 (1991).

75. K. I. Goldberg, J.-Y. Yan, and E. M. Breitung, *J. Am. Chem. Soc.* **117**, 6889 (1995).

76. A. Pedersen and M. Tilset, *Organometallics* **12**, 56 (1993).

77. G. Smith and J. K. Kochi, *J. Organomet. Chem.* **198**, 199 (1980).

78. H. W. Sternberg, I. Wender, R. A. Friedel, and M. Orchin, *J. Am. Chem. Soc.* **75**, 2717 (1953).

79. J. J. Eisch and R. B. King, *Organometallic Syntheses* **1**, 158 (1965).

80. R. J. Hoxmeier, J. R. Blickensderfer, and H. D. Kaesz, *Inorg. Chem.* **18**, 3453 (1979).

81. W. D. Jones and R. G. Bergman, *J. Am. Chem. Soc.* **101**, 5447 (1979).

82. M. J. Nappa, R. Santi, S. P. Diefenbach, and J. Halpern, *J. Am. Chem. Soc.* **104**, 619 (1982).

83. R. J. Ruszczyk, B. L. Huang, and J. D. Atwood, *J. Organomet. Chem.* **299**, 205 (1986).

84. R. M. Bullock and B. J. Rappoli, *J. Am. Chem. Soc.* **113**, 1659 (1991).

85. J. R. Norton, *Acc. Chem. Res.* **12**, 139 (1979).

86. R. J. Cross and F. Glockling, *J. Chem. Soc.*, 5422 (1965).

87. U. Belluco, M. Giustiniani, and M. Graziani, *J. Am. Chem. Soc.* **89**, 6494 (1967).

88. G. W. Parshall, *J. Am. Chem. Soc.* **90**, 1669 (1968).

5.6. Problems

5.1. Identify which of the following are oxidative addition and reductive elimination reactions

a. $2HCo(CO)_3PPh_3 \rightarrow Co_2(CO)_6(PPh_3)_2 + H_2$

b. $CH_3Mn(CO)_5 + PMe_3 \rightarrow CH_3C(O)Mn(CO)_4PMe_3$

c. $HCo(N_2)(PPh_3)_3 + C_2H_4 \rightarrow HCo(C_2H_4)(PPh_3)_3 + N_2$

d. $HCo(N_2)(PPh_3)_3 + H_2 \rightarrow H_3Co(PPh_3)_3 + N_2$

e. $2Co(CN)_5^{3-} + CH_3I \rightarrow CH_3Co(CN)_5^{3-} + Co(CN)_5I^{3-}$

f. *trans*-$HPt(PMe_3)_2Cl + CH_3I \rightarrow$ *trans*-$Pt(PMe_3)_2(I)(Cl) + CH_4$

5.2. The following reaction has been investigated in each direction.

$$(Me)_n(EtO)_{3-n}SiH + Ir(dppe)_2^+ + \underset{k_r}{\overset{k_f}{\rightleftharpoons}} HIr(dppe)_2SiMe_n(OEt)_{3-n}^+$$

Interpret the following activation parameters data:

n	$\Delta H_f^{\ddagger}$, kcal/mol	$\Delta S_f^{\ddagger}$, eu	$\Delta H_r^{\ddagger}$, kcal/mol	$\Delta S_r^{\ddagger}$, eu
0	5.56	−47.3	25.3	8.2
1	5.62	−46.5	22.7	1.2
2	5.59	−48.2	19.2	0.1

5.3. Given the following data determine the rate law for oxidative addition of $C_6H_5CH_2I$ to $CpRh(CO)PPh_3$.

$[C_6H_5CH_2I][M]$	$10^4k_{obsd}(s^{-1})$
0.166	7.37
0.237	10.7
0.402	16.0

[A. J. Hart-Davis and W.A.G. Graham, *Inorg. Chem.* **10**, 1653 (1971).]

5.4. The dependence on the halide is as follows

Cl < Br < I

for $C_6H_5CH_2X$ in the reaction of Problem 5.3. Interpret this in terms of the mechanism.

5.5. Since oxidation number is a formalism that may or may not have any relationship with reality, there has been some discussion as to whether additions are oxidative in character. Discuss the significance of the following data for addition to Ir(*o*-phen)(cod)$^+$.

Adding Group	Oxidation Number
SCN$^-$	0.26
I$^-$	0.37
PPh$_3$	0.47
CH$_3$I	2.58
HCl	3.00
Cl$_2$	3.00

[W. J. Louw, D.J.A. deWaal, T.I.A. Gerber, C. M. Demanet, and R. G. Copperthwaite, *Inorg. Chem.* **21**, 1667 (1982).]

5.6.

a. The complex $CH_3Rh(PPh_3)_3$ is cleaved by H_2 to give methane and $HRh(PPh_3)_3$. If $CH_3Rh(PPh_3)_3$ is heated by itself methane is again evolved and an orange complex isolated which analyzes approximately as

1Rh:3PPh$_3$—Suggest the structure and the mechanism of CH_4 formation

b. The orange complex from (a) reacts with CO to give a new complex that analyzes as

1Rh:3PPh$_3$:2CO

and has IR peaks at 1965 and 1620 cm^{-1}. Suggest structures for this product.

5.7. a. On treating Ir(CO)CpPPh$_3$ with CH$_3$I, a white solid is obtained, which is soluble in fairly polar solvents, completely insoluble in benzene, has a ν_{CO} at 2050 cm^{-1}, and analyzes as IrC$_{25}$H$_{23}$IOP. Suggest a structure of the product.

b. Similar reactions of the cobalt and rhodium analogs M(CO)Cp(PPh$_3$) with CH$_3$I yield products that are colored, somewhat soluble in nonpolar solvents, and have

ν_{CO} between 1650 and 1666 cm^{-1}. However, the analyses are the same as they are in (a). Suggest structures.

c. Compare Ir(CO)CpPPh$_3$ with Vaska's compound, *trans*-IrCl(CO)(PPh$_3$)$_2$. What is the key difference between them? How does this affect the nature of their reactions with CH$_3$I? (The activation parameters are very similar).

5.8. Reaction of a macrocyclic rhodium complex with various alkyl halides, showed the following dependence on the alkyl group methyl > ethyl > secondary > cyclohexyl. In addition, reaction of *n*-butyl bromide in the presence of LiCl led to only the *trans*-chloro rhodium alkyl complex under conditions where the *trans*-bromo Rh alkyl does not exchange with LiCl. Present a mechanism that accounts for these data.

5.9. Reaction of *cis*-Me$_2$Pd(PPh$_3$)$_2$ with PhCH$_2$Br leads primarily to *trans*-MePd(PPh$_3$)$_2$Br and PhCH$_2$CH$_3$. Starting with α-deuteriobenzylbromide leads to α-deuterioethylbenzene with overall inversion of configuration at the benzylic carbon. Discuss the implications of this work. [D. Milstein and J. K. Stille, *J. Am. Chem. Soc.* **101**, 4981 (1979).]

5.10. Reaction of Ni(Et)$_2$(dppe) with CO leads to ethylene, ethane, 3-pentanone and propionaldehyde. Account for these products. [T. Yamamoto, T. Kohara, and A. Yamamoto, *Chem. Lett.* 1217 (1976).]

5.11. Similar reactions have been observed on Fe(II) complexes.

Fe(Et)$_2$(bipy) → C$_2$H$_6$ + C$_2$H$_4$ + [Fe(bipy)]

$$\text{Fe(Et)}_2\text{(bipy)}^+ \rightarrow \underset{45\%}{\text{C}_4\text{H}_{10}} + \underset{48\%}{\text{C}_2\text{H}_6} + \underset{8\%}{\text{C}_2\text{H}_4} + [\text{Fe(bipy)}]^+$$

Fe(Et)$_2$(bipy)$^{2+}$ → C$_4$H$_{10}$ + [Fe(bipy)]$^{2+}$

The rates of the reactions vary with half-lives of 15 days, 38 min, and 10^{-3}s for Fe(II), Fe(III), and Fe(IV), respectively. Account for the product distribution and rates of reaction.

5.12. A stereoselective synthetic route to alkenes has been presented as shown in the following reaction:

$$\text{R}'\text{R}''\text{C}{=}\text{CR}'''\text{X} \xrightarrow[\text{Pd(PPh}_3)_4]{\text{RLi}} \text{R}'\text{R}''\text{C}{=}\text{CR}'''\text{R}$$

Suggest a scheme.

5.13. The complex Cp$_2$Re(H)(Me) reacts in CD$_2$Cl$_2$ to give CH$_4$ and Cp$_2$ReCl.

$$\text{Cp}_2\text{Re(H)(Me)} \xrightarrow{\text{CD}_2\text{Cl}_2} \text{Cp}_2\text{ReCl} + \text{CH}_4$$

Methyl hydrogens and the hydride undergo site exchange with very similar activation parameters to the reductive elimination of CH$_4$. Suggest the key intermediate for these observations. [G. L. Gould and D. M. Heinekey, *J. Am. Chem. Soc.* **111**, 5502 (1989).]

6

Homogeneous Catalysis by Transition Metal Complexes

Homogeneous catalysis provides an important application fueling the explosive growth in organometallic chemistry. Homogeneous catalysis, with advantages in selectivity, milder conditions for reactions, and economy in reagents, will become more important as resources diminish. As shown in Table 6.1 a number of industrial processes use homogeneous catalysts.[1,2] While the processes in Table 6.1 are volume commodities, stereospecific syntheses are beginning to have a major impact on the pharmaceutic industry with products such as L-dopa.[3–7]

In this chapter the mechanisms of homogeneous hydrogenation, including asymmetric hydrogenation, hydroformylation, hydrocyanation, polymerization, alkene metathesis, ethylene oxidation to acetaldehyde, and methanol homologation are presented. As we discuss these mechanisms and the supporting data, it is important to remember that a catalytic cycle is composed of many reactions that may be equilibria and are usually impossible to investigate completely. Thus, the mechanisms suggested will be the simplest that satisfactorily describe the data. Books[1,2] and a nice series of articles in the *Journal of Chemical Education* provide other sources for information on homogeneous catalysis.[3]

6.1. General Considerations

A catalyst accelerates the rate of a reaction without changing the thermodynamic stability of the reactants or products. The catalyst lowers the activation energy and thereby speeds the reaction (Fig. 6.1). By lowering the activation energy, a catalyst increases the rate of reaction in both directions; that is, it increases the rate of

Table 6.1. SOME PROCESSES
CATALYZED HOMOGENEOUSLY[1]

Reaction	1990 Production[a]
hydroformylation	1818
hydrocyanation (adiponitrile)	420
alkene polymerizations	10,000
methanol carbonylation	1,164

[a] Thousands of metric tons from reference 1b.

attainment of equilibrium. Therefore, a catalyst is useful only for reactions with favorable equilibria.

Reactions may be catalyzed heterogeneously or homogeneously. Heterogeneous catalysts operate in a different phase than the reactants. For the reactions that we shall discuss, heterogeneous catalysts are most often metals or metal oxides. Homogeneous catalysts function in the same phase as the reactants; for our considerations, the homogeneous catalyst is usually a metal complex in solution. Heterogeneous catalysts, which are more commonly used in industry, have advantages in stability, generality for different reactions, and ease of separation from the products. Homogeneous catalysts often require lower temperature and pressure conditions, and they are more selective for a given product.[8] Advantages of homogeneous catalysts have made them very attractive as raw materials become scarce. An additional advantage of homogeneous catalysts is that they can be studied and the reaction mechanism can be investigated by spectroscopic techniques.[9]

In order to function as a catalyst a complex must have open coordination sites or be capable of easily generating coordination sites. Most catalysts involve square-planar complexes or complexes with readily dissociable ligands.[1,2,8,9] Saturated complexes with tightly bound ligands do not function in catalytic cycles. The transition metal also must hold the substrate molecules within the coordination sphere in a position appropriate for reaction. In some cases the catalyst holds the substrate molecules in position for stereospecific reactions. This is termed a *temp-*

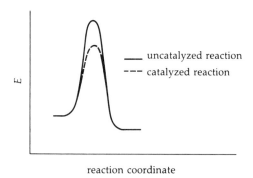

uncatalyzed reaction
--- catalyzed reaction

E

reaction coordinate

Figure 6.1. The effect of a catalyst on the reaction profile.

late effect.[10] A structure illustrating the potential of metal complexes for a template effect is shown in Figure 6.2.[11] The tetranickel cluster is a catalyst for the cyclotrimerization of alkynes.[11] In the structure the alkynes are held along the basal nickel atoms in a position for trimerization; thus, the catalyst brings the three acetylenes together as a template.[11] It has been suggested that four acetylene molecules are simultaneously coordinated to nickel in the Ni(II) catalyzed cyclotetramerization of acetylene forming cyclooctatetraene,[12]

$$HC \equiv CH \xrightarrow{\text{Ni(II)}} \text{(cyclooctatetraene)} \tag{6.1}$$

and successive coordination of alkynes was suggested in cyclotrimerization on Ir(I).[13] The mechanistic distinction between simultaneous and successive coordination is not always straightforward. The important feature is that the metal complex must be able to coordinate the substrate(s) in the proper position for further reaction.

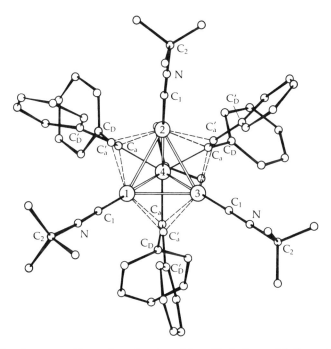

Figure 6.2. Structure illustrating the template effect for catalytic reactions. The $Ni_4[CNC(CH_3)_3]_4[\mu_3(\eta^2)\text{-}C_6H_5C{\equiv}CC_6H_5]_3$ molecule is viewed normal to the triangular base along the idealized threefold axis. Nickel atoms are represented by large numbered open circles and carbon and nitrogen atoms by small open circles. [Reprinted with permission from M. G. Thomas, E. L. Muetterties, R. O. Day, and V. W. Day, *J. Am. Chem. Soc.* **98**, 4645 (1976). Copyright 1976 American Chemical Society.]

(6.2)

Studying catalytic reactions is different from studying the stoichiometric reactions we have discussed in earlier chapters. For catalytic reactions, interpretation of product distribution is very important; kinetic studies are less significant. Identification of intermediates is usually impossible. In fact, the postulate has been advanced that if a complex is reactive enough to participate in a catalytic cycle, then it is not stable enough to be observed.[9] This postulate suggests the difficulty of trying to define a catalytic cycle. Evidence cited in support of this postulate is the observation of the intermediate shown in Figure 6.3, which was intercepted from the catalytic cycle shown in Figure 6.4, but which has the wrong configuration for the primary product.[9,14] The mechanisms described in this chapter are those accepted for the different catalytic reactions.

6.2. Homogeneous Hydrogenation of Alkenes

A catalyst must accomplish four things in homogeneous hydrogenations: (1) it must coordinate the alkene; (2) it must coordinate hydrogen; (3) it must allow the hydrogen to add to the alkene; and (4) it must eliminate the hydrogenated product.

$$\text{alkene} + H_2 \xrightarrow{\text{catalyst}} \text{alkane} \tag{6.3}$$

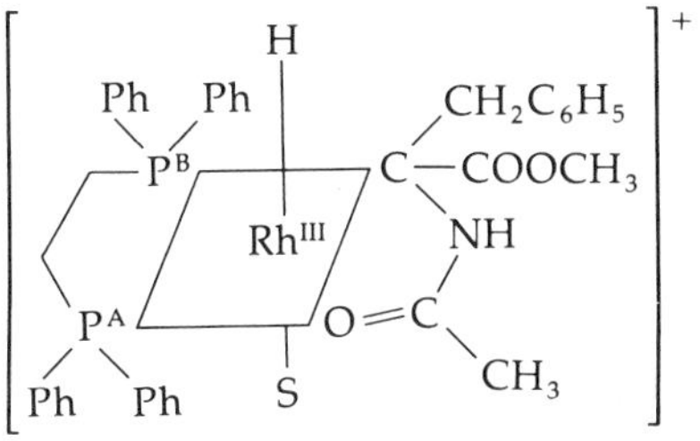

$$S = CH_3OH \text{ or } CH_3CN$$

Figure 6.3. An alkyl hydride intermediate that was intercepted from a cycle for catalytic hydrogenation. [Reprinted with permission from A.S.C. Chan and J. Halpern, *J. Am. Chem. Soc.* **102,** 838 (1980). Copyright 1980 American Chemical Society.]

Coordination of alkenes has been studied often, beginning in the early nineteenth century with Zeise's salt, $KPtCl_3(C_2H_4)$.[15] Many alkene complexes have been prepared, and the stability of the alkene–metal bond evaluated.[16,17] Stability constants determined for the reaction

$$Ag^+ + \text{alkene} \rightleftharpoons Ag^+(\text{alkene}) \qquad (6.4)$$

are reported in Table 6.2.[16] The data in Table 6.2 indicate that increasing the hydrocarbon chain length or increasing substitution on the olefinic carbons de-

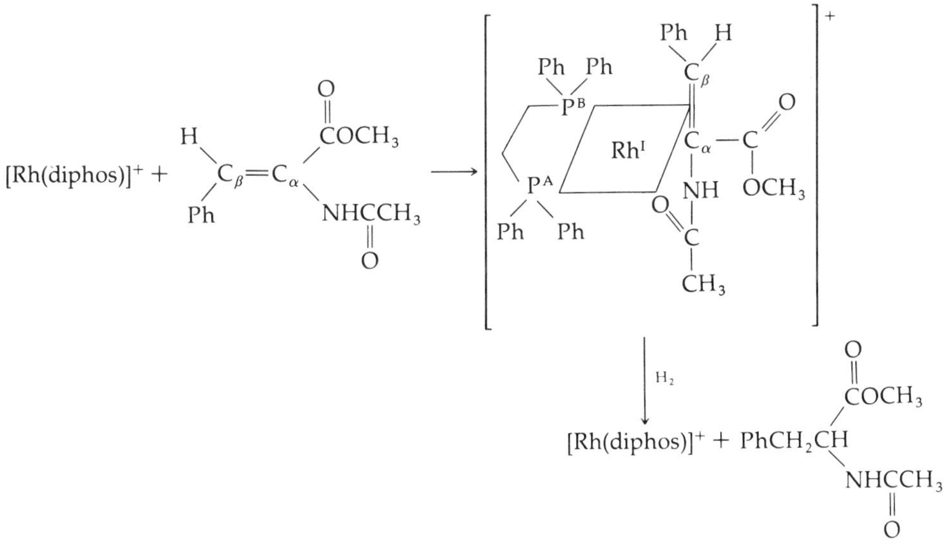

Figure 6.4. Cycle for the catalytic hydrogenation of methyl(Z)-α-acetamidocinnamate by Rh(diphos)$^+$.[14]

Table 6.2. STABILITY CONSTANTS
FOR ALKENE COMPLEXATION

Alkene	K^a
C_2H_4	17.5
$CH_3CH{=}CH_2$	7.5
$C_2H_5CH{=}CH_2$	8.8
$C_3H_7CH{=}CH_2$	6.7
$C_4H_9CH{=}CH_2$	4.3
$C_5H_{11}CH{=}CH_2$	3.2
$cis\text{-}CH_3CH{=}CHCH_3$	4.9
$trans\text{-}CH_3CH{=}CHCH_3$	1.6
$(CH_3)_2C{=}CHCH_3$	1.01
$(CH_3)_2C{=}C(CH_3)_2$	0.34

^{a} For reaction with Ag^+ in ethylene glycol at 25°C.

$$K = \frac{[Ag(alkene)^+]}{[Ag^+][alkene]}.$$

creases the stability of the alkene–metal bond. In general *cis* isomers are more stable than *trans*.[16]

Hydrogen activation has also been commonly observed (see Chapter 5). Heterolytic activation has sometimes been observed,[18] although oxidative-addition is the most commonly observed means of activation (Chapter 5). Transfer of hydrogen to a coordinated alkene has not been as thoroughly investigated. The transfer must occur in two steps.

$$H_2M(alkene) \rightleftharpoons HM(alkyl) \qquad (6.5)$$

$$HM(alkyl) \longrightarrow M + alkane \qquad (6.6)$$

The first step would be alkene insertion into an M–H bond (or hydride migration). It has proven extremely difficult to study this step. Two olefin hydride complexes that have been observed illustrate the reaction:

$$CpRh(C_2H_4)PMe_3 + HBF_4 \longrightarrow [CpRhH(C_2H_4)PMe_3]BF_4 \qquad (6.7)$$

$$\Updownarrow$$

$$[CpRh(C_2H_5)PMe_3]BF_4$$

$$HCo(N_2)(PPh_3)_3 + alkene \longrightarrow HCo(alkene)(PPh_3)_3 + N_2 \qquad (6.8)$$

In Reaction 6.7 the olefin insertion to the alkyl can be observed[10]; in Reaction 6.8 the alkene and the hydride are in *trans* positions, and the alkyl is not formed.[20] Insertion of ethylene into a Pt–H bond has been investigated in some detail.[21]

$$trans\text{-}Pt(H)(acetone)(PEt_3)_2^+ + C_2H_4 \longrightarrow \qquad (6.9)$$

$$trans\text{-}Pt(C_2H_5)(acetone)(PEt_3)_2^+$$

The rate expression precluded coordination of the ethylene and insertion into the Pt–H bond in the five-coordinate intermediate. The following mechanism was suggested[21]:

$$trans\text{-}HPt(PEt_3)_2S + C_2H_4 \rightleftharpoons trans\text{-}HPt(C_2H_4)(PEt_3)_2 \qquad (6.10)$$

$$trans\text{-}HPt(PEt_3)_2(C_2H_4) \rightleftharpoons cis\text{-}HPt(PEt_3)_2(C_2H_4) \qquad (6.11)$$

$$cis\text{-}HPt(PEt_3)_2(C_2H_4) \rightleftharpoons cis\text{-}Pt(C_2H_5)(PEt_3)_2S \qquad (6.12)$$

$$cis\text{-}Pt(C_2H_5)(PEt_3)_2S \rightleftharpoons trans\text{-}Pt(C_2H_5)(PEt_3)_2S \qquad (6.13)$$

S = acetone

The intermediate, $trans\text{-}HPt(C_2H_4)(PEt_3)_2$, was observed by NMR. The reverse of Reaction 6.5 is considered to occur by β-elimination, a general reaction of transition metal alkyl complexes (see Chapter 5).

The stoichiometric hydrogenation of activated alkenes by Cp_2MoH_2 showed the stereochemistry of the alkene insertion to be *cis* by reaction of Cp_2MoD_2 with dimethyl fumarate, which leads to *threo*-$Cp_2MoD[CH(CO_2Me)CHD(CO_2Me)]$, or with dimethylmaleate, which leads to the *erythro* isomer. The isomers were identified by NMR.[22] Since the stereochemistry of Cp_2MoRH was known, the stereochemistry of the reductive-elimination could also be investigated and proceeded with retention of configuration at the σ-bonded carbon.[22] The stereochemistry is shown in Figure 6.5. This was in agreement with the results on reductive-elimination described in Chapter 5.

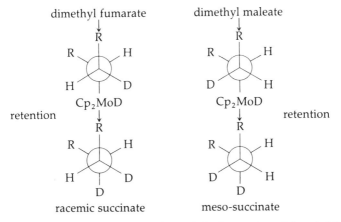

Figure 6.5. Stereochemistry of the stoichiometric hydrogenation of dimethyl fumarate or dimethyl maleate by Cp_2MoH_2.[20]

6.2.1. Rh(PPh$_3$)$_3$Cl

Tris(triphenylphosphine)chlororhodium is one of the most active homogeneous hydrogenation catalysts known and has been often studied.[23-30] It is a square-planar, 16-electron complex that readily undergoes oxidative-addition reactions and functions as a hydrogenation catalyst at ambient conditions.[23] The rates of hydrogenation of several alkenes are reported in Table 6.3.[25] The data show that less hindered olefins are hydrogenated more rapidly.[25] This order is very similar to that expected for olefin coordination stability.[19,20]

Oxidative addition of H$_2$ to Rh(PPh$_3$)$_3$Cl has been studied by proton and P-31 NMR.[28] Addition of H$_2$ at $-25°$C led quantitatively to H$_2$Rh(PPh$_3$)$_3$Cl with the stereochemistry shown in Figure 6.6. Upon warming to 30°C, the coupling from Rh to P$_1$ and from P$_1$ to P$_2$ is lost, while the Rh to P$_2$ coupling is retained. This is consistent with rapid dissociation of the PPh$_3$ represented as P$_1$. The rate constant for the reaction

$$H_2Rh(PPh_3)_3Cl \rightleftharpoons H_2Rh(PPh_3)_2Cl + PPh_3 \qquad (6.14)$$

was estimated as 200 s^{-1} at 30°C by line-shape analysis.[28] Although the equilibrium is rapidly established, it must greatly favor the six-coordinate species since there was no evidence for free PPh$_3$ or the five-coordinate complex. Studies of the kinetics of these reactions are complicated by the following reactions:

$$Rh(PPh_3)_3Cl \rightleftharpoons Rh(PPh_3)_2Cl + PPh_3 \qquad (6.15)$$

$$Rh(PPh_3)_2 Cl \rightleftharpoons Rh_2Cl_2(PPh_3)_4 \qquad (6.16)$$

Both products react with H$_2$ and can participate in the catalytic cycle. An excess of PPh$_3$ inhibits Reactions 6.15 and 6.16, allowing the kinetics of the hydrogen addition to Rh(PPh$_3$)$_3$Cl to be studied.[29] The results are consistent with the following reactions:

$$Rh(PPh_3)_3Cl + H_2 \xrightarrow{k_1} Rh(PPh_3)_3Cl(H_2) \qquad (6.17)$$

Table 6.3. HYDROGENATION OF ALKENES WITH Rh(PPh$_3$)$_3$Cl AT 25°C IN BENZENE SOLUTION

Alkene	$k(10^2 M^{-1}s^{-1})$
1-hexene	29.1
2-methylpent-1-ene	26.6
cis-4-methylpent-2-ene	9.9
trans-4-methylpent-2-ene	1.8
cyclohexene	31.6
1-methylcyclohexene	0.6
styrene	93.0

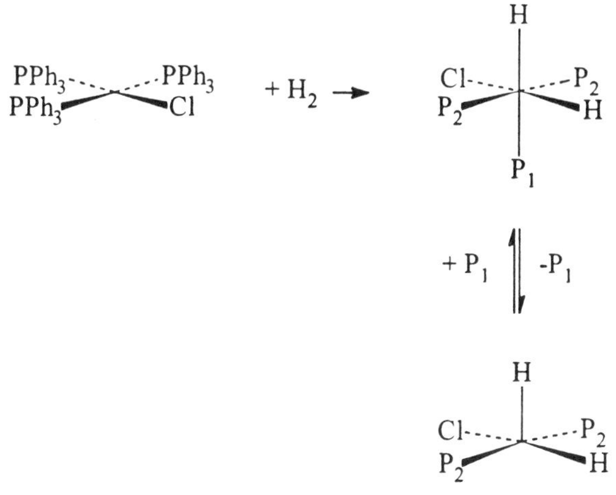

$$P_1 = P_2 = PPh_3$$

Figure 6.6.   Stereochemistry of H_2 addition to $Rh(PPh_3)_3Cl$.[28]

$$Rh(PPh_3)_3Cl \underset{k_{-2}}{\overset{k_2}{\rightleftharpoons}} Rh(PPh_3)_2Cl + PPh_3 \qquad (6.18)$$

$$Rh(PPh_3)_2 Cl + H_2 \overset{k_3}{\longrightarrow} Rh(PPh_3)_2Cl(H_2) \qquad (6.19)$$

$$Rh(PPh_3)_2Cl(H_2) + PPh_3 \overset{fast}{\rightleftharpoons} Rh(PPh_3)_3Cl(H_2) \qquad (6.20)$$

The rate law was derived

$$\frac{-d[Rh(PPh_3)_3Cl]}{dt} =$$

$$\left(k_1 + \frac{k_2 k_3}{k_{-2}[PPh_3] + k_3[H_2]} \right) [H_2][Rh(PPh_3)_3Cl] \quad (6.21)$$

with the rate constants, $k_1 = 4.8\ M^{-1}s^{-1}$, $k_2 = 0.71\ s^{-1}$, and $k_{-2}/k_3 = 1.1$.[29] The remainder of the catalytic reaction, the transfer of H_2 from $H_2Rh(PPh_3)_3Cl$ to an alkene, was studied for cyclohexene.[30] The rate law is shown in Equation 6.22

$$\frac{-d[H_2Rh(PPh_3)_3Cl]}{dt} = \left(\frac{K_5 k_6 [C_6H_{10}]}{[PPh_3] + K_5[C_6H_{10}]} \right) [H_2Rh(PPh_3)_3Cl] \quad (6.22)$$

where the constants are defined by the following equations:

$$H_2Rh(PPh_3)_3Cl + C_6H_{10} \overset{K_5}{\rightleftharpoons} H_2Rh(PPh_3)_2Cl(C_6H_{10}) + PPh_3 \qquad (6.23)$$

$$H_2Rh(PPh_3)_2Cl(C_6H_{10}) \xrightarrow{k_6} [\quad] \xrightarrow[\text{rapid}]{+PPh_3} \qquad (6.24)$$

$$Rh(PPh_3)_3Cl + C_6H_{12}$$

Equation 6.23 would be a rapid preequilibrium described by the equilibrium constant K_5. The reaction represented by k_6 is the rate-determining step. A complete catalytic cycle is represented by reactions 6.25–6.29; the stereochemistries are shown in Figure 6.7.

$$Rh(PPh_3)_3Cl + H_2 \rightleftharpoons H_2Rh(PPh_3)_3Cl \qquad (6.25)$$

$$H_2Rh(PPh_3)_3Cl \rightleftharpoons H_2Rh(PPh_3)_2Cl + PPh_3 \qquad (6.26)$$

$$H_2Rh(PPh_3)_2Cl + \text{alkene} \rightleftharpoons H_2Rh(PPh_3)_2Cl(\text{alkene}) \qquad (6.27)$$

$$H_2Rh(PPh_3)_2Cl(\text{alkene}) \rightleftharpoons H(\text{alkyl})Rh(PPh_3)_2Cl \qquad (6.28)$$

$$H(\text{alkyl})Rh(PPh_3)_3Cl + PPh_3 \longrightarrow Rh(PPh_3)_3Cl + \text{alkane} \qquad (6.29)$$

These reactions may be termed oxidative-addition (Eq. 6.25) ligand substitution (Eqs. 6.26 and 6.27), alkene insertion into the Rh–H bond (Eq. 6.28), and reductive elimination of alkane. With the exception of the final step each reaction is reversible, and isomerization of alkenes does occur. The following three reactions may replace Reactions 6.25–6.27 at low concentrations of PPh_3.[29]

$$Rh(PPh_3)_3Cl \rightleftharpoons Rh(PPh_3)_2Cl(S) + PPh_3 \qquad (6.30)$$

$$Rh(PPh_3)_2Cl(S) + H_2 \rightleftharpoons H_2Rh(PPh_3)_2Cl(S) \qquad (6.31)$$

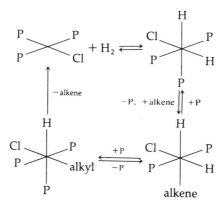

Figure 6.7. Stereochemistry at Rh for catalytic hydrogenation by $Rh(PPh_3)_3Cl$.

$$H_2Rh(PPh_3)_2Cl(S) + alkene \rightleftharpoons H_2Rh(PPh_3)_2(alkene)Cl \qquad (6.32)$$

S = solvent

Although $Rh(PPh_3)_2Cl(S)$ is of such low concentration that it cannot be detected in the catalytic system, the very rapid reaction with H_2 allows $Rh(PPh_3)_2Cl(S)$ to contribute to the catalytic reaction. There has been active discussion about whether PPh_3 dissociation occurs before or after oxidative-addition of hydrogen.[27] This is an extremely difficult question to resolve since the two products $H_2Rh(PPh_3)_3Cl$ and $H_2Rh(PPh_3)_2Cl$ are both present in solution. The kinetics indicated that both reactions contribute to formation of $H_2Rh(PPh_3)_3Cl$.

6.2.2. HRh(PPh₃)₃CO

Just as $Rh(PPh_3)_3Cl$ serves as a model for catalytic hydrogenations by square-planar, 16-electron complexes, $HRh(PPh_3)_3CO$ serves as a model for hydrogenations by 18-electron hydride complexes.[31] The geometry is trigonal bipyramidal with equatorial PPh_3 ligands as shown in Figure 6.8.[32,33] An 18-electron complex must open a coordination site for the alkene as an initial step; PPh_3 readily dissociates from $HRh(PPh_3)_3(CO)$ under mild conditions.[34,35] The proposed mechanism is:

$$HRh(PPh_3)_3CO \rightleftharpoons HRh(PPh_3)_2CO + PPh_3 \qquad (6.33)$$

$$HRh(PPh_3)_2CO + alkene \xrightarrow{K} Rh(PPh_3)_2(CO)(alkyl) \qquad (6.34)$$

$$Rh(PPh_3)_2(CO)(alkyl) \xrightarrow{k, H_2} HRh(CO)(PPh_3)_2 + alkane \qquad (6.35)$$

The rate shows a first-order dependence on the concentration of hydrogen and rhodium complex and a dependence between zero and first order in alkene.[36] The rate law for the mechanism of Reactions 6.33–6.35 shows these characteristics.

$$rate = \frac{kK[Rh][alkene][H_2]}{1 + K[alkene]} \qquad (6.36)$$

The presence of PPh_3 inhibits the reaction by reversing Reaction 6.33, leaving no coordination site for the alkene. None of the intermediates has been isolated or

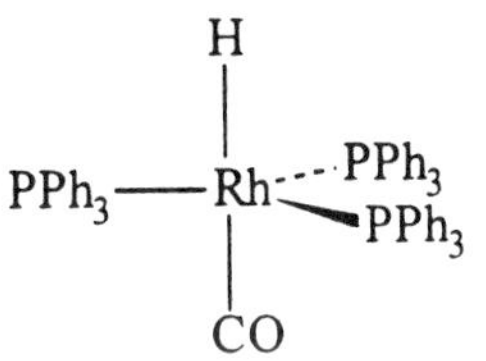

Figure 6.8.   Structure of $HRh(PPh_3)_3(CO)$.

spectroscopically characterized. Evidence for the first two steps is shown by the exchange of H from alkene to rhodium in the absence of H_2.[37]

$$DRh(PPh_3)_3CO + alkene \longrightarrow hydrogen\ exchange \qquad (6.37)$$

The scheme shown in Equations 6.33–6.35 includes reactions that probably occur in more than one step. The sequence can be described as: PPh_3 dissociation (Eq. 6.33); addition of alkene followed by insertion of the alkene into a Rh–H bond (Eq. 6.34); and oxidative-addition of H_2 followed by reductive-elimination of alkane (Eq. 6.35). The key differences mechanistically between $Rh(PPh_3)_3Cl$ and $HRh(PPh_3)_3CO$ lie in the electron count; the 16-electron $Rh(PPh_3)_3Cl$ may undergo oxidative-addition, but the 18-electron $HRh(PPh_3)_3CO$ complex must initially undergo PPh_3 dissociation. The presence of the hydride ligand in $HRh(PPh_3)_3CO$ allows the alkene to add and insert into the Rh–H bond retaining the coordination site (no further dissociation is necessary); $H_2Rh(PPh_3)_3Cl$ must undergo PPh_3 dissociation to open a site for alkene.

The carbonyl complex $HRh(PPh_3)_3CO$ shows considerably more selectivity toward terminal alkenes, but is only about half as active as $Rh(PPh_3)_3Cl$. The results for several olefins are shown in Table 6.4[36] and may be compared with the data in Table 6.3.

6.2.3. $HCo(CN)_5^{3-}$

Much of the earliest research on mechanisms of homogeneous hydrogenation of alkenes by transition metal complexes derived from studies of $HCo(CN)_5^{3-}$, which hydrogenates activated C–C double bonds.[38] Two mechanisms have been suggested, depending on the substrate. For hydrogenation of conjugated dienes to monoenes, illustrated here by 1,3-butadiene, the mechanism involves an alkyl complex similar to the mechanisms considered earlier. The products of the hydrogenation of 1,3-butadiene are 1-butene and cis- and trans-2-butene with 1-butene dominant at high concentrations of CN^- and trans-2-butene dominant at low concentrations of CN^-.[38] A detailed mechanism has been suggested for the hydrogenation of 1,3-butadiene by $HCo(CN)_5^{3-}$.

Table 6.4. RATE OF HYDROGENATION OF SEVERAL OLFINS BY $HRh(PPh_3)_3CO$ AT 25°C

Alkene	Rate (ml min^{-1})
1-hexene	16.7
ethylene	35
styrene	1.4
allyl benzene	11.1
cyclohexene	<0.1
cis-2-heptene	<0.1

$$HCo(CN)_5^{3-} + C_4H_6 \rightleftharpoons \left[(CN)_5Co-\overset{\displaystyle Me}{\underset{\displaystyle CH=CH_2}{\overset{|}{\underset{|}{CH}}}} \right]^{3-} \quad (6.38)$$

$$(CN)_5CoCHMeCH = CH_2^{3-} \underset{+CN^-}{\overset{-CN^-}{\rightleftharpoons}} (CN)_4Co(\eta^3\text{-}C_4H_7)^{2-} \quad (6.39)$$

$$(CN)_5CoCHMeCH = CH_2^{3-} \rightleftharpoons \quad (6.40)$$
$$(CN)_5CoCH_2CH = CHCH_3^{3-}$$

$$(CN)_5CoCHMeCH = CH_2^{3-} + HCo(CN)_5^{3-} \longrightarrow \quad (6.41)$$
$$2Co(CN)_5^{3-} + 1\text{-}C_4H_8$$

$$(CN)_5CoCH_2CH = CHCH_3^{3-} + HCo(CN)_5^{3-} \longrightarrow \quad (6.42)$$
$$2Co(CN)_5^{3-} + 2\text{-}C_4H_8 \ (cis \text{ and } trans)$$

$$(CN)_4Co(\eta^3\text{-}C_4H_7)^{2-} + HCo(CN)_5^{3-} \longrightarrow 2Co(CN)_5^{3-} + 1\text{-}C_4H_8 \quad (6.43)$$

$$(CN)_4Co(\eta^3\text{-}C_4H_7)^{2-} + HCo(CN)_5^{3-} \longrightarrow \quad (6.44)$$
$$2Co(CN)_5^{3-} + 2\text{-}C_4H_8(trans)$$

The rate constants were calculated for each step of this catalytic cycle [the cycle is completed by $2Co(CN)_5^{3-} + H_2 \rightarrow 2HCo(CN)_5^{3-}$]. The products of the hydrogenation of several conjugated dienes are shown in Table 6.5.[38] The hydrogenation of α,β-unsaturated acids by $HCo(CN)_5^{3-}$ has been suggested to occur by a free radical mechanism.[38] The reduction of cinnamic acid to β-phenylpropionic acid can serve as an example:

$$(CN)_5CoH^{3-} + Ph(H)C = CHCOOH \longrightarrow \quad (6.45)$$
$$Co(CN)_5^{3-} + PhCH_2\overset{\bullet}{C}HCOOH$$

Table 6.5. PRODUCTS OF THE HYDROGENATION OF CONJUGATED DIENES BY $HCo(CN)_5^{3-}$

Alkene	Products
butadiene	1-butene and 2-butene
1,3-pentadiene	1-pentene and *trans*-2-pentene
2,4-hexadiene	2-hexenes
norbornadiene	nortricyclene and norbornene
cyclopentadiene	cyclopentene
1,3-cyclohexadiene	cyclohexene

$$PhCH_2\overset{\bullet}{C}HCOOH + HCo(CN)_5^{3-} \longrightarrow \tag{6.46}$$

$$PhCH_2CH_2COOH + Co(CN)_5^{3-}$$

$$PhCH_2\overset{\bullet}{C}HCOOH + Co(CN)_5^{3-} \rightleftharpoons PhCH_2CHCOOH \tag{6.47}$$
$$\overset{|}{Co(CN)_5^{3-}}$$

Kinetic analysis rules out the mechanism involved in hydrogenation of conjugated dienes (Reactions 6.38–6.44) for hydrogenation of α,β-unsaturated acids. The complex $HCo(CN)_5^{3-}$ is an active hydrogenation catalyst for many activated C–C double bonds. It serves as a model for a free-radical mechanism in homogeneous catalysis by transition metal complexes.

6.2.4. $CH_3Co(CO)_2[P(OMe)_3]_2$

The most important requirement for an active catalyst is the ability to coordinate both alkene and hydrogen. An alkyl carbonyl complex may open two coordination sites if the carbonylation–decarbonylation reaction is a rapid equilibrium:

$$R\overset{}{\underset{\underset{CO}{|}}{-\!\!\!-M}}\rightleftharpoons RC(O)M \tag{6.48}$$

One coordination site is opened by the migration of the alkyl to a CO and the second by dissociative loss of a ligand from the acyl complex. A catalytic system for hydrogenation of alkenes has been developed using this concept of carbonylation–decarbonylation to open the coordination sites necessary.[39] The catalyst, $CH_3Co(CO)_2[P(OMe)_3]_2$, is active for the hydrogenation of terminal alkenes at 450 turnovers per hour. In both rate and selectivity this alkyl complex is very similar to $HRh(PPh_3)_3(CO)$. The rates of hydrogenation of several alkenes are shown in Table 6.6.[39] The following cycle has been suggested for these catalytic reactions[39]:

$$CH_3Co(CO)_2[P(OMe)_3]_2 + alkene \rightleftharpoons \tag{6.49}$$
$$CH_3C(O)Co(CO)(alkene)[P(OMe)_3]_2$$

$$CH_3C(O)Co(CO)(alkene)[P(OMe)_3]_2 \rightleftharpoons \tag{6.50}$$
$$CH_3Co(CO)(alkene)[P(OMe)_3]_2 + CO$$

$$CH_3Co(CO)(alkene)[P(OMe)_3]_2 + H_2 \rightleftharpoons \tag{6.51}$$
$$CH_3C(O)Co(alkene)(H_2)[P(OMe)_3]_2$$

$$CH_3C(O)Co(alkene)(H_2)[P(OMe)_3]_2 \rightleftharpoons \tag{6.52}$$
$$CH_3Co(CO)(alkyl)(H)[P(OMe)_3]_2$$

$$CH_3Co(CO)(alkyl)(H)[P(OMe)_3]_2 + CO \rightleftharpoons \tag{6.53}$$
$$CH_3Co(CO)_2[P(OMe)_3]_2 + RH$$

Table 6.6. THE RATES OF
HYDROGENATION OF ALKENES
BY $CH_3Co(CO)_2[P(OMe)_3]_2$

Alkene	Turnovers/Hour
1-hexene	450
cyclohexene	11
2-hexene	4
1-octene	>100
styrene	>100
1,7-octadiene	>100
3-hexene	<1
cyclohexenone	<1

Each of the intermediates in this scheme contains 18 electrons. An understanding of why alkane instead of CH_4 is reductively eliminated in the last step can only come from consideration of the possible geometries. The suggested geometries for Reaction 6.52 are shown in Figure 6.9. The CH_3 group is *trans* to H while the alkyl group is *cis*, thus the selectivity for reductive-elimination of alkane. If the phosphite donors are constrained to be *cis* as in the complex $CH_3Co(CO)_2$(Pom-Pom), the catalyst has a much shorter lifetime, and the rate of CH_4 elimination is increased [Pom-Pom = $(CH_3O)_2PCH_2CH_2P(OCH_3)_2$].[40]

6.3. Asymmetric Hydrogenation

One of the most promising areas of application for homogeneous hydrogenation catalysis is in the use of chiral catalysts to effect the asymmetric hydrogenation of prochiral alkenes with high optical yields.[1b,3,7] The processes for L-DOPA (for treatment of Parkinson's disease, reaction 6.54), L-phenylalanine (for conversion to aspartame, reaction 6.55), and Naproxen® (anti-inflammatory, reaction 6.56) indicate the tremendous potential of asymmetric hydrogenation.

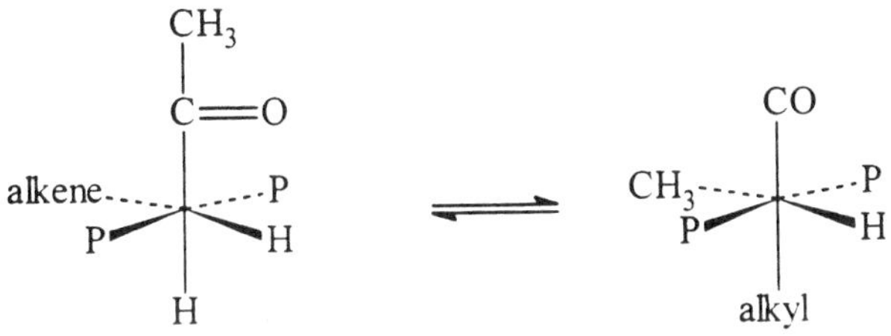

Figure 6.9. Stereochemistry of Reaction 6.52.

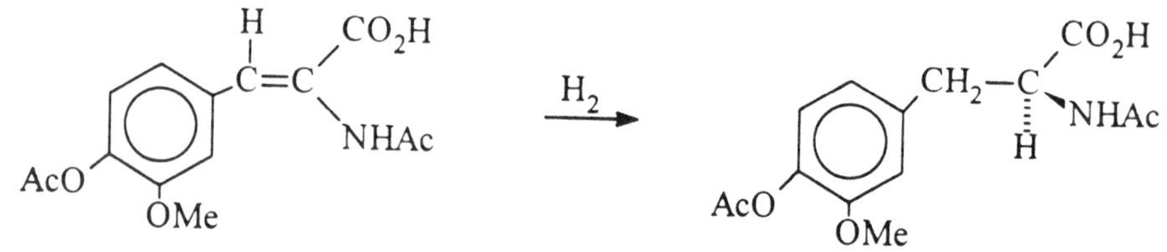

(6.54)

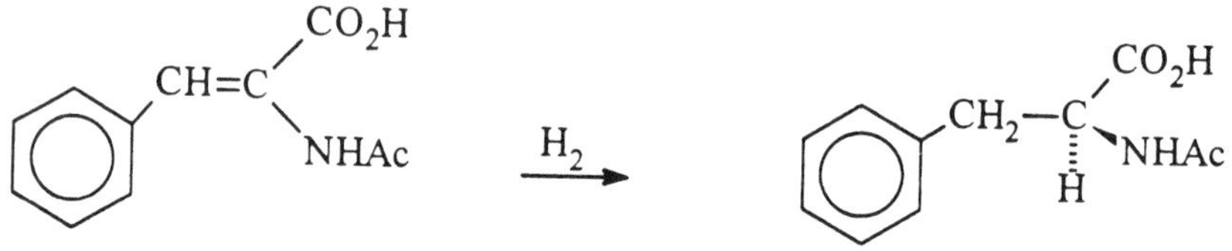

(6.55)

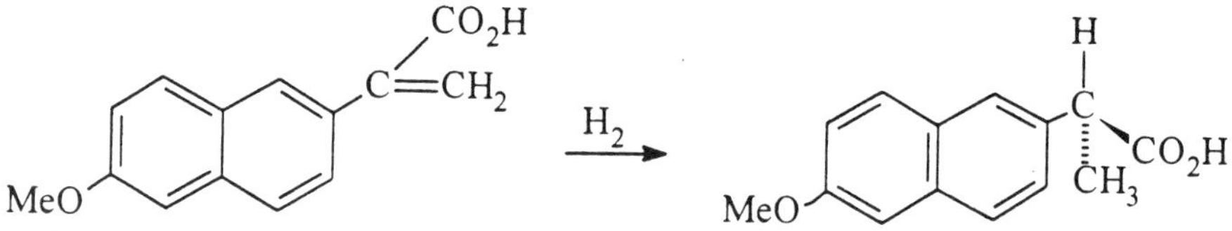

(6.56)

Asymmetric hydrogenation prevents costly and difficult resolution experiments.

The chirality of the catalyst derives from a chiral ligand coordinated to the catalytic center. A few of the ligands used are shown in the following; a thorough review describes the chiral phosphine ligands used.[42]

P(Cy)(Me)(o-C$_6$H$_4$OMe)

CAMP

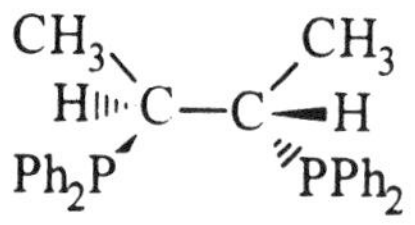

chiraphos

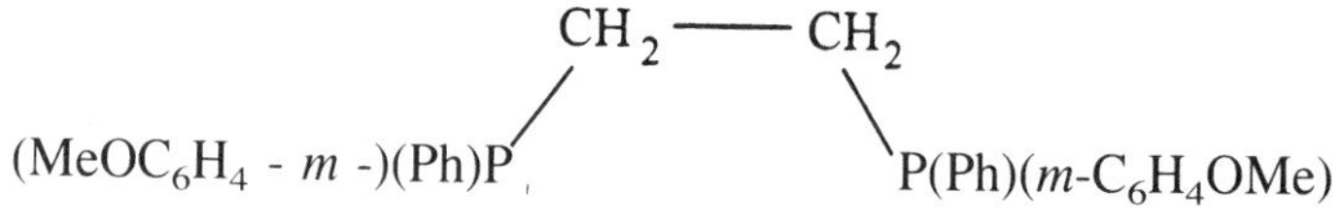

(MeOC$_6$H$_4$ - m -)(Ph)P

P(Ph)(m-C$_6$H$_4$OMe)

DIPAMP

The enantiomeric excesses found for several chiral phosphines in asymmetric hydrogenations leading to L-DOPA are shown in Figure 6.10.[41]

The catalysts used are usually formed by addition of two equivalents of phosphine (or one equivalent of bidentate phosphine ligand) ligand to one equivalent of [Rh(cod)Cl]$_2$. Conditions and enantiomeric excesses are reported in Table 6.7 for CAMP and DIPAMP.[41] The synthesis of L-DOPA is shown in Figure 6.11. Using a

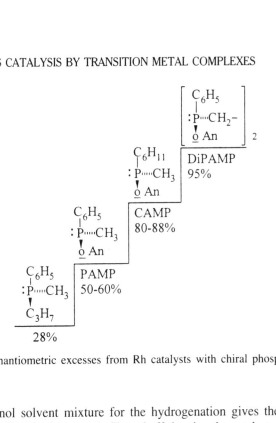

Figure 6.10. Enantiometric excesses from Rh catalysts with chiral phosphine ligands.[41]

water–isopropanol solvent mixture for the hydrogenation gives the product as a slurry. Upon cooling the L-isomer is filtered off, leaving the catalyst and byproducts in solution. The mechanism of the hydrogenation reaction,

(6.57)

$$\underset{Ph}{\overset{H}{}}C=C\overset{\overset{O}{\parallel}}{\underset{NHCCH_3}{\overset{}{\underset{\parallel}{O}}}}\overset{C-OC_2H_5}{} \xrightarrow{Rh^I,\ H_2} C_6H_5CH_2C\overset{\overset{O}{\parallel}}{\underset{NHCCH_3}{\overset{\overset{COC_2H_5}{}}{\underset{\parallel}{O}}}}H$$

Table 6.7. HYDROGENATION OF Z-ACETAMIDOCINNAMIC ACID[41]

	Temperature (°C)	Pressure (atm)	% ee	
			CAMP	DiPAMP
Free Acid	25	27	54	78
	25	3	80	94
	25	0.7	87	—
	50	3	79	94
Anion	50	3	56	95
	25	3	79	96
	0	3	88	—
	25	27	71	96
	68	3	—	94

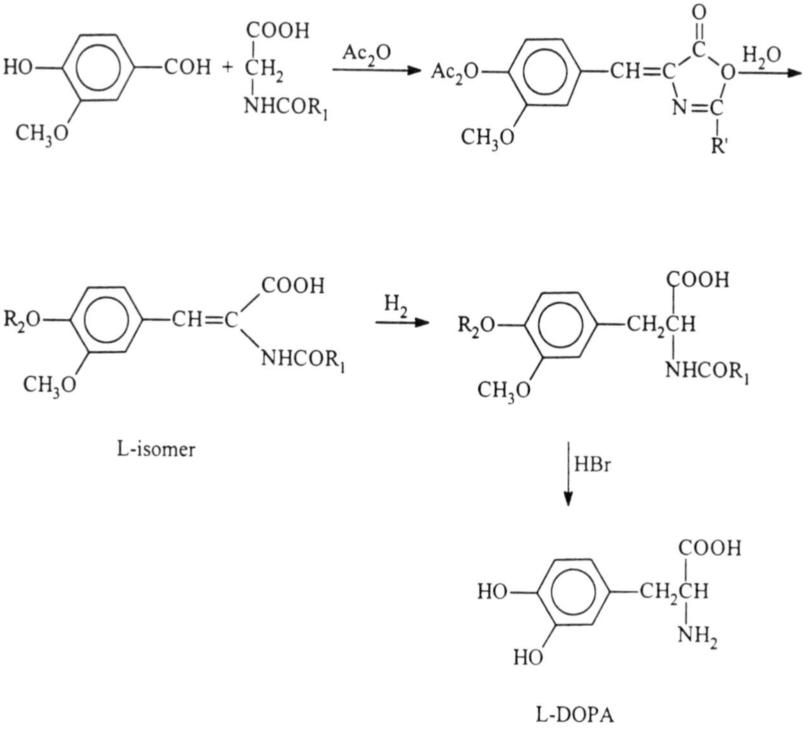

Figure 6.11. L-DOPA synthesis using an asymmetric hydrogenation (R_1 = CH_3 and R_2 = Ac).[40]

has been examined and the following scheme has been suggested,[9]

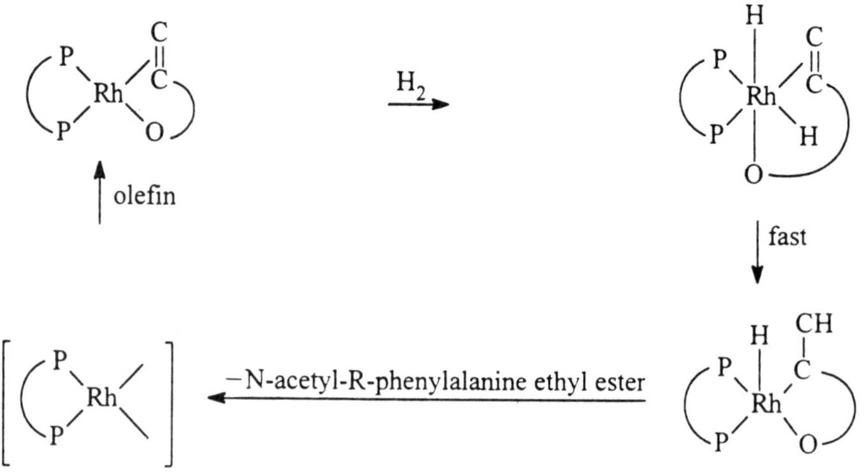

(6.58)

where P–P is a bidentate phosphine ligand and the ethyl-(Z)-α-acetamidocinnamate is functioning as a bidentate ligand through the alkene and an oxygen. At low temperature the monohydride alkyl complex could be characterized.[14] From this reaction it was suggested that the enantiomeric selectivity in hydrogenation does not arise from a preferred mode of bonding of the prochiral olefin, but rather that it arises from differences in the rates of the subsequent reactions.[9]

Asymmetric syntheses using chiral catalysts are relatively new, but hydrogenations are important in several processes. As the field matures and the benefits of chemical efficiency (less waste generated and less raw materials required) become more important, direct chemical synthesis of enantiomerically pure materials must continue to grow.

6.4. Hydroformylation Reaction

Hydroformylation involves conversion of an alkene to an aldehyde with one additional C atom in the presence of synthesis gas, a mixture of CO and H_2.

$$C_nH_{2n} + CO + H_2 \longrightarrow C_nH_{2n+1}CHO \tag{6.59}$$

This reaction is of considerable industrial importance as a means to functionalize alkenes.[43] The traditional catalyst is $HCo(CO)_4$, which is derived from $Co_2(CO)_8$ and H_2 under the reaction conditions. The extra selectivity to straight-chain aldehydes exhibited by $HRh(PPh_3)_3CO$ has led industrial processes to shift to this catalyst.[44] A nice review summarizes the catalysts and products important to hydroformylation.[44] Table 6.8 provides a comparison of three systems for hydroformylation.

Hydroformylation by the Co system is first order in [Co] and first order in alkene. The rate increases with increasing H_2 pressure and decreases with increasing CO pressure.[45] The conditions required are rather strenuous: over 100°C and over 200 atm pressure.[44,45] Based on these observations and a number of other reactions, the following sequence was proposed for hydroformylation of alkenes by $Co_2(CO)_8$[46]:

$$Co_2(CO)_8 + H_2 \rightleftharpoons 2HCo(CO)_4 \tag{6.60}$$

Table 6.8. COMPARISON OF THREE CATALYST SYSTEMS FOR HYDROFORMYLATION OF ALKENES[44b]

Catalyst	Temperature (°C)	Pressure (atm)	Product(s)	N/I[a]
$Co_2(CO)_8$	140–180	200–300	aldehyde/alcohol	3–4:1
$Co_2(CO)_8 + PR_3$	160–200	50–100	alcohol/aldehyde	8–9:1
$HRh(CO)_2(PPh_3)_2$	90–110	10–20	aldehyde	12–15:1

[a] N/I refers to the normal to iso ratio, which is a measure of the selectivity for the desired linear product.

$$HCo(CO)_4 \rightleftharpoons HCo(CO)_3 + CO \qquad (6.61)$$

$$HCo(CO)_3 + \text{alkene} \rightleftharpoons HCo(\text{alkene})(CO)_3 \qquad (6.62)$$

$$HCo(\text{alkene})(CO)_3 + CO \rightleftharpoons RCo(CO)_4 \qquad (6.63)$$

$$RCo(CO)_4 \rightleftharpoons RC(O)Co(CO)_3 \qquad (6.64)$$

$$RC(O)Co(CO)_3 + H_2 \rightleftharpoons RC(O)Co(CO)_3 (H_2) \qquad (6.65)$$

$$RC(O)Co(CO)_3(H_2) \rightleftharpoons RC(O)H + HCo(CO)_3 \qquad (6.66)$$

Primary side products are alkane, isomerized alkene, nonlinear aldehydes, and alcohols, which are all logical products of the proposed scheme.[43] Each reaction involves two-electron transitions, either 16- or 18-electron complexes. Infrared studies of a hydroformylation reaction have suggested that Reactions 6.65 and 6.66 should be replaced by a binuclear reaction (see Chapter 5).[47]

$$RC(O)Co(CO)_3 + HCo(CO)_4 \xrightarrow{\;CO\;} RC(O)H + Co_2(CO)_8 \qquad (6.67)$$

Considering the relative ease of $Co(CO)_4$ formation from $HCo(CO)_4$, it is surprising that radical mechanisms have not received more attention. Photolysis has been shown to inhibit the hydroformylation reaction, which is a difficult observation to justify in terms of a radical mechanism.[48]

Terminal alkenes are hydroformylated more rapidly than are more highly substituted alkenes.[45] Pertinent data (Table 6.9)[45] show that the degree of substitution at the olefinic carbons is the most important consideration in rates of hydroformylation. This would correspond to the ease of coordination of the alkene as discussed earlier. Both associative and dissociative mechanisms have been suggested for hydroformylation by $HRh(CO)(PPh_3)_3$.[44,49] Schemes are shown in Figure 6.12.

Table 6.9. RATES OF HYDROFORMYLATION OF ALKENES BY $HCo(CO)_4$ AT 110°C

Alkene	$k(\times 10^5 s^{-1})$
1-hexene	110
1-octene	109
2-hexene	30
2-octene	31
2-methyl-1-pentene	13
4-methyl-1-pentene	107
2-methyl-2-pentene	8.1
cyclohexene	9.7

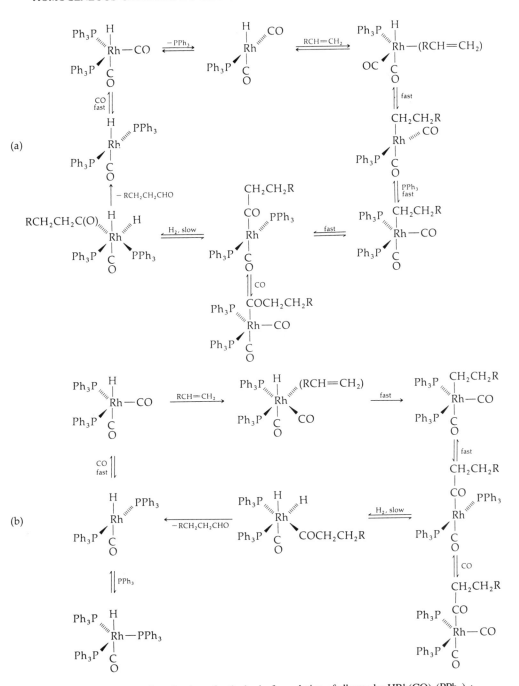

Figure 6.12. Proposed mechanisms for the hydroformylation of alkenes by HRh(CO)$_2$(PPh$_3$)$_2$: (a) dissociative; (b) associative.

The active species in either scheme is $HRh(CO)_2(PPh_3)_2$, which is formed by reaction of $HRh(CO)(PPh_3)_3$ with CO.[49]

$$HRh(CO)(PPh_3)_3 + CO \longrightarrow HRh(CO)_2(PPh_3)_2 + PPh_3 \qquad (6.68)$$

The relative rates of hydroformylation of several alkenes by $HRh(CO)(PPh_3)_3$ are shown in Table 6.10.[49] Comparison of the data in Tables 6.6 and 6.7 indicates that the Rh complex is considerably more selective for terminal alkenes and reacts at milder conditions.[44,49] The Rh system also produces fewer side products, as shown in Table 6.11.[44] The advantages of Rh catalysts in mild conditions and greater specificity (hydroformylation of l-hexene produces heptanal in $>90\%$ yield on Rh while Co produces a lower yield of the stright-chain aldehyde) are very attractive for industrial processes and offset the extra expense of Rh over Co.

Other metal carbonyl species that have been investigated show reactivity that is quite dependent on the metal center.[44]

metal:	Rh	Co	Ru	Mn	Fe	Cr, Mo, W, Ni
relative rate:	10^3- 10^4	1	10^{-2}	10^{-4}	10^{-6}	0

In 1984 a new process for hydroformylation of propene was brought on line.[50] This process incorporates an organorhodium complex in water functioning as the catalyst. The aldehyde product forms a layer on top of the catalytic solution that can be decanted to remove the product from the catalytic mixture. The catalyst involves water-soluble phosphine ligands, $P(m\text{-}C_6H_4SO_3Na)_3$, such that the Rh catalyst dissolves in water. The propylene, H_2, and CO are introduced, and the n-butanal is collected. This water-soluble Rh system requires slightly higher conditions of temperature and pressure, but the ease of separation of the product has led to this process being used for 100,000 tons of n-butanal annually.[50]

6.5. Wacker Acetaldehyde Synthesis

The Wacker process, which was one of the first industrial applications of homogeneous catalysis, led to the displacement of acetylene as the starting material for

Table 6.10. RATES OF HYDROFORMYLATION OF ALKENES BY $HRh(CO)(PPh_3)_3$ IN BENZENE AT 25°C[49]

Alkene	Rate (gas uptake in ml/min)
1-hexene	3.52
1-pentene	3.74
1-heptene	3.50
2-pentene	0.15
cis-2-heptene	0.12
2-methyl-1-pentene	0.06

Table 6.11. PRODUCTS FROM THE
HYDROFORMYLATION OF PROPYLENE
FOR Co AND Rh-BASED CATALYSTS[44]

Product	Co (%)	Rh
aldehyde	80	97%
alcohol	11	trace
butylformate	2	trace
propane	2	2%
others	5	<1%

manufacture of organic chemicals. The efficient conversion of ethylene to acet-aldehyde using $PdCl_2$ in a catalytic sequence was accomplished in the mid-1950s.[1,51] The stoichiometric reaction

$$C_2H_4 + PdCl_2 + H_2O \longrightarrow CH_3CHO + Pd(0) + 2HCl \qquad (6.69)$$

had been known since 1894. Combination with two reactions to regenerate the palladium chloride allowed a catalytic synthesis of acetaldehyde from ethylene.

$$Pd(0) + 2CuCl_2 \longrightarrow PdCl_2 + 2CuCl \qquad (6.70)$$

$$2CuCl + 2HCl + 1/2O_2 \longrightarrow 2CuCl_2 + H_2O \qquad (6.71)$$

Addition of the three reactions gives the net reaction

$$C_2H_4 + 1/2O_2 \longrightarrow CH_3CHO \qquad (6.72)$$

in which ethylene is converted by oxygen to acetaldehyde. While Reactions 6.70 and 6.71 are essential to the catalytic sequence, they are electron-transfer processes, which will be discussed in Chapter 8. The conversion of ethylene into acetaldehyde occurs in Reaction 6.69, which we shall examine in more detail.

The rate law for product formation in the Wacker process

$$\frac{-d[C_2H_4]}{dt} = \frac{k[PdCl_4^{2-}][C_2H_4]}{[H^+][Cl^-]^2} \qquad (6.73)$$

has led to the following mechanism for acetaldehyde formation.[1,51,52]

$$PdCl_4^{2-} + (C_2H_4) \rightleftharpoons [(C_2H_4)PdCl_3]^- + Cl^- \qquad (6.74)$$

$$(6.75)$$

$$
H_2O + \quad \underset{H_2O}{\overset{Cl}{\diagdown}} Pd \underset{Cl}{\overset{}{\diagup}} \overset{CH_2}{\underset{CH_2}{\|}} \rightleftharpoons \tag{6.76}
$$

$$
\underset{H_2O}{\overset{Cl}{\diagdown}} Pd^{=} \underset{Cl}{\overset{}{\diagup}} CH_2-CH_2\overset{+}{O}H_2
$$

$$
\underset{H_2O}{\overset{Cl}{\diagdown}} Pd \underset{Cl}{\overset{}{\diagup}} \overset{CH_2}{\underset{CH_2}{\|}} + OH^- \xrightarrow{\text{slow}} \tag{6.77}
$$

$$
\left[\underset{H_2O}{\overset{Cl}{\diagdown}} Pd \underset{Cl}{\overset{}{\diagup}} CH_2-CH_2OH \right]^-
$$

$$
H_2O + \underset{H_2O}{\overset{Cl}{\diagdown}} Pd^{=} \underset{Cl}{\overset{}{\diagup}} CH_2-CH_2\overset{+}{O}H_2 \rightleftharpoons \tag{6.78}
$$

$$
\left[\underset{H_2O}{\overset{Cl}{\diagdown}} Pd \underset{Cl}{\overset{}{\diagup}} CH_2-CH_2OH \right]^- + H_3O^+
$$

$$
\left[\underset{H_2O}{\overset{Cl}{\diagdown}} Pd \underset{Cl}{\overset{}{\diagup}} CH_2-CH_2OH \right]^- \xrightarrow{\text{slow}} \tag{6.79}
$$

$$
\left[\underset{H_2O}{\overset{Cl}{\diagdown}} Pd-CH_2-CH_2OH \right] + Cl^-
$$

$$
\left[\underset{H_2O}{\overset{Cl}{\diagdown}} Pd-CH_2-CH_2OH \right] \xrightarrow{\text{fast}} \underset{H_2O}{\overset{Cl}{\diagdown}} Pd \underset{H}{\overset{}{\diagup}} \overset{CHOH}{\underset{CH_2}{\|}} \tag{6.80}
$$

(6.81)

The first two reactions are ligand substitutions at a square-planar center (Chapter 2). Note that the strong *trans* effect of the C_2H_4 leads to the *trans* product.[52,53] Reaction 6.77 and reactions 6.76 and 6.78 involve attack of OH^- or H_2O on coordinated ethylene in what has been called a hydroxypalladation. The mechanism of this step has been demonstrated to be attack of hydroxide or water from outside the coordination sphere on coordinated ethylene by using specifically deuterium-labeled ethylenes.[52,53] β-Hydrogen elimination (Reaction 6.80) leads to the olefin complex, which may rearrange to a second σ-bonded isomer (Reaction 6.81). Acetaldehyde is formed by β-elimination of the hydrogen on oxygen. The palladium hydride decomposes to palladium metal and HCl.

The primary discussion has centered on the hydroxypalladation reaction that converts the coordinated ethylene into the σ alkyl (Pd-CH_2CH_2OH).[52,53] The earlier suggestion of ethylene insertion into a Pd-OH bond was shown to be inconsistent with two separate stereochemical studies. The hydroxypalladation of *cis*-1,2-dideuterioethylene under CO allowed the β-hydroxyethylpalladium complex to be trapped and the stereochemistry to be determined.[52]

(6.82)

Since the insertion of CO into a C-Pd σ bond occurs with retention of configuration at C, the hydroxypalladation proceeds with *anti* stereochemistry, which indicates that attack on the coordinated ethylene takes place outside of the coordination sphere of the complex.[52] A similar conclusion was obtained from studies of *trans*-1,2-dideuterioethylene with the stereochemistry determined for chloroethanol and the corresponding epoxide.[53]

threo   97% *cis*

(6.83)

Formation of the epoxide is known to proceed with inversion of configuration at one carbon. Thus, external attack on the coordinated olefin is confirmed.[53]

The Wacker chemistry can also be used for other alkenes. Terminal alkenes produce methylketones, and acetone can be readily produced from propene. The rates and yields decrease with increasing chain length of the alkene. Cyclic alkenes produce good yields of cyclic ketones, but other internal alkenes give mixtures of products.

The Wacker process has been a very important reaction in transition metal homogeneous catalysis for the production of acetic acid as a building block of industrial organic chemistry. This process may soon be replaced by synthesis gas (CO and H_2) chemistry in industry, but it will remain a model for the adaptation of known chemistry into a useful catalytic cycle.[1]

6.6. Hydrocyanation of 1,3-Butadiene

The synthesis of 6,6-nylon involves an elimination polymerization that requires 1,6-hexanediamine. 1,6-Hexanediamine can be readily produced by hydrogenation of adiponitrile using heterogeneous catalysts.

$$N\equiv CCH_2CH_2CH_2CH_2C\equiv N \xrightarrow{H_2} \qquad (6.84)$$

$$H_2NCH_2CH_2CH_2CH_2CH_2CH_2NH_2$$

Thus, the convenient production of adiponitrile has been an active area of research.[1,54] A commercial route using homogeneous hydrocyanation of 1,3-butadiene has been developed and investigated mechanistically.[54]

$$(6.85)$$

These reactions are catalyzed by Ni(0) complexes. The catalytic cycle is composed of three separate reactions: hydrocyanation of an activated olefin, isomerization, and hydrocyanation. Each of these reactions will be considered independently, and then the cycle in its entirety.

The catalyst is a tetrahedral nickel complex, NiL_4, where L is a phosphite ligand.[54] Since NiL_4 is an 18-electron complex, ligand dissociation is important in initiating the catalytic reaction.

$$NiL_4 \rightleftharpoons NiL_3 + L \qquad (6.86)$$

The equilibrium constants for ligand dissociation of L from NiL_4, given in Table 6.12, indicate the importance of steric interactions. Hydrogen cyanide oxidatively adds to the NiL_3 complex.

$$NiL_3 + HCN \rightleftharpoons HNiL_3(CN) \tag{6.87}$$

The hydridocyanonickel complexes could be spectroscopically characterized.[54] The hydrocyanation of ethylene can be used as a model for the mechanism of a nonactivated olefin.

$$NiL_4 \rightleftharpoons NiL_3 + L \tag{6.88}$$

$$NiL_3 + HCN \rightleftharpoons HNiL_3(CN) \tag{6.89}$$

$$HNiL_3CN \rightleftharpoons HNiL_2CN + L \tag{6.90}$$

$$HNiL_2CN + C_2H_4 \rightleftharpoons HNiL_2(C_2H_4)CN \tag{6.91}$$

$$HNiL_2(C_2H_4)CN \rightleftharpoons C_2H_5NiL_2CN \tag{6.92}$$

$$C_2H_5NiL_2CN + L \rightleftharpoons C_2H_5NiL_3CN \tag{6.93}$$

$$C_2H_5NiL_3CN \longrightarrow C_2H_5CN + NiL_3 \tag{6.94}$$

$$L = P(O\text{-}o\text{-tolyl})_3$$

The hydridocyano (Reaction 6.89) and ethylcyano (Reaction 6.92) complexes could be spectroscopically identified in the catalytic cycle.[54] This scheme involves a sequence of 16- and 18-electron conversions.

Hydrocyanation of a conjugated alkene is considered to be slightly different. Addition of the first molecule of HCN to 1,3-butadiene has been studied for $Ni[P(OEt)_3]_4$.

$$NiL_4 \rightleftharpoons NiL_3 + L \tag{6.95}$$

$$NiL_3 + HCN \rightleftharpoons HNiL_3CN \tag{6.96}$$

$$HNiL_3CN \rightleftharpoons HNiL_2CN + L \tag{6.97}$$

Table 6.12. EQUILIBRIUM CONSTANTS FOR LIGAND DISSOCIATION FROM NiL_4 IN BENZENE AT 25°C[54]

L	$K(M)$
$P(OEt)_3$	$<10^{-10}$ (at 70°C)
$P(O\text{-}p\text{-tolyl})_3$	6×10^{-10}
$P(O\text{-}i\text{-Pr})_3$	2.7×10^{-5}
$P(O\text{-}o\text{-tolyl})_3$	4.0×10^{-2}

$$HNiL_2CN + C_4H_6 \rightleftharpoons (\eta^3\text{-}C_4H_7)NiL_2CN \qquad (6.98)$$

$$L = P(OEt)_3$$

The η^3-allyl group combines with the cyanide in two ways to form the desired linear 3-pentenenitrile and the branched 2-methyl-3-butenenitrile. The path leading to linear product was favored by a small rate factor.

The catalytic conversion of 1,3-butadiene to adiponitrile with HCN in the presence of nickel(0) complexes is completed by the addition of BPh$_3$, a Lewis acid promoter. The BPh$_3$ coordinates to the N of the coordinated cyanide and has three beneficial effects on the production of adiponitrile: (1) the rate of hydrocyanation of inactivated olefins is enhanced; (2) the amount of linear product is increased; and (3) the catalyst lifetime is enhanced, presumably because a second oxidative-addition of HCN to form NiL$_2$CN$_2$ is prevented.[54]

$$HNiL_2CN + HCN \longrightarrow H_2 + NiL_2CN_2 \qquad (6.100)$$

$$RNiL_2CN + HCN \longrightarrow RH + NiL_2CN_2 \qquad (6.101)$$

Each of the proposed steps in the hydrocyanation of 1,3-butadiene with NiL$_4$ (L = phosphite ligand) is suggested to occur by 16- and 18-electron intermediates.

6.7. Olefin Metathesis

The olefin metathesis reaction involves transfer of alkylidene units between two alkenes.[55-58]

$$(R_1)_2C{=}C(R_1)_2 \qquad (6.102)$$

$$+ \qquad \rightleftharpoons 2(R_1)_2C{=}C(R_2)_2$$

$$(R_2)_2C{=}C(R_2)_2$$

The olefin metathesis reaction is of considerable industrial importance.[57] Propylene is converted to ethylene and 2-butene in a commercial process, and cycloolefins polymerize to linear unsaturated polyalkenamers.[57]

$$(6.103)$$

The importance of olefin metathesis led to extensive investigation that produced mechanistic understanding only 15 years after first reports on the reaction. A primary difference from the other catalytic systems that we have examined is that the catalyst for olefin metathesis is not well defined. The most active catalysts are combinations (WCl_6, EtOH, $EtAlCl_2$; and $Mo(PPh_3)_2(NO)_2Cl_2$, $Me_3Al_2Cl_3$). Despite this uncertainty in the coordination around the active metal center, the primary steps are now understood.

A number of experiments have been devised to differentiate between the two suggested mechanisms.[55,59,60]

$$(6.104)$$

$$(6.105)$$

The first mechanism (Eq. 6.104) is the cyclobutane mechanism, and the second is the metallocyclobutane or carbene mechanism. Reactions in which a mixture of a cycloalkene and an alkene undergo metathesis allow these two mechanisms to be distinguished and have shown the cyclobutane mechanism to be unsatisfactory.[55] Metathesis of cyclooctene with 2-hexene was examined for the relative amounts of C_{12}, C_{14}, and C_{16} dialkenes. Only C_{14} would be expected for the cyclobutane mechanism since both alkenes would be coordinated to the metal.[55]

$$(6.106)$$

For the metallocyclobutane mechanism more possibilities are available. One reaction is as follows:

$$(6.107)$$

With this mechanism one expects a statistical distribution, 1:2:1 for C_{12}, C_{14}, and C_{16} dialkenes.[55] Since the alkenes formed could undergo further metathesis, the significant observation to distinguish the mechanism must be made early in the reaction. The ratios, $[C_{12}]/[C_{14}]$ and $[C_{16}]/[C_{14}]$, extrapolated to $t = 0$ are nonzero in confirmation of the metallocyclobutane mechanism. Similar experiments with three alkenes, cyclooctene, 2-butene, and 4-octene were performed.[55]

$$(6.108)$$

For the cyclobutane mechanism no C_{14} dialkene should form initially, but one would again expect a statistical ration of $C_{12}:C_{14}:C_{16}$, 1:2:1 for the metallocyclobutane. In plots extrapolated to $t = 0$ the amount of C_{14} was invariably larger than the amount of C_{12} or C_{16}. A sample plot is shown in Figure 6.13.[55]

Another experiment to distinguish between the metallocyclobutane and the cyclobutane mechanism involved metathesis of 1,7-octadiene.[60]

$$(6.109)$$

The ratios of deuterated ethylenes expected for a cyclobutane mechanism with a 1:1 ratio of d^0 to d^4 1,4-octadiene was calculated to be 1:1.6:1, and for the carbene or metallocyclobutane mechanism the ratio was calculated to be 1:2:1.[60] The experimental ratio of 1:2.1:1 is most consistent with the metallocyclobutane mechanism.

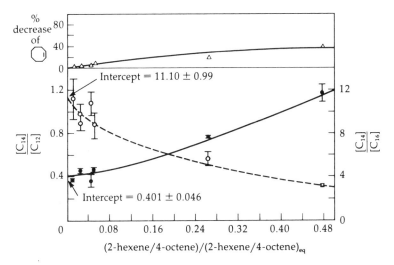

Figure 6.13. Plot showing the formation of products from the metathesis of cyclooctene with 2-butene and 4-octene. The ordinate is labeled at the left for the solid curve and at the right for the dashed curve. [Reprinted with permission from T. J. Katz and J. McGinnis, *J. Am. Chem. Soc.* **99,** 1903 (1977). Copyright 1977 American Chemical Society.]

The rate of olefin metathesis depends on the degree of substitution of the alkene, decreasing in the following order: the degenerate exchange of methylene units between terminal alkenes > the cross metathesis of terminal and internal alkenes > the metathesis of internal alkenes > the metathesis of terminal alkenes to give ethylene and internal alkenes.[59,61]

The exchange of methylene units between an alkylidene complex and an olefin has been demonstrated stoichiometrically,

$$(CO)_5W{=}CPh_2 + CH_2{=}CPh(OMe) \longrightarrow \qquad (6.110)$$

$$(CO)_5W{=}C\overset{\displaystyle Ph}{\underset{\displaystyle OMe}{\big\langle}} + CH_2{=}CPh_2$$

as required in the metallocyclobutane mechanism.[62] Exploring the reactions of $(CO)_5W = C(tol)_2$ with alkenes allowed the relative reactivities to be quantitatively determined (Table 6.13).[59] In each case the new alkene formed corresponds to the least substituted possibility or the more highly substituted alkylidene (carbene). The scheme in Figure 6.14 illustrates these reactions.[59]

In the discussion thus far we have been concerned with the nature of the alkene products and have shown that the metallocyclobutane mechanism is consistent with the products obtained. One needs an alkylidene metal complex for the metallocyclobutane mechanism to initiate the reaction. A theoretical study of the metal

Table 6.13.   THE REACTIONS OF ALKENES WITH $(CO)_5W=C(tol)_2$[59]

Alkene	Alkene Products	
$CH_2=CH(CH_2)_2CH_3$	$(tol)_2C=CH_2$	$(tol)_2C=CH(CH_2)_2CH_3$
	36%	0.06%
$CH_2=C(CH_3)_2$	$(tol)_2C=CH_2$	$(tol)_2C=C(CH_3)_2$
	73%	<0.06%
$CH_2=CHC_6H_5$	$(tol)_2C=CH_2$	$(tol)_2C=CHC_6H_5$
	61%	<0.2%

and ligand environment that facilitates olefin metathesis has appeared.[63] Many metals have been used, but the most active catalysts are based on W, Mo, and Re.[58] The alkylidene complex $(CO)_5W = C(Ph)_2$, functions as a catalyst but much more slowly than other catalysts.[59] It has been suggested that the slowness of this alkylidene is attributable to the saturation of the coordination around W; CO dissociation is required to open a coordination site before the alkene can be coordinated.[59] The most active catalysts are based on transition metal halides and alkylaluminum species, and they probably proceed initially to a transition metal alkyl complex. In a few cases α-hydrogen abstraction has been observed from an alkyl to form an alkylidene.[64] A W alkylidene that may serve as a model for alkylidenes formed when W oxides or chlorides are treated with alkylaluminum compounds has been isolated by the following reaction.[65]

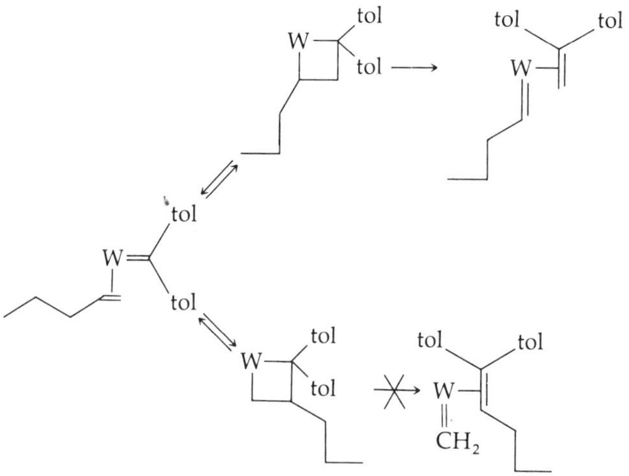

Figure 6.14.   Preferred reaction of an alkene with an alkylidene leading to the most highly substituted alkylidene product.

$$Ta(CHCMe_3)(PEt_3)_2Cl_3 + W(O)(OCMe_3)_4 \longrightarrow \qquad (6.111)$$

The oxoalkylidene complex is an active catalyst for metathesis of alkenes. Reaction with 1-butene in the presence of $AlCl_3$ for 1–2 hours led to 3,3-dimethyl-1-butene (1 equiv.), 3-hexenes (1–2 equiv.), and an undetermined amount of C_2H_4. The alkylidene product was predominantly the propylidene product, although some methylidene product was observed. It was suggested that the metallocyclobutane *A* is more stable than *B*.

In several active catalyst systems—such as $Mo(PPh_3)_2Cl_2(NO)_2$, $Me_3Al_2Cl_3$; WCl_6, $SnMe_4$; and WCl_6, $ZnMe_2$—methane and ethylene are eliminated in the absence of alkene, which is evidence for methylidene formation.[58] Further evidence for methylidene intermediacy is shown by initial formation of propylene in the metathesis of 2-alkenes.[58] Labeling showed that the methylidene of the propylene arose from the cocatalyst, and the ethylidene arose from the 2-alkene.[58]

Research on olefin metathesis supports the metallocyclobutane mechanism. The nature of the alkene products and the rates of alkene reactions are fully consistent with the mechanism. The nature of the catalyst centers is not clearly defined, but there is evidence of alkylidene formation in the catalytic systems in several cases.

6.8. Polymerization of Alkenes

Polymerization remains one of the most important industrial applications of homogeneous catalysis.[1,2] The polymerization of simple derivatives of ethylene leads to a number of important products, as shown in Table 6.14. The discovery of alkene polymerization by Ziegler-Natta catalysts is one of the milestones of organometallic chemistry. Ziegler-Natta catalyst systems, composed of a transition metal halide and a main group alkyl (usually an aluminum alkyl), are poorly defined in terms of

Table 6.14. PRODUCTS FROM THE POLYMERIZATION OF ETHYLENE DERIVATIVES

Alkene	Polymer (Application)
$CH_2{=}CH_2$	Polyethylene (plastic bags)
$CF_2{=}CF_2$	Teflon (pan coating)
$CH_2{=}CHCl$	Polyvinylchloride (phonograph records)
$CH_2{=}CHCN$	Orlon (rug fibers)
$CH_2{=}CHPh$	Polystyrene (combs)

the coordination sphere of the transition metal. In this respect they are similar to olefin metathesis catalysts.

At the current time the two most plausible mechanisms for polymer growth on a transition metal cannot be differentiated on the basis of experimental data on the catalytic polymerization reaction. One proposed scheme involves insertion of an alkene into a metal alkyl bond, similar to insertion of an alkene into a metal–hydride bond as discussed for hydrogenation and hydroformylation.

$$\text{(6.112)}$$

The second involves alkene insertion into a metal-alkylidene bond and subsequent hydrogen transfer.

$$\text{(6.113)}$$

Each of these proposed chain growth steps will be briefly examined.

The insertion of an alkene into a transition metal alkyl bond is analogous to the suggestions for the mechanism of the growth reaction in alkyl aluminum chemistry, which was the precursor to polymerization. Although a number of alkyl–alkene transition metal complexes are known, the direct insertion of an alkene into a transition metal alkyl bond has not been observed.[66-69] The insertion of alkenes into a Lu–C bond has been observed.[66]

$$\text{(6.114)}$$

$$Cp'_2LuCD_3 + CD_2{=}CDCH_3 \longrightarrow Cp'_2LuCD_2CD(CH_3)CD_3 \qquad (6.115)$$

$$Cp' = \eta^5\text{-}C_5Me_5$$

Deuterium-labeling (Reaction 6.115) demonstrated that the propene inserted into the lutetium–methyl bond.[66] Further insertion of propene occurred, although the details were not reported. Evidence for ethylene insertion into a Co–methyl bond has also been obtained from deuterium-labeling experiments on $CpCo(PPh_3)(CH_3)_2$.[67]

$$CpCo(PPh_3)(CH_3)_2 + C_2H_4 \longrightarrow \qquad (6.116)$$
$$CpCo(PPh_3)(C_2H_4) + CH_4 + CH_3CH{=}CH_2$$

When the completely deuterated starting complex (deuterated phosphine and methyl groups) reacted with C_2H_4, the products were CD_3H and d_3-propene.[67] This labeling strongly suggests that reaction proceeds by PPh_3 loss, addition and insertion of C_2H_4, β-hydrogen elimination, and reductive-elimination of CD_3H:

$$CpCo(CD_3)_2PPh_3 \rightleftharpoons CpCo(CD_3)_2 + PPh_3 \qquad (6.117)$$

$$CpCo(CD_3)_2 + C_2H_4 \rightleftharpoons CpCo(CD_3)_2(C_2H_4) \qquad (6.118)$$

$$CpCo(CD_3)_2(C_2H_4) \rightleftharpoons CpCo(CD_3)(CH_2CH_2CD_3) \qquad (6.119)$$

$$CpCo(CD_3)(CH_2CH_2CD_3) \rightleftharpoons CpCo(CD_3)(H)(\eta^2\text{-}CH_2{=}CHCD_3) \quad (6.120)$$

$$CpCo(CD_3)(H)(\eta^2\text{-}CH_2{=}CHCD_3) \longrightarrow \qquad (6.121)$$
$$CpCo(\eta^2\text{-}CH_2{=}CHCD_3) + CD_3H$$

$$CpCo(\eta^2\text{-}CH_2{=}CHCD_3) + C_2H_4 + PPh_3 \longrightarrow \qquad (6.122)$$
$$CpCo(C_2H_4)(PPh_3) + CH_2{=}CHCD_3$$

This scheme nicely accounts for the observed products.

The apparent difficulty of inserting a coordinated alkene into a metal–alkyl bond and the similarity of the polymerization catalyst systems to the metathesis catalyst systems led to a second postulated mechanism for chain growth.[68,69] This mechanism is based on the known reaction between an alkylidene and an alkene to form a metallocyclobutane, as in the metathesis reaction. The chain growth for propylene is described as:

(6.123)

$$M-C\overset{H}{\underset{Me}{\diagup}}R \rightleftharpoons M=C\overset{R}{\underset{Me}{\diagdown}} \xrightarrow{C_3H_6} \quad M=C\overset{R}{\underset{Me}{\diagdown}}$$

$$H-\overset{M}{\underset{Me}{C}}-CH_2-CH(Me)R \leftarrow H-\overset{H}{\underset{Me}{C}}-\overset{R}{\underset{CH_2}{C}}\diagdown\overset{R}{Me}$$

The first step is α-abstraction, converting the alkyl into an alkylidene, followed by reaction of the alkylidene with alkene to form a metallocyclobutane. The final step is reductive elimination of one end of the metallocyclobutane. The similarity to metathesis catalyst systems would indicate that alkylidene complexes could form in systems active for polymerization of alkenes, and that the other reactions in this mechanism are well documented in stoichiometric reactions.

There is no evidence at this point that allows either of these postulated chain growth steps for polymerization to be excluded. The separation and identification of the active species in a polymerization catalytic cycle would be extremely valuable to further understanding of the mechanism of this important reaction.

6.9. Methanol Carbonylation

The conversion of methanol to acetic acid is a reaction of considerable importance.[70]

$$CH_3OH + CO \longrightarrow CH_3C(O)OH \qquad (6.124)$$

Commercial processes based on cobalt and rhodium complexes exist. Mechanistic work has been accomplished on the Rh system, and the cycle is understood.[70] Use of any of a number of Rh complexes as starting compounds produces equivalent catalytic behavior, suggesting that the catalyst is formed under the reaction conditions, 180°C and 30–40 atm of CO. The iodide promoter that is also required may be in any of several different forms, CH_3I, HI, or I_2. Kinetic studies show a first-order dependence on Rh concentration and on the concentration of iodide promoter. The rate was independent of the partial pressure of CO, the concentration of CH_3OH, and the concentration of $CH_3C(O)OH$.

Carbonylation of RhX_3 leads to the carbonyl anions, $Rh(CO)_2X_2^-$, at lower temperatures.

$$RhX_3 + 3CO + H_2O \longrightarrow Rh(CO)_2X_2^- + CO_2 + 2H^+ + X^- \qquad (6.125)$$

The same anionic species forms under the reaction conditions for a number of Rh precursors.[70] A sample withdrawn from a reactor initially charged with $Rh(PPh_3)_2(CO)Cl$ showed $Rh(CO)_2I_2^-$ to be present. A catalytic cycle can be constructed for the carbonylation of CH_3OH based on $Rh(CO)_2I_2^-$ from the reactions we have previously discussed. The scheme is shown in Figure 6.15. The first step is one expected in reactions of square-planar, 16-electron complexes, oxidative-addition of CH_3I. The methyl iodide is present as the promoter, or it is formed from methanol.

$$CH_3OH + HI \longrightarrow CH_3I + H_2O \qquad (6.126)$$

The next step is an alkyl migration to coordinated CO with CO taking up the open coordination site. Reductive elimination of acetyl iodide regenerates the catalyst, $Rh(CO)_2I_2^-$. Reaction of acetyl iodide with H_2O leads to $CH_3C(O)OH$ and HI, which completes the catalytic reaction.

Room temperature reaction of $Rh(CO)_2I_2^-$ with CH_3I leads directly to an acetyl complex that was isolated and characterized as a dimer.[70]

$$Rh(CO)_2I_2^- + CH_3I \longrightarrow \qquad (6.127)$$

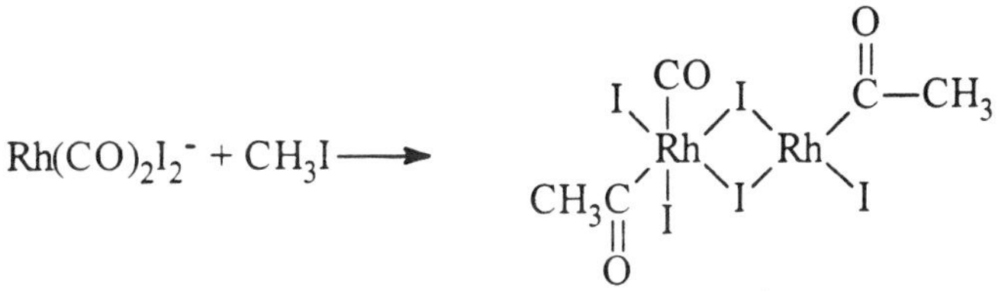

The dimer readily reacts with CO to form an acetyl carbonyl complex that is unstable toward reductive elimination of $CH_3C(O)I$. The kinetic dependencies indi-

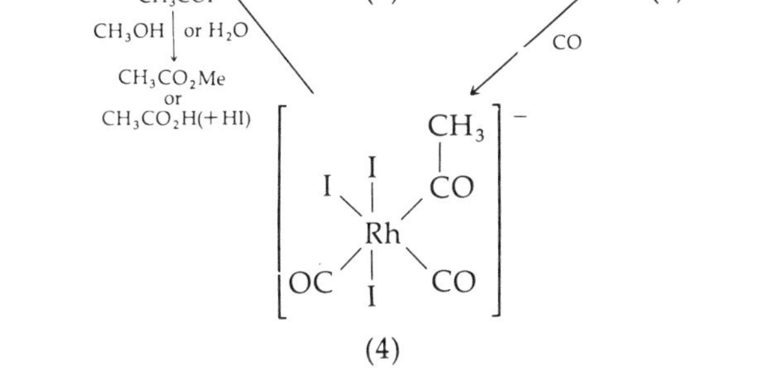

Figure 6.15. Proposed catalytic cycle for the carbonylation of methanol.[70]

cate that oxidative addition of methyl iodide to $Rh(CO)_2I_2^-$ is the rate-determining step. This is also indicated since $Rh(CO)_2I_2^-$ is the predominant Rh species present under catalytic conditions.

Thus, the proposed cycle accounts for the products and the observed kinetics. In addition, each step of the catalytic cycle can be independently shown to occur at low temperature and pressure, so that the cycle is unusually well characterized, especially for a reaction that occurs at high temperature and pressure.[70]

6.10. Summary

We have examined the proposed mechanisms of several homogeneously catalyzed reactions. Although the reactions are different, the basic catalytic steps at the metal center are composed of steps that we have examined in earlier chapters: oxidative-addition, ligand dissociation, ligand addition to 16-electron complexes, insertion reactions, and reductive elimination.

6.11. References

1. (a) G. W. Parshall, *Homogeneous Catalysis* (New York: John Wiley and Sons, Inc., NY, 1980). (b) G. W. Parshall and S. D. Ittel, *Homogeneous Catalysis,* 2nd Ed. (John Wiley and Sons, Inc., NY, 1992).

2. C. Masters, *Homogeneous Transition-Metal Catalysis: A Gentle Art* (London: Chapman and Hall, 1980).

3. *J. Chem. Ed.* **63,** March, 1986, is devoted to industrial applications of organometallic chemistry and catalysis. Specific manuscripts important to this chapter are G. W. Parshall and R. E. Putscher, general considerations, p. 189; B. Goodell, polymerization, p. 191; R. L. Pruett, hydroformylation, p. 196; C. A. Tolman, hydrocyanation, p. 199; D. Forster and T. W. DeKleva, carbonylation of alcohols, p. 204; and W. S. Knowles, L-DOPA, p. 222.

4. E. L. Muetterties and J. Stein, *Chem. Rev.* **79,** 479 (1979).

5. C. Masters, *Adv. Organomet. Chem.* **17,** 61 (1979).

6. M. D. Fryzuk and B. Bosnich, *J. Am. Chem. Soc.* **100,** 5491 (1978).

7. K. E. Koenig and W. S. Knowles, *J. Am. Chem. Soc.* **100,** 7561 (1978).

8. E. L. Muetterties, *Inorg. Chim. Acta* **50,** 1 (1981).

9. J. Halpern, *Inorg. Chim. Acta* **50,** 11 (1981).

10. R. E. Harmon, S. K. Gupta, and D. J. Brown, *Chem. Rev.* **73,** 21 (1973).

11. M. G. Thomas, E. L. Muetterties, R. O. Day, and V. W. Day, *J. Am. Chem. Soc.* **98,** 4645 (1976).

12. S. Eichler, *Chem. Ber.* **95,** 550 (1962).

13. J. P. Collman, J. W. Kang, W. F. Little, and M. F. Sullivan, *Inorg. Chem.* **7,** 1298 (1968).

14. A. S. C. Chan and J. Halpern, *J. Am. Chem. Soc.* **102,** 838 (1980).

15. J. S. Thayer, *J. Chem. Ed.* **46,** 442 (1969) and references therein.

16. F. R. Hartley, *Chem. Rev.* **73,** 163 (1973).

17. F. R. Hartley, *Angew. Chem. Int. Ed. Engl.* **11**, 596 (1972).

18. J. Halpern, *Pure and Appl. Chem.* **51**, 2171 (1979).

19. H. Werner and R. Feser, *Angew. Chem. Int. Ed. Engl.* **18**, 157 (1979).

20. Y. Kubo, A. Yamamoto, and S. Ikeda, *J. Organomet. Chem.* **60**, 165 (1973).

21. H. C. Clark and C. R. Jablonski, *Inorg. Chem.* **13**, 2213 (1974).

22. A. Nakamura and S. Otsuka, *J. Am. Chem. Soc.* **95**, 7262 (1973).

23. J. A. Osborn, F. H. Jardine, J. F. Young, and G. Wilkinson, *J. Chem. Soc.* (A), 1711 (1966).

24. J. F. Mague and G. Wilkinson, *J. Chem. Soc.* (A), 1736 (1966).

25. F. H. Jardine, J. A. Osborn, and G. Wilkinson, *J. Chem. Soc.* (A), 1574 (1967).

26. S. Montelatici, A. van der Ent, J. A. Osborn, and G. Wilkinson, *J. Chem. Soc.* (A), 1054 (1968).

27. F. H. Jardine, *Prog. Inorg. Chem.* **28**, 63 (1981).

28. P. Meakin, J. P. Jesson, and C. A. Tolman, *J. Am. Chem. Soc.* **94**, 3240 (1972).

29. J. Halpern and C. S. Wong, *J. Chem. Soc., Chem. Commun.* 629 (1973).

30. J. Halpern, T. Okamato, and A. Zakhariev, *J. Mol. Catal.* **2**, 65 (1976).

31. B. R. James, *Homogeneous Hydrogenation* (New York: John Wiley and Sons, 1973).

32. L. Vaska, *J. Am. Chem. Soc.* **88**, 4100 (1966).

33. S. J. LaPlaca and J. A. Ibers, *Acta Cryst.* **18**, 511 (1965).

34. L. Vaska, *Inorg. Nucl. Chem. Lett.* **1**, 89 (1965).

35. D. Evans, G. Yagupsky, and G. Wilkinson, *J. Chem. Soc.* (A), 2660 (1968).

36. C. O'Connor and G. Wilkinson, *J. Chem. Soc.* (A), 2665 (1968).

37. M. Yagupsky and G. Wilkinson, *J. Chem. Soc.* (A), 941 (1970).

38. B. R. James, *Homogeneous Hydrogenation* (New York: John Wiley and Sons, 1973) p. 106.

39. T. S. Janik, M. F. Pyszczek, and J. D. Atwood, *J. Mol. Catal.* **11**, 33 (1981).

40. P. Sullivan, M. F. Pyszczek, T. S. Janik, and J. D. Atwood, *Annals of N.Y. Academy of Sciences* **415**, 259 (1983).

41. W. S. Knowles, *J. Chem. Ed.* **63**, 222 (1986).

42. W. S. Knowles, *Acc. Chem. Res.* **16**, 106 (1983).

43. A. J. Chalk and J. F. Harrod, *Adv. Organomet. Chem.* **6**, 119 (1968).

44. (a) R. L. Pruett, *Adv. Organomet. Chem.* **17**, 1 (1979). (b) R. L. Pruett, *J. Chem. Ed.* **63**, 196 (1986).

45. I. Wender, S. Metlin, S. Ergun, H. W. Sternberg, and H. Greenfield, *J. Am. Chem. Soc.* **78**, 5401 (1956).

46. R. F. Heck and D. S. Breslow, *J. Am. Chem. Soc.* **83**, 4023 (1961).

47. N. H. Alemdarogly, J. L. M. Penniger, and E. Ostay, *Monatschift für Chemie* **107**, 1153 (1976).

48. M. J. Mirbach, M. F. Mirbach, A. Saus, N. Topalsavoglou, and T. N. Phu, *J. Am. Chem. Soc.* **103**, 7590 (1981).

49. C. K. Brown and G. Wilkinson, *J. Chem. Soc.* (A), 2753 (1970).

50. E. G. Kuntz, *Chem. Tech.* 570 (1987).

51. G. Henrici-Olivé and S. Olivé, *Coordination and Catalysis* (Weinheim, Verlag Chemie, 1977).

52. J. K. Stille and R. Divakaruni, *J. Organomet. Chem.* **169**, 239 (1979).

53. J. E. Backvall, B. Akermark, and S. O. Ljunggren, *J. Am. Chem. Soc.* **101**, 2411 (1979).

54. (a) W. C. Seidel and C. A. Tolman, *Annals of New York Academy of Sciences* **415**, 201 (1983). (b) C. A. Tolman, R. J. McKinney, W. C. Seidel, J. D. Druliner, and W. R. Stevens, *Adv. Catal.* **33**, 1 (1985).

55. T. J. Katz, *Adv. Organomet. Chem.* **16**, 283 (1978).

56. N. Calderon, *Acc. Chem. Res.* **5**, 127 (1972).

57. R. J. Haines and G. J. Leigh, *Chem. Soc. Rev.* **4**, 155 (1975).

58. N. Calderon, J. P. Lawrence, and E. A. Ofstead, *Adv. Organomet. Chem.* **17**, 449 (1979).

59. C. P. Casey, H. E. Tuinstra, and M. C. Sueman, *J. Am. Chem. Soc.* **98**, 3478 (1976).

60. R. H. Grubbs, D. D. Carr, C. Hoppin, and P. L. Burk, *J. Am. Chem. Soc.* **98**, 3478 (1976).

61. C. P. Casey, L. D. Albin, and T. J. Burkhardt, *J. Am. Chem. Soc.* **99**, 2533 (1977).

62. C. P. Casey and T. J. Burkhardt, *J. Am. Chem. Soc.* **96**, 7808 (1974).

63. A. K. Rappe and W. A. Goddard III, *J. Am. Chem. Soc.* **104**, 448 (1982).

64. R. R. Schrock, *Acc. Chem. Res.* **12**, 98 (1979).

65. J. H. Wengrovius, R. R. Schrock, M. R. Churchill, J. R. Missert, and W. J. Youngs, *J. Am. Chem. Soc.* **102**, 4515 (1980).

66. P. L. Watson, *J. Am. Chem. Soc.* **104**, 337 (1982).

67. E. R. Evitt and R. G. Bergman, *J. Am. Chem. Soc.* **101**, 3973 (1979).

68. M. L. H. Green, *Pure and Appl. Chem.* **50**, 27 (1978).

69. K. J. Ivin, J. J. Rooney, C. D. Stewart, M.L.H. Green, and R. Mahtab, *J. Chem. Soc., Chem. Commun.* 604 (1978).

70. (a) D. Forster, *Adv. Organomet. Chem.* **17**, 255 (1978). (b) D. Forster and T. W. DeKleva, *J. Chem. Ed.* **63**, 204 (1986).

6.12. Problems

6.1. Write the balanced equation for each of the following homogeneously catalyzed reactions:

a. hydrogenation
b. asymmetric hydrogenation
c. hydroformylation
d. hydrocyanation of 1,3-butadiene

e. olefin metathesis
f. methanol carbonylation
g. polymerization of an alkene

6.2. Suggest a catalyst and appropriate reaction conditions for each reaction in Problem 1.

6.3. Replacing PPh_3 by PBu_3 in $Rh(PPh_3)_3Cl$ causes a loss of catalytic activity. Combining the information from Chapter 5, suggest a reason for this loss of activity.

6.4. Reaction of diphenylacetylene with Cp_2MoH_2 led to cis-1,2-diphenylethylene and $Cp_2Mo(PhCCPh)$. Suggest a mechanism. [A. Nakamura and S. Otsuka, *J. Am. Chem. Soc.* **94**, 1886 (1972).]

6.5. Reaction of Cp_2NbCl_2 with $(n\text{-}C_3H_7)MgCl$ gives an 18-electron mononuclear complex that adds one equivalent of CO to form $Cp_2Nb(CO)C_3H_7$. Present a scheme to account for these observations. [A. H. Klazimga and J. H. Teuben, *J. Organomet. Chem.* **194**, 309 (1980).]

6.6. Most homogeneous catalysts show a selectivity in hydrogenation of alkenes (Tables 6.2, 6.3, and 6.4). Suggest a reason for this.

6.7. The complex $H_3Co(PPh_3)_3$ is an active hydrogenation catalyst. Suggest the mechanism for the reaction. (Refer to Reaction 6.4 and the relevant discussion.)

6.8. Asymmetric hydrogenations are accomplished by an optically active center on the metal complex. Suggest at least two different ways to generate an optically active center on a square-planar rhodium complex. [See J. M. Brown and D. Parker, *Organometallics* **1**, 950 (1982) and references therein.]

6.9. Look at the hydroformylation mechanism catalyzed by $HCo(CO)_4$ (Eqs. 6.59–6.66). Discuss the assumptions made for this mechanism, and account for the formation of the major impurities (alkane, alcohol, etc.).

6.10. The addition of PBu_3 to a $Co_2(CO)_8$ hydroformylation mixture leads to a catalyst that is more selective for 1-alkenes to n-aldehydes, that produces alcohols, and that is active at more moderate conditions. Account for these observations. [L. H. Slaugh and R. D. Mullineaux, *J. Organomet. Chem.* **13**, 469 (1968).]

6.11. There are several similarities between the mechanisms of alkene hydrogenation by $CH_3Co(CO)_2(P(OMe)_3)_2$, hydroformylation by $HCo(CO)_4$, and hydroformylation by $HRh(CO)(PPh_3)_3$. Compare and contrast these mechanisms.

6.12. Explain the evidence that implicates metal alkylidenes (carbenes) in olefin metathesis.

6.13. Explain in some detail the steric interactions in the metallocyclobutane intermediate that lead to specificity in the metathesis reactions.

6.14. How does the mechanism suggested for methanol carbonylation account for the observed dependencies on the added reagents?

6.15. The major advances in homogeneous polymerization catalysis have been in metallocene-based catalysts. Choose one of the catalysts (Exxon and Dow are major players) and describe the catalyst, emphasizing the differences to Ziegler-Natta catalysts. (A. M. Thayer, *Chem. Eng. News*, Sept. 11, 1995, p. 15).

6.16. When $Cp_2Ta(\mu\text{-}CH_2)_2IrCp^*(H)$ is placed in solution with C_2H_4 and H_2, ethane is formed. However, $Cp_2Ta(\mu\text{-}CH_2)_2IrCp^*(H)$ is not the active catalyst. Suggest experiments that would demonstrate that $Cp_2Ta(\mu\text{-}CH_2)_2IrCp^*(H)$ is *not* the catalyst. [M. D. Butts and R. G. Bergman, *Organometallics* **13**, 2668 (1994).]

6.17. Polymerization of styrene using $Cp*Ti(^{13}CH_3)_3$ produces polystyrene with a terminal $^{13}CH_3$ group that has been used to indicate a polyinsertion mechanism and to exclude a carbocationic mechanism. Explain. [C. Pellecchia, D. Pappalardo, L. Oliva, and A. Zambelli, *J. Am. Chem. Soc.* **117**, 6593 (1995)].

6.18. A metathesis catalyst is made from $WOCl_4$ with two equivalents of 2,6-dibromophenol and two equivalents of $PbEt_4$. What is the likely composition of the active catalyst? [W. A. Nugent, J. Feldman, and J. C. Calabrese, *J. Am. Chem. Soc.* **117**, 8992 (1995).]

6.19. Deuterioformylation (D_2 and CO) of 1-hexene using $HRh(CO)_2L_2$ complexes gave $CH_3(CH_2)_3CHDCH_2CDO$ and $CH_3(CH_2)_3CH(CH_2D)CDO$ with very little deuterium in recovered hexenes. What are the mechanistic implications of this observation? [C. P. Casey and L. M. Petrovich, *J. Am. Chem. Soc.* **117**, 6007 (1995).]

CHAPTER

7

Stereochemical Nonrigidity

Molecules that interchange rapidly between different structures are said to be stereo-chemically nonrigid. If the conversion(s) leads to a chemically distinguishable species, then the process is *isomerization;* if the conversion(s) leads to a chemically indistinguishable configuration, then the process is *fluxionality*. The primary technique for studying stereochemically nonrigid molecules is NMR spectroscopy, which is convenient for rates from 10^2 to $10^5 s^{-1}$. (See Chapter 1 for a discussion of kinetic studies by NMR.) In this chapter three types of stereochemical nonrigidity will be considered: coordination number isomerization, "ring-whizzing" or migration of a metal around an unsaturated ring, and rearrangements in metal clusters.

7.1. Coordination Number Isomerization

Transition metal complexes of coordination number four to nine have at least two possible geometries.[1,2] These geometries may interconvert (leading to either isomerization or fluxional behavior) at quite different rates depending on the coordination number.[1,3] Rearrangement mechanisms have been defined by consideration of solid-state structures that lie on the reaction coordinate between two or more distinct geometries.[3–5] The rearrangements often involve movement along the normal vibrational modes. The interconversions for each coordination number will be considered separately.

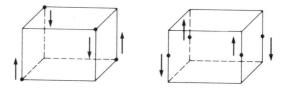

Figure 7.1. Molecular motion involved in conversion of tetrahedral to square geometry.

7.1.1. Four-Coordinate Complexes

Four-coordinate complexes, in which the metal contains eight d-electrons, may be either square planar or tetrahedral. Pt(II) and Pd(II) complexes are invariably square planar, but the smaller Ni(II) may be tetrahedral, square-planar, or an intermediate along the reaction coordinate between these two primary geometries.[1,2] Electronic factors (CFSE) favor square-planar complexes, but steric interactions favor tetrahedral ones. Our discussion will center on complexes of Ni(II) for which the energy differences between square-planar and tetrahedral are relatively small and interconversions are possible.

The interconversion between tetrahedral and square-planar geometries lies along a vibrational mode and involves relatively little molecular motion (Fig. 7.1). The small molecular motion required suggests rapid interconversion. However, a spin change is also involved since tetrahedral complexes are paramagnetic with two unpaired electrons and square-planar complexes are diamagnetic. Thus, the interconversion between tetrahedral and square-planar complexes is relatively slow and can be conveniently studied.

Complexes of formula NiL_2X_2, where L is a phosphorus donor and X is a halide, have been observed as tetrahedral, as square planar, and as mixtures of the two isomers.[1,2,6,7] Table 7.1 lists the percentage of tetrahedral isomer for several differ-

Table 7.1. STRUCTURES OF SEVERAL NiL_2Cl_2 COMPLEXES GIVEN AS THE PERCENTAGE OF TETRAHEDRAL ISOMER

Complex	Percentage Tetrahedral at 0°C
$Ni(PEt_3)_2Cl_2$	0
$Ni(PPh_3)_2Cl_2$	100
$Ni(PPh_2Me)_2Cl_2$	31
$Ni(PPh_2Et)_2Cl_2$	40
$Ni(PPh_2Pr)_2Cl_2$	47
$Ni[(p\text{-}ClC_6H_4)_2PMe]_2Cl_2$	10
$Ni[(p\text{-}MeC_6H_4)_2PMe]_2Cl_2$	35
$Ni(PPh_2Me)_2Br_2$	59
$Ni(PPh_2Me)_2I_2$	73

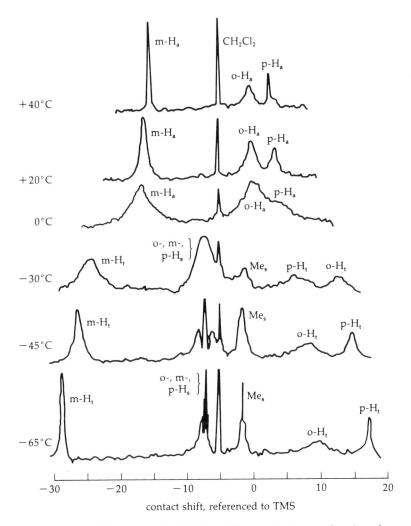

Figure 7.2. Proton NMR traces of [MeP(Ph)$_2$]$_2$NiBr$_2$ in CD$_2$Cl$_2$ as a function of temperature. The H$_a$ refer to averaged resonances, the H$_t$ to tetrahedral, and the H$_s$ to square planar. [Reprinted with permission from G. N. LaMar and E. O. Sherman, *J. Am. Chem. Soc.* **92,** 2691 (1970). Copyright 1970 American Chemical Society.]

ent complexes.[6,7] The data indicate that the nature of the ligands on Ni affects the geometry in a similar set of complexes.[6,7] Many of the data suggest the preference of larger ligands for tetrahedral geometry in which the ligand–ligand interactions are reduced. For Ni(PPh$_2$R)$_2$Cl$_2$, as R changes from Me to Et, Pr, and Ph, the structures become 31, 40, 47, and 100% tetrahedral, respectively. The halide dependence also indicates the preference of larger ligands for tetrahedral geometry with 31, 59, and 73% tetrahedral for the chloro, bromo, and iodo complexes,

Table 7.2. RATE CONSTANTS AND ACTIVATION
PARAMETERS FOR THE CONVERSION FROM
TETRAHEDRAL TO SQUARE-PLANAR GEOMETRY
FOR $Ni(PPh_2R)_2X_2$

R	X	$k(10^{-5}s^{-1})$	$\Delta H^{\ddagger}$	$\Delta S^{\ddagger}$
Me	Cl	2.6	9.0	-4
	Br	0.45	9.3	-6.1
	I	29	8.5	-0.6
Et	Cl	9.2	9.0	-1
	Br	1.6	9.0	-4.7
	I	>10	7.0	-5
Pr	Cl	6.6	9.4	-0.6
	Br	1.3	9.8	-2.4
	I	>10	8.0	-3

respectively. The results on *para* substituted phenylphosphines indicate that electronic properties are also important.

The kinetics of these reactions are studied by NMR techniques (as described in Chapter 1). A typical set of data is shown in Figure 7.2 for $Ni(PPh_2Me)_2Br_2$.[7] The resonances for the paramagnetic tetrahedral complex are contact shifted; those for the diamagnetic square-planar complex occur in the normal aromatic region. The kinetic parameters for the interconversion of $Ni(PPh_2R)_2X_2$ are shown in Table 7.2.[7] The rate of isomerization depends on both R and X, although not in a readily explainable manner. For fixed R the rate depends on the halide (Br < Cl < I); for constant X, the R group dependence is Me < Pr < Et. The factors determining the kinetics are apparently complex and involve a combination of steric and electronic effects. This is in contrast to the thermodynamics where the preference for tetrahedral (X = Cl < Br < I and L = R_3P < ArR_2P < Ar_2RP < Ar_3P) is clearly understandable in terms of steric effects.

The interconversion from tetrahedral to square planar has been shown to be important in ligand substitution of Ni(II) chelate complexes.[8] For the *bis*(N,N′-dialkyl-2-aminotropane iminato)Ni(II), shown in Figure 7.3, the thermodynamic parameters for the square-planar $\rightleftarrows$ tetrahedral interconversion were evaluated. The data in Table 7.3 show that the planar configuration is favored, but that steric

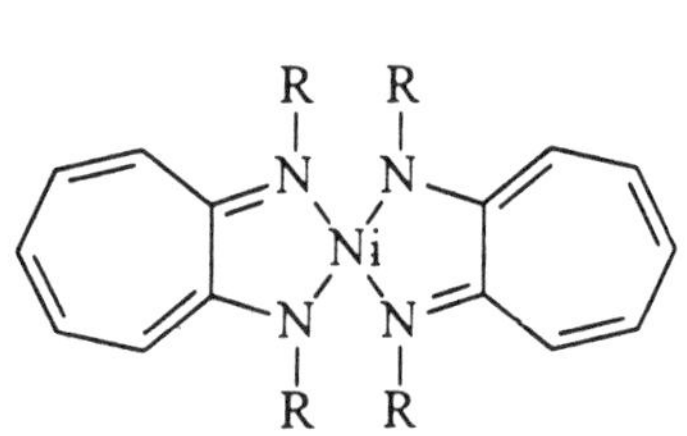

Figure 7.3. Bis(N,N′-dialkyl-2-aminotropone iminato)Ni(II).

Table 7.3. THERMODYNAMIC DATA FOR THE PLANAR
$\rightleftharpoons$ TETRAHEDRAL CONVERSION OF
bis(*N*,*N'*-DIALKYL-2-AMINOTROPANE IMINATO)Ni(II)[8]

R	ΔH(kJ/mole)	ΔS(kJ/mole K)	ΔG(kJ/mole)	K
Me	30.9 ± 0.8	70 ± 2	10 ± 1	0.018
Et	8 ± 2	70 ± 10	-12 ± 5	100
i-Pr	no square planar observed			
Ph	6.9 ± 0.6	37 ± 3	-4 ± 1	5 ± 3

interactions of the alkyl groups are very important in driving toward a tetrahedral geometry.[8] Substitution reactions with pyridine proceeded through the square-planar geometry; for those complexes dominated by tetrahedral the substitution was preceded by rearrangement to square planar.[8]

7.1.2. Five-Coordinate Complexes

Five-coordinate rearrangements are important for five-coordinate complexes and for intermediates in dissociative reactions of six-coordinate complexes.[9-11] Two geometries are observed in five coordination, trigonal bipyramidal, and square pyramidal. Both geometries and the interconversion between them are shown in Figure 7.4. Five-coordinate complexes are usually trigonal bipyramidal; but five-coordinate intermediates generated by dissociation from six-coordinate complexes are often square pyramidal.

For five-coordinate complexes of trigonal bipyramidal geometry there are two symmetry-distinct coordination positions; the three equatorial positions are equivalent (1, 2, and 3 of Figure 7.4) but distinct from the two axial positions (4 and 5 in Figure 7.4). The two different environments should be detectable by experimental measurements. X-ray diffraction experiments do confirm the geometry and show

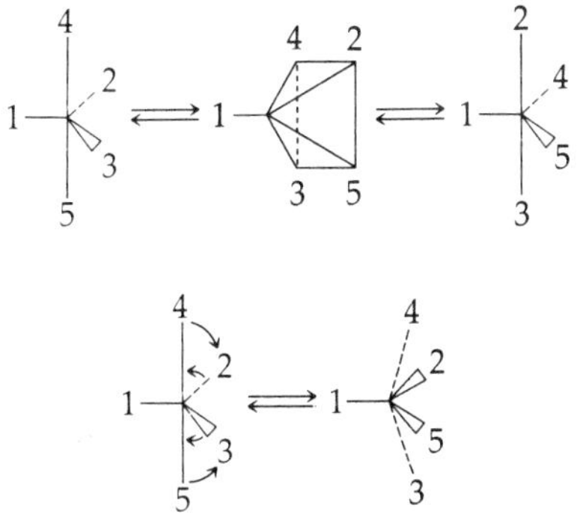

Figure 7.4. The interconversion between trigonal-bipyramidal and square-pyramidal geometries.

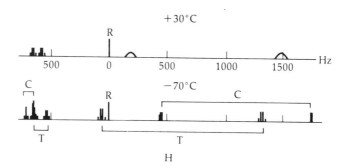

Figure 7.5. Fluorine NMR spectra of $CF_3Co(CO)_3(PF_3)$ at 30°C and −70°C (reference, $CFCl_3$). [Reprinted with permission from C. A. Udovich and R. J. Clark, *J. Am. Chem. Soc.* **91**, 526 (1969). Copyright 1969 American Chemical Society.]

bond differences between the axial and equatorial ligands. NMR studies of molecules such as PF_5 and $Fe(CO)_5$ (^{19}F and ^{13}C, respectively) show only one resonance.[9–11] This indicates that there is a chemical process that averages the environments of the axial and equatorial positions on the NMR time scale but not on the X-ray time scale.

There are two primary sources of five-coordinate complexes: Group 15 species and d^8 transition metal complexes. Species of both types undergo rapid intramolecular rearrangements. Since in many cases the exchange cannot be stopped (on the NMR time scale) at temperatures as low as −150°C, the activation barrier must be quite low (see Chapter 1). A mechanism involving pseudorotation along the normal coordinates as illustrated in the bottom of Figure 7.4 has been suggested.[12] The procedure may be viewed as holding one equatorial position of the trigonal bipyramid fixed and bending the other two equatorial positions apart. The two axial positions bend away from the one stationary equatorial site, leading to the square pyramid. The reaction may proceed to a new trigonal bipyramidal configuration or return to the starting configuration.

For many five-coordinate complexes the exchange cannot be stopped on the NMR time scale.[9–12] One of the first complexes that was investigated was $CF_3Co(CO)_3PF_3$.[12] The room temperature ^{19}F NMR spectrum shows an averaged environment for the complex, as shown schematically in Figure 7.5, where the

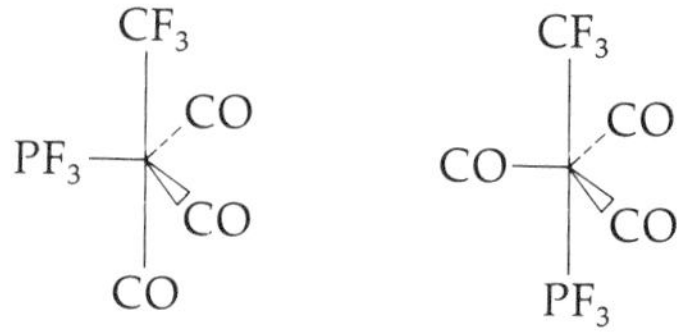

Figure 7.6. Suggested structures of the two low-temperature isomers of $CF_3Co(CO)_3PF_3$.

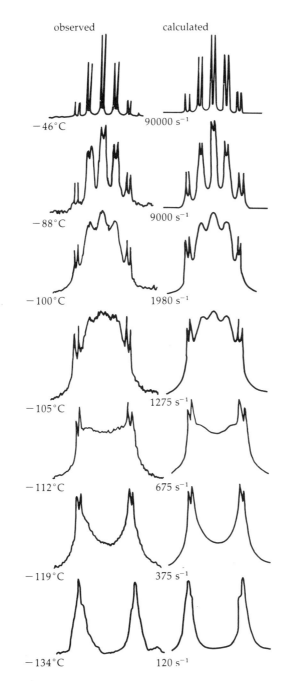

observed   calculated

−46°C   90000 s⁻¹

−88°C   9000 s⁻¹

−100°C   1980 s⁻¹

−105°C   1275 s⁻¹

−112°C   675 s⁻¹

−119°C   375 s⁻¹

−134°C   120 s⁻¹

Figure 7.7. Observed and calculated hydride region proton (90 MHz) NMR spectra for $HRh[P(OC_2H_5)_3]_4$ in $CHClF_2$. [Reprinted with permission from P. Meakin, E. L. Muetterties, and J. P. Jesson, *J. Am. Chem. Soc.* **94**, 5271 (1972). Copyright 1972 American Chemical Society.]

methyl fluorines are downfield from the reference and the PF_3 resonances are broad and upfield. Lowering the temperature to $-70°C$ stops the exchange and leads to the two isomers shown in Figure 7.6.[13]

A number of HML_4 complexes have also been investigated; the temperature-dependent NMR spectra are shown in Figure 7.7 for $HRh(P(OEt)_3)_4$.[14] The $-134°C$ spectrum does not represent a low-temperature limit spectrum, but it does indicate a C_{3v} ground state. The calculated lineshapes shown in Figure 7.7 were obtained assuming a first-order, low-temperature limit corresponding to a C_{3v} geometry with the NMR parameters $J_{HRh} = 9$ Hz, $J_{HP_{ax}} = 152$ Hz, and $J_{HP_{eq}} = 5$ Hz.[14] The kinetic parameters ($\Delta H^{\ddagger} = 5.2$ kcal/mole and $\Delta S^{\ddagger} = -11.9$ eu) were also obtained. By similar techniques the activation barriers of the five-coordinate HML_4 species shown in Table 7.4 were obtained.[14] These data illustrate one feature that is commonly observed: The rate of intramolecular interchange is first row > second row > third row. These hydrido complexes, HML_4, are best described as a tetrahedral arrangement of phosphorus ligands with the hydride capping a tetrahedral face, and a hydride tunneling mechanism is favored.[14]

In cases where steric interactions may be involved, $M[P(OMe)_3]_5$, the order of dependence on the metal is altered as shown by the data in Table 7.5.[15] The Ru complex has the lowest activation barrier in contrast to the trends shown in Table 7.4 where steric interactions are not as important. The steric interaction is shown nicely by the increase in activation barrier for $Fe[P(OEt)_3]_5$ over that of $Fe[P(OMe)_3]_5$.

The rate of exchange in $Fe(CO)_5$ is too rapid to be investigated directly by NMR. However, the exchange frequency was calculated from time-averaged chemical shifts and weighting factors obtained from infrared data.[16] The value determined, $1.1 \times 10^{10}s^{-1}$, is very rapid and suggests an exchange mechanism based on normal vibrational modes.

Table 7.4. ACTIVATION BARRIERS FOR A SERIES OF HML_4 COMPLEXES OBTAINED FROM DYNAMIC NMR INVESTIGATIONS

Compound	$\Delta G^{\ddagger}$ (kcal/mole)
$HRh[P(OEt)_3]_4$	7.25
$HIr(CO)_2(PPh_3)_2$	8.4
$Fe(PF_3)_5$	<5
$Ru(PF_3)_5$	<5
$Os(PF_3)_5$	<5
$HFe(PF_3)_4^-$	<5
$HRu(PF_3)_4^-$	7.0
$HOs(PF_3)_4^-$	8.0
$HCo(PF_3)_4$	5.5
$HRh(PF_3)_4$	9.0
$HIr(PF_3)_4$	10.0

Table 7.5. ACTIVATION PARAMETERS FOR
INTRAMOLECULAR REARRANGEMENTS
IN ML_5 SYSTEMS

Complex	$\Delta G^{\ddagger}$
$Fe[P(OMe)_3]_5$	8.8
$Ru[P(OMe)_3]_5$	7.2
$Os[P(OMe)_3]_5$	7.6
$Fe[P(OEt)_3]_5$	11.3

Most of the studies of five-coordinate complexes have centered on 18-electron species because these are readily accessible. A few studies have been accomplished on intermediates generated by ligand dissociation from six-coordinate complexes. For these intermediates the important question is whether the intermediate exists long enough to undergo rearrangement. The best estimate of the lifetime for an intermediate from ligand dissociation is 10^{-5}s. This is about the rate observed for intramolecular rearrangement in 18-electron five-coordinate complexes. As discussed in Chapter 4, CO dissociation from $Mn(CO)_5Br$ leads to $Mn(CO)_4Br$, which undergoes intramolecular rearrangement.[17] This is in contrast to $Mo(CO)_4PPh_3$, which is apparently rigid.[18] The only conclusion possible for 16-electron, five-coordinate complexes is that one should be aware of the possibility for intramolecular rearrangement in intermediates generated by dissociation from six-coordinate complexes. Too little information exists to make further generalizations at this point.

7.1.3. Six-Coordinate Complexes

Six-coordinate complexes differ from four-coordinate and five-coordinate complexes in that they have one predominant isomer.[1,2] Complexes that are octahedral do not rearrange to trigonal prismatic species with any finite lifetime. Isomerization and racemization reactions of octahedral complexes provide information on octahedral $\rightleftarrows$ trigonal prismatic interconversions. These processes often occur by dissociation of a ligand from an octahedral complex. Because this process is not distinct from ligand substitution, as discussed in Chapter 3, we will not consider it here. The reactions to be considered in this section involve rearrangement without complete dissociation of a ligand from the coordination sphere. Two types of mechanisms will be considered: one involving dissociation of one end of chelating ligand; the second involving twists of the intact octahedral complex. In both cases the most informative examples involve *tris*-chelate complexes that are substitutionally inert.[19–21]

The bond-rupture mechanism involves dissociation of one end of a chelate ligand leading to a five-coordinate intermediate followed by reattachment of the chelate. The five-coordinate intermediate may be either a trigonal bipyramid or a square pyramid, and it may be either rigid or fluxional. These possibilities lead to racemization for a symmetrical chelate or to racemization and isomerization for an unsymmetrical one. The case for an unsymmetrical chelate is shown in Figure 7.8. The analysis is further complicated by the possibility of mixtures of intermediates.

The twist mechanism can be considered as rotation of one trigonal face of an octahedron while the opposite face remains in its original position.[19] Two possibilities are shown in Figure 7.9. Rotation about the real C_3 axis, which leads to inversion of configuration without isomerization, is called a *trigonal twist*. The twist mechanisms can be thought of as

$$\text{octahedral} \rightleftarrows \text{[trigonal prismatic]} \rightleftarrows \text{octahedral}$$

reactions in which the trigonal prismatic structure represents the transition state.[19]

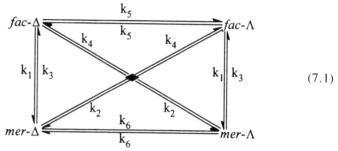

$$(7.1)$$

Sorting out the different intramolecular rearrangements for complexes is not simple.[19-21] The isomerization and racemization of unsymmetrical *tris*-chelate

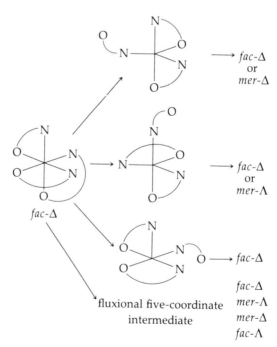

Figure 7.8. Possible isomerizations and racemization of a facial unsymmetrical *tris*–chelate complex undergoing bond rupture.

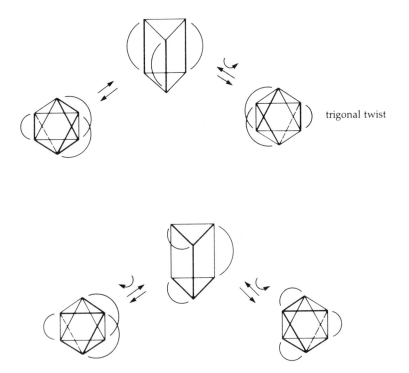

trigonal twist

Figure 7.9. Possible twist mechanisms in isomerization and racemization of *tris*–chelate complexes. [Based on J. C. Bailer, Jr. *J. Inorg. Nucl. Chem.* **8**, 165 (1958).]

complexes can be described in general for the four possible isomers shown in Figure 7.10.[21] The rate constants have been evaluated for two Co(III) complexes, *tris(S*-methylhexane-2,4-dionato)Co(III), Co(mhd)$_3$, and *tris*(benzoylacetonate)Co(III), Co(bzac)$_3$ (Table 7.6).[22,23] The data show differences for the *cis* and *trans* isomers,

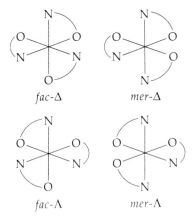

fac-Δ   *mer*-Δ

fac-Λ   *mer*-Λ

Figure 7.10. Four possible isomers of unsymmetrical *tris*–chelate complexes.

Table 7.6. RATE CONSTANTS FOR ISOMERIZATION AND RACEMIZATION OF TWO UNSYMMETRICAL $tris$-CHELATE COMPLEXES[a]

$k(10^4 s^{-1})$	Co(mhd)$_3$	Co(bzac)$_3$
k_1	0.11	0.02
k_2	0.96	1.81
k_3	0.05	0.01
k_4	0.46	0.67
k_5	~0	0.44
k_6	0.86	1.29

[a] The rate constants are as defined in Scheme 7.1[22,23]

with the cis isomers undergoing primarily isomerization and inversion ($k_2 \gg k_1$, k_5) and the $trans$ isomer undergoing inversion and inversion with isomerization at comparable rates ($k_6 > k_4 \gg k_3$). These rate data cannot be readily accommodated into one mechanistic path. It has been concluded on the basis of these rate constants that Co(mhd)$_3$ follows a bond-rupture mechanism primarily through a trigonal bipyramidal transition state with the unidentate chelate occupying an axial site. However, some of the transition states are trigonal bipyramidal with the unidentate chelate in an equatorial site (10%) and some are square pyramidal with the unidentate chelate in the axial site (20%).[21] The other complex Co(bzac)$_3$ could be accommodated equally well by bond rupture with 80% axially substituted trigonal bipyramidal and 20% axially substituted square pyramidal intermediates and by a twist mechanism with 20% trigonal twist.[21] The best way to differentiate between the two mechanisms is by comparing rates and activation parameters for the isomerization (racemization) to those for dissociation of the chelate. The oxalate complex of Rh^{3+}, Rh(ox)$_3^{3-}$, illustrates this comparison. The similarity in activation parameters for aquation, for ^{18}O exchange of the bound oxygen of oxalate, and for racemization strongly suggests that bond rupture is the mechanism. The comparison is shown in Table 7.7.[21] The activation parameters indicate a similar rate-determining step for the three reactions, probably a bond rupture. Bond rupture must be the mechanism for the exchange of the coordinated oxygen of oxalate for free ^{18}O, and

Table 7.7. COMPARISON OF RATE DATA FOR THREE REACTIONS OF Rh(ox)$_3$ $^{3-}$[21]

Reaction	$k_{obs}(10^6 s^{-1})$	$\Delta H^{\ddagger}$(kcal/mole)	$\Delta S^{\ddagger}$(eu)
^{18}O exchange[a]	14.7	23.4	−6.3
aquation	0.9	25.5	−7.8
racemization	19.7	23.3	−8.2

[a] This is exchange of the coordinated oxygen on oxalate for free ^{18}O.

it can be involved in the replacement of oxalate (aquation). The comparison suggests that racemization also occurs by bond rupture.

Racemization of $Ni(en)_3^{2+}$ was investigated

$$\Lambda(\lambda\lambda\lambda) \rightleftharpoons \Delta(\delta\delta\delta)$$

and found to occur at a rate of $5.5 \times 10^3 s^{-1}$ ($\Delta G^{\ddagger} = 15.7$ kcal/mole) at 100°C.[21] Comparison to N,N'-dimethylethylenediamine, where bond rupture would lead to an interconversion of the *dl* and meso isomers that is not observed, suggests that the racemization occurs by a twist mechanism.

Intramolecular racemization and isomerization of octahedral complexes with

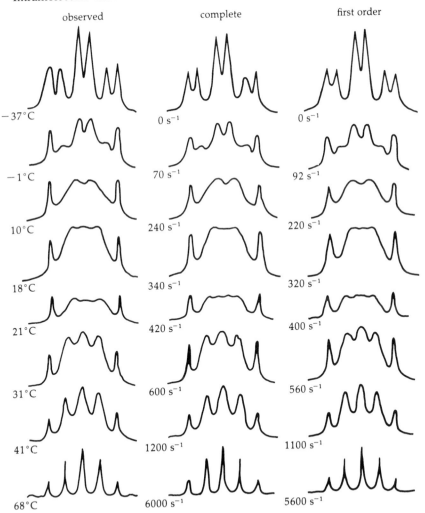

Figure 7.11. Calculated and observed NMR spectra of $H_2Fe[P(OEt)_3]_4$ at various temperatures. [Reprinted with permission from P. Meakin, E. L. Muetterties, F. N. Tebbe, and J. P. Jesson, *J. Am. Chem. Soc.* **93**, 4701 (1971). Copyright 1971 American Chemical Society.]

bidentate ligands occur relatively slowly either by dissociation to five-coordinate intermediates or by a twist mechanism (rotation of one octahedral face with respect to the other).

The complex $H_2Fe(P(OEt)_3)_4$ has been investigated by variable-temperature NMR.[24] The changes in the spectrum with temperature are shown in Figure 7.11. The high-temperature spectrum (68°C) clearly shows the equivalence of the phosphorus nuclei; the low-temperature spectrum indicates the *cis* stereochemistry of the rigid molecule. A number of mechanisms including twists were investigated, but the only mechanism giving satisfactory agreement involves a hydrogen migrating from one face of a pseudotetrahedral arrangement of $P(OEt)_3$ ligands to another.[24] It is significant that the octahedral complex undergoes rearrangement with a substantially higher barrier (13.7 kcal/mole) than do the HMP_4 complexes (Table 7.4). This can also be seen in the higher temperature required for the fast-exchange limit.

7.1.4. Seven-Coordinate Complexes

Seven-coordinate complexes are not often observed, but they are of considerable interest as intermediates in associative reactions of six-coordinate complexes and in dissociative reactions of eight-coordinate complexes.[1,2] Three geometries are commonly found: a pentagonal bipyramid derived from the octahedral arrangement by addition to an edge, a capped octahedron with the seventh ligand occupying a face position of an octahedron, and a capped trigonal prism with a square face capped.[25] These possibilities are shown in Figure 7.12. The energy differences between these geometries are small, as only minor movement is required to interconvert them (Fig. 7.13). Calculations show barriers to interconversion between these three isomers of 0–4 kcal/mole.[25] Values for several complexes are shown in Table 7.8. Thus, under most experimental conditions seven-coordinate complexes are stereochemically nonrigid.

The Cr complex $Cr[P(OMe)_3]_5H_2$ has a pentagonal bipyramidal structure with the two hydrogens in the pentagonal plane as shown by ^{31}P NMR at $-130°C$. The high-temperature limit spectra show a triplet for the ^{31}P and a sextet for the ^1H. Detailed analysis of the mechanism for the intramolecular exchange was not offered, but a scheme involving simultaneous exchange of two axial P ligands with two equatorial ligands would be consistent with the variable-temperature NMR.[26]

The reaction dynamics of *bis*[1,2-*bis*(dimethylphosphino)ethane]tricarbonyltitanium, $Ti(Me_2PCH_2CH_2PMe_2)_2(CO)_3$, have been investigated by variable-temperature NMR.[27] The structure of this analogue of $Ti(CO)_7$ was shown to be a capped octahedron (structure B in Fig. 7.12) in which a CO is the capping group. Both ^{13}C and ^{31}P spectra were investigated at variable temperatures, and the exchange was stopped at $-115°C$, where the ^{31}P spectrum showed four distinct, equally intense resonances. This is one of the few seven-coordinate complexes in which the exchange can be stopped and investigated. The change in the resonances as the temperature was raised to $-60°C$ allowed determination of the activation parameters for the dynamic behavior ($\Delta G^{\ddagger} = 7.8$ kcal/mole; $\Delta H^{\ddagger} = 8.6$ kcal/mole; $\Delta S^{\ddagger} = 4$ eu).[27] A trigonal twist of the face of the octahedron occupied by three

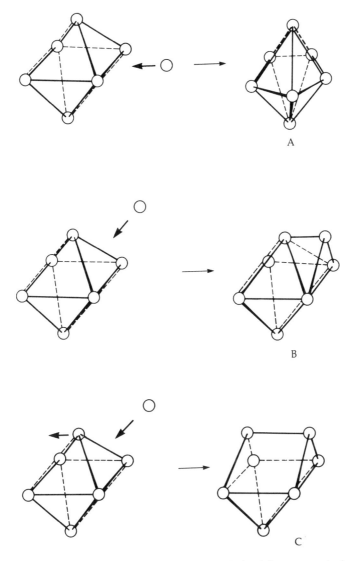

Figure 7.12. Seven-coordinate geometries as derived from an octahedron.

phosphorus ligands was suggested as the simplest mechanism consistent with the NMR data.

7.1.5. Eight-Coordinate Complexes

Eight-coordinate complexes are commonly observed for the high oxidation states (+4, +5), of the early transition metals.[1,2] Two geometrical forms are observed:

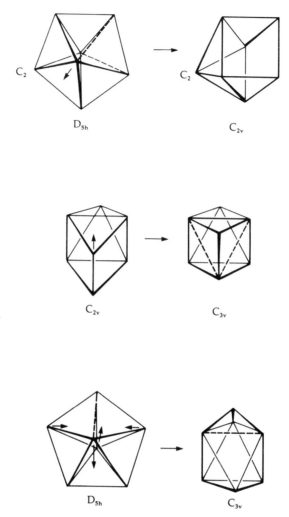

Figure 7.13. Rearrangements of the idealized seven-coordinate geometries. [Based on R. Hoffman, B. F. Beier, E. L. Muetterties, and A. R. Rossi, *Inorg. Chem.* **16,** 311 (1977).]

the square antiprism and the dodecahedron (Fig. 7.14). Each of these structures can be readily derived from a cube—the square antiprism by rotating one face by 45° relative to the opposite face and the dodecahedron by closing the cube along opposite faces. Conversion between the two occurs relatively rapidly along vibrational modes.[28] A possible scheme is shown in Figure 7.15. The $Mo(CN)_8^{4-}$ complex shows a dodecahedral arrangement of cyanides around molybdenum that would give two different environments for the two CN^- groups. Only one ^{13}C NMR signal is seen in solution. This is most readily explained by a rapid equilibration, as shown in Figure 7.15.[28]

Table 7.8. CALCULATED RELATIVE VALUES OF THE ENERGY (kcal/mole) OF THE PENTAGONAL BIPYRAMID, CAPPED OCTAHEDRON, AND CAPPED TRIGONAL PRISM FOR d^4-CONFIGURATIONS[25]

Geometry	MCl_7	$M(CO)_7$
pentagonal bipyramid	0	0
capped octahedron	1	5
capped trigonal prism	2	4

7.1.6. Nine-Coordinate Complexes

Nine-coordinate complexes have not been extensively investigated.[1,2] The geometry observed for almost all nine-coordinate complexes is the symmetrically tricapped trigonal prism.[29] This geometry is related by a stretching motion to the monocapped square antiprism. The rearrangement is shown in Figure 7.16. The barrier to this intramolecular rearrangement is calculated to be very small, but experimental data demonstrating the rearrangement have not been presented.[29]

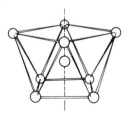

Figure 7.14. The square antiprism and the dodecahedron for eight-coordinate complexes. [Reprinted with permission from S. J. Lippard, *Progress Inorganic Chemistry* **8**, 115 (1976). Copyright 1976 S. J. Lippard.]

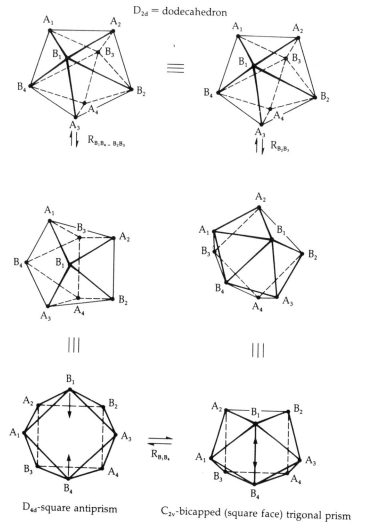

Figure 7.15. Rearrangements in eight-coordination. [Based on E. L. Muetterties, *Inorg. Chem.* **4**, 769 (1965).]

7.2. Metal Migration in an Unsaturated Ring—Ring Whizzers

When a metal coordinates to a ring with more than one equivalent position, characterization by NMR often indicates an equilibration process. The cyclopentadienyl
ring, η^1-C_5H_5, serves as an example (Fig. 7.17). Only one resonance is seen in the
proton NMR despite the existence of three nonequivalent sites in the static struc-

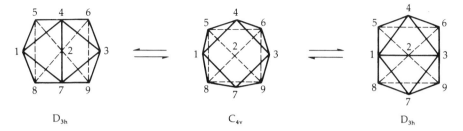

Figure 7.16. The suggested intramolecular path for stereochemical nonrigidity in nine-coordinate complexes. [Reprinted with permission from L. J. Guggenberger and E. L. Muetterties, *J. Am. Chem. Soc.* **98,** 7221 (1976). Copyright 1976 American Chemical Society.]

ture.[29] The process that equilibrates the three sites is metal migration around the ring, and complexes that show such behavior are commonly termed *ring whizzers*. We shall examine several examples that illustrate different ring whizzers.

7.2.1. $(\eta^5\text{-}C_5H_5)(\eta^1\text{-}C_5H_5)Fe(CO)_2$

One of the first complexes that demonstrated fluxional behavior for a ring complex was $(\eta^5\text{-}C_5H_5)(\eta^1\text{-}C_5H_5)Fe(CO)_2$,

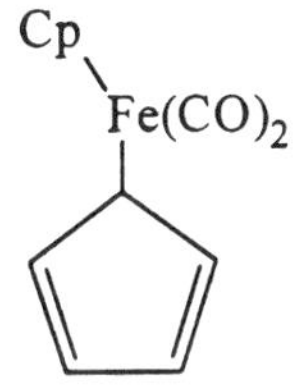

Cp
 \
  Fe(CO)$_2$

which shows two singlet resonances of equal intensity at room temperature.[30] The NMR spectra at several temperatures are shown in Figure 7.18. At $-80°C$ the spectrum shows a sharp single resonance from $\eta^5\text{-}C_5H_5$ and a singlet from the unique hydrogen on $\eta^1\text{-}C_5H_5$. The assignment of the two multiplets is crucial to the mechanistic interpretation; the downfield signal is assigned to the H_b hydrogens as defined in Figure 7.17.[30] As the spectra in Figure 7.18 show, collapse of the low-

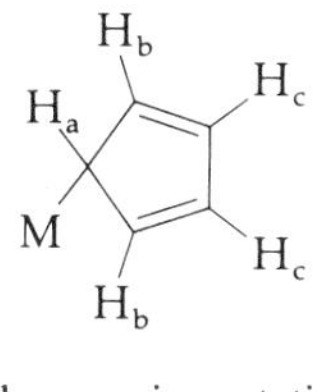

H_b
H_a — H_c
M
H_c
H_b

Figure 7.17. Nonequivalent hydrogens in a static structure on an $\eta^1\text{-}C_5H_5$ complex.

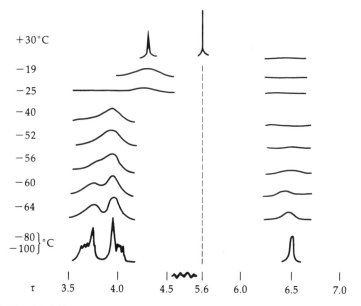

Figure 7.18. Variable-temperature proton NMR spectra of (η^5-C$_5$H$_5$)(η^1-C$_5$H$_5$)Fe(CO)$_2$ in CS$_2$. [Reprinted with permission from F. A. Cotton, *Acc. Chem. Res.* **1**, 257 (1968). Copyright 1968 American Chemical Society.]

temperature spectrum is unsymmetrical (the H$_b$ hydrogens are exchanging more rapidly than are the H$_c$s). This rules out a random process or any process in which all sites become equivalent and limits the reasonable mechanistic possibilities to a 1,2 metal migration or a 1,3 metal migration, as shown in Figure 7.19. The effect of a 1,2 shift on the hydrogen positions can be illustrated as follows.

$$
\begin{bmatrix} H_a \\ H_b \\ H_c \\ H_c \\ H_b \end{bmatrix} \xrightarrow{\text{1,2 shift}} \begin{bmatrix} H_b \\ H_a \\ H_b \\ H_c \\ H_c \end{bmatrix}
$$

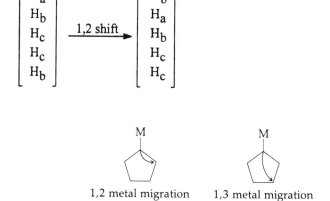

1,2 metal migration   1,3 metal migration

Figure 7.19. The definition of 1,2 and 1,3 metal migration for fluxional isomerism in η^1-C$_5$H$_5$ complexes.

With the 1,2 shift both H_bs are equilibrated in a single migration while only one H_c is equilibrated. Thus, H_b would exchange more rapidly than would H_c as observed in the NMR spectrum. The 1,3 shift would lead to more rapid equilibration of H_c than H_b.

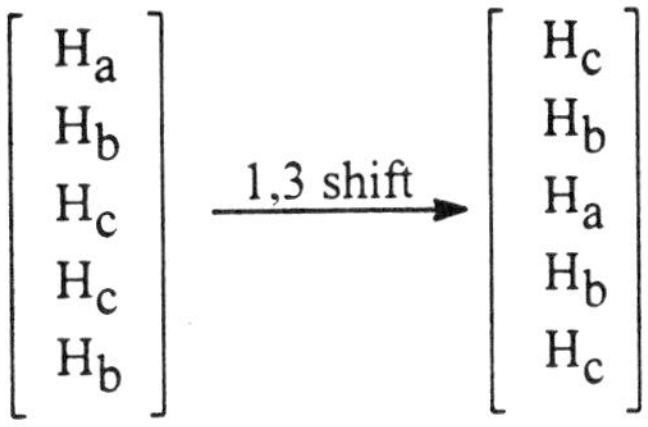

$$
\begin{bmatrix} H_a \\ H_b \\ H_c \\ H_c \\ H_b \end{bmatrix}
\xrightarrow{\text{1,3 shift}}
\begin{bmatrix} H_c \\ H_b \\ H_a \\ H_b \\ H_c \end{bmatrix}
$$

The 1,2 shift is thus demonstrated for the metal migration in the η^1-cyclopentadienyl ring of $(\eta^5\text{-}C_5H_5)(\eta^1\text{-}C_5H_5)Fe(CO)_2$.[30]

7.2.2. $(\eta^4\text{-}C_8H_8)Ru(CO)_3$

Cyclooctatetraene complexes also allow for metal migration around the unsaturated ring. The complex $(\eta^4\text{-}C_8H_8)Ru(CO)_3$ (Fig. 7.20) serves as an example.[30,31] The NMR spectrum at low temperature ($-147°C$) shows four distinct resonances of equal intensity, but at room temperature there is only one sharp resonance, indicating complete equilibration.[31] The spectra at different temperatures are shown in Figure 7.21. An unsymmetrical collapse like that of $(\eta^5\text{-}C_5H_5)(\eta^1\text{-}C_5H_5)Fe(CO)_2$ is observed. After assignment of the spectra and interpretation of the possibilities, a 1,2 shift was again suggested. The activation energy of 9.4 kcal/mole indicates the ease of metal migration along the unsaturated C_8H_8 ring.

7.2.3. Cycloheptatriene Complexes

The predominant mechanism for metal migration in unsaturated ring systems is the 1,2 shift. This mechanism is observed for $\eta^1\text{-}C_5H_5$ complexes, $(\eta^1\text{-}C_7H_7)Re(CO)_5$,

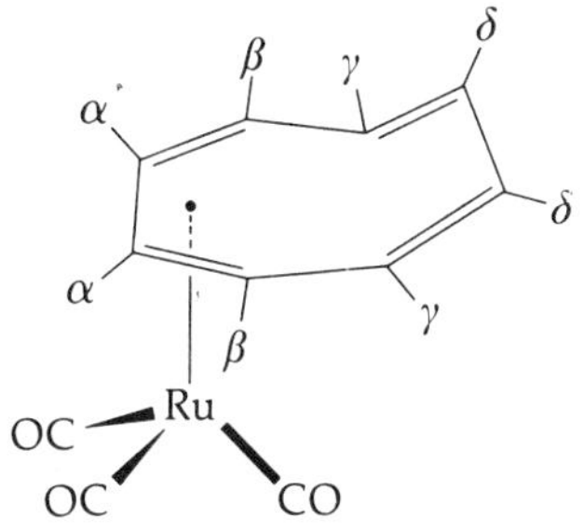

Figure 7.20. Static structure of $(\eta^4\text{-}C_8H_8)Ru(CO)_3$ showing the four different hydrogen positions. [Reprinted with permission from F. A. Cotton, A. Davison, T. J. Marks, and A. Musco, *J. Am. Chem. Soc.* **91**, 6598 (1969). Copyright 1969 American Chemical Society.]

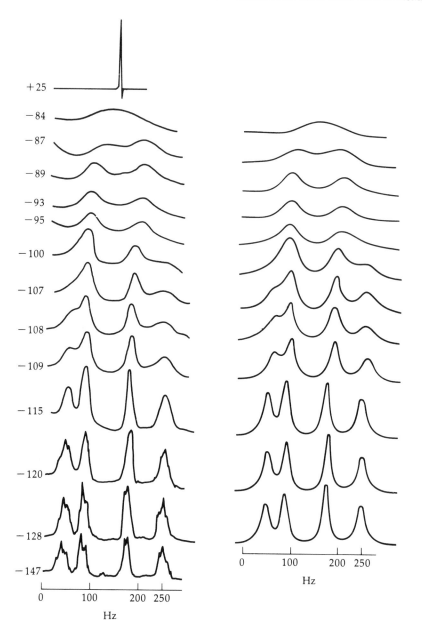

Figure 7.21.   The variable-temperature NMR spectra of $(\eta^4\text{-}C_8H_8)Ru(CO)_3$ with computer simulation for a 1,2 shift at the right. [Reprinted with permission from F. A. Cotton, A. Davison, T. J. Marks, and A. Musco, *J. Am. Chem. Soc.* **91,** 6598 (1969). Copyright 1969 American Chemical Society.

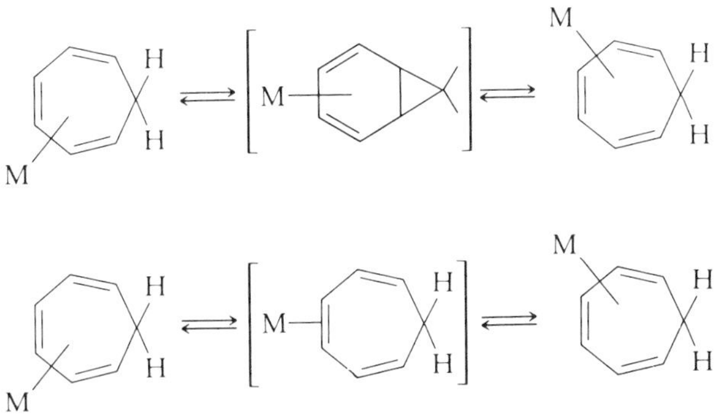

Figure 7.22. Possible mechanisms for rearrangement of cycloheptatriene complexes.

η^3-C_7H_7 complexes, η^4-C_6R_6 Rh complexes, η^4-C_8H_8 complexes, η^5-C_7H_7 complexes, and some η^6-$C_8H_4Me_4$ complexes. In this section two cycloheptatrienyl complexes in which 1,3 and 1,4 shifts are suggested will be examined.[32,33] For a cycloheptatriene complex such as (η^4-C_7H_8)Fe(CO)$_3$, a simple 1,2 shift is not possible and the proton NMR spectrum shows no line broadening up to 100°C. However, a high-energy process does exchange the environments.[32] Two possible mechanisms are shown in Figure 7.22. The 1,2 shift must be accompanied by ring closure to the norcaradiene intermediate. The alternative is a 1,3 shift through an unsaturated intermediate. The mechanism of rearrangement was investigated by spin saturation transfer of the proton NMR.[32] A number of mechanisms can be ruled out based on the lack of scrambling of the exo and endo substituents in the d_7 complex. The free energy of activation is 22.3 kcal/mol.[32] The selective nature of the averaging process was only consistent with a 1,3 shift. The results of an extended Hückel molecular orbital calculation suggested an η^2 transition state/intermediate with the iron shifted toward the interior of the cycloheptatriene.[32]

The rearrangement of the cycloheptatrienyl complex CpRu(CO)$_2$(η^1-C_7H_7) has been interpreted in terms of two concurrent modes: a 1,2 shift and a 1,4 shift.[33] The structure of CpRu(CO)$_2$(η^1-C_7H_7) is shown in Figure 7.23. The spin saturation transfer technique was used to explore the possible migration pathways.[33] Rate constants for the different possibilities were calculated from ^{13}C NMR data; for the

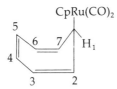

Figure 7.23. The structure of CpRu(CO)$_2$(η^1-C_7H_7) with the numbering scheme.

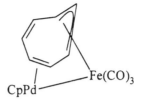

Figure 7.24. $(CO)_3Fe(\mu\text{-}\eta^3\colon\eta^2\text{-}(C_7H_7))PdCp$.

1,2 shift $k = 1.7 \times 10^{-2}\text{s}^{-1}$, for the 1,3 shift $k \approx 0$, and for the 1,4 shift $k = 0.3$ s^{-1}.[33] This suggests that the 1,2 and 1,4 shifts are of comparable energy ($\Delta G^{\ddagger} \approx 19$ kcal/mol).

Another interesting example is $(CO)_3Fe(\mu\text{-}\eta^3\colon\eta^2\text{-}C_7H_7)PdCp$ which shows a singlet for the cycloheptyl ring at room temperature.[34] The low temperature ^{1}H and ^{13}C NMR spectra indicate the structure shown in Figure 7.24. The interesting ring-whizzing involving two metals almost certainly uses the free double bond, but detailed mechanistic studies were not reported.[34]

7.2.4. Other Complexes and Summary

An allene complex (this is not formally an unsaturated ring system) also undergoes fluxional behavior.[35] The structure of $(\eta^2\text{-}Me_2C=C=CMe_2)Fe(CO)_4$ is shown in Figure 7.25. The proton NMR spectrum at room temperature shows only one resonance, although the spectrum is resolved at low temperature.[35] The rearrangement must involve migration of the iron from coordination to one double bond to coordination to the orthogonal double bond. This is illustrated in Figure 7.25. The activation energy for the process was determined as 9.0 kcal/mole.[35]

The examples illustrate the ease of metal migration in unsaturated systems, where there are equivalent structural forms. The most common mechanism is a 1,2 shift, but others may be possible, especially if the 1,2 shift is energetically unfavorable. The metal remains coordinated during the process and must migrate along the electron density from one energy minimum to another.

7.3.   M–M Bonded Systems

In systems that involve more than one metal center, the exchange of ligands between metal centers is commonly observed. This type of stereochemical nonrigidity

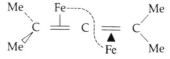

Figure 7.25.   The structure of $(\eta^2\text{-}Me_2C=C=CMe_2)Fe(CO)_4$ and migration in its fluxional behavior.[33]

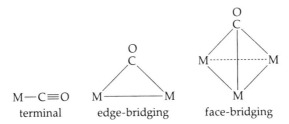

Figure 7.26. Bonding modes of terminal, edge-bridging and face-bridging CO in metal clusters.

has been investigated extensively as a possible model to surface mobility of chemisorbed species.[36] Fluxional behavior of metal carbonyl clusters has dominated studies of M–M bonded complexes. Bonding of carbon monoxide in metal clusters may take different forms, as shown in Figure 7.26. These structures are models for carbon monoxide mobility on clusters and for the rearrangements of other ligands, such as nitric oxide, isocyanides, and hydrides. Fluxional processes in a few cluster compounds will be examined in the following sections.

7.3.1. Metal Carbonyl Dimers

Dimers are not clusters, but the intramolecular rearrangements of dimers provide data on exchange of ligands between two metal centers. For $Co_2(CO)_8$ in solution, infrared spectroscopy reveals three forms, one bridged and two nonbridged[37]; the two primary ones are shown in Figure 7.27. This molecule is fluxional on the NMR time scale below $-90°C$, but the mechanism has not been established.[36] Conversion of the bridged isomer to the unbridged and back would be logical steps.

The iron dimer $Cp_2Fe_2(CO)_4$ has been investigated extensively by a number of techniques; of these, NMR is the most informative.[36,38] The complex exhibits only one resonance in the ^1H spectrum down to $-48°C$, but the peak splits to two singlets at lower temperatures.[38,39] These two singlets are assigned to *cis* and *trans* isomers, as shown in Figure 7.28. The NMR spectra at different temperatures are shown in Figure 7.29. Note the slight preference for the *cis* isomer. The interconversion between the two isomers occurs with a rather low activation barrier ($\Delta G^\ddagger = 10.4$ kcal/mole). ^{13}C NMR has shown that bridge-terminal CO exchange in the *cis* complex occurs with an activation barrier very similar to that for isomerization, but the bridge–terminal CO exchange of the *trans* complex was considerably more facile. These data suggest a simultaneous breaking of the two bridge bonds with a

Figure 7.27. The solution structures of $Co_2(CO)_8$.

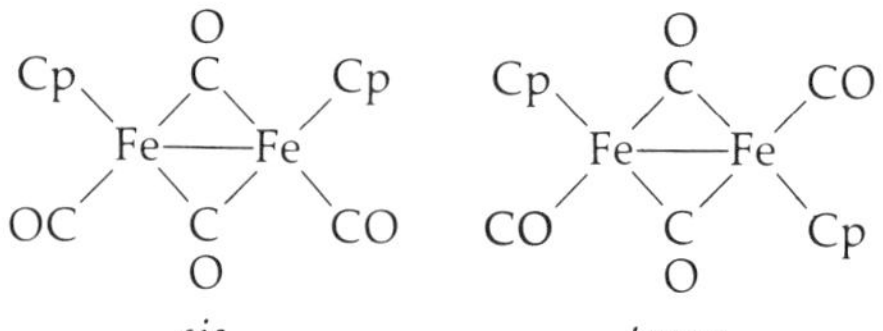

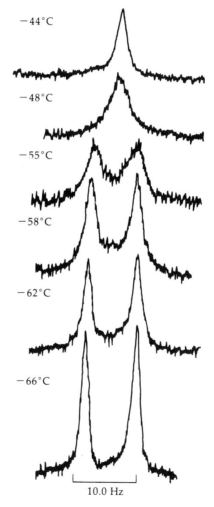

Figure 7.28. The *cis* and *trans* isomers of $Cp_2Fe_2(CO)_4$.

CO moving to a terminal position on each iron. Bridge–terminal exchange in the *trans* isomer would occur by just this process, leading to an intermediate as shown in the top part of Figure 7.30, where all four COs are equivalent. *Cis–trans* isomerization and bridge–terminal exchange in the *cis* isomer require simultaneous break-

Figure 7.29. The proton NMR spectrum of $Cp_2Fe_2(CO)_4$ shows the interconversion between *cis* and *trans* isomers that equilibrates the environment of the hydrogens. [Reprinted with permission from J. G. Bullitt, F. A. Cotton and T. J. Marks, *Inorg. Chem.* **11**, 671 (1972). Copyright 1972 American Chemical Society.]

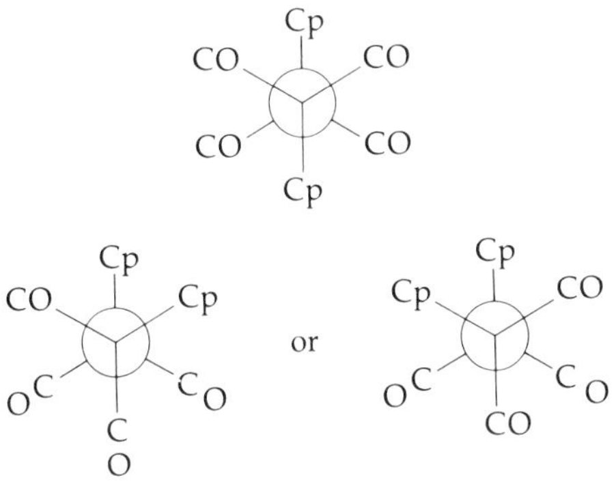

Figure 7.30. The nonbridged intermediates in the fluxional behavior of $Cp_2Fe_2(CO)_4$. The *trans* isomer leads only to the top possibility in which the four COs are equivalent. The *cis* isomer may lead to either of the lower possibilities. Since the COs are nonequivalent in the *cis* isomer, rotation must occur for bridge-terminal exchange.[36]

ing of the two bridge bonds and rotation of one iron by 120° with respect to the other. This rotation requires an additional activation barrier and slows these processes relative to the bridge–terminal exchange in the *trans* isomer. In the larger Ru analogue, the *cis–trans* isomerization could not be stopped at the lowest temperature investigated.[39] In a complex where the cyclopentadienyls are linked and cannot rotate, $(\eta^{10}\text{-}C_{10}H_8\text{-}C_{10}H_8)Fe_2(CO)_4$ (Fig. 7.31), exchange of bridge and terminal COs occurs only above 80°C.[36]

7.3.2. Trinuclear Clusters

The dodecacarbonyl trimetal complexes are diverse in structure and fluxional behavior.[36] The iron complex adopts the bridged structure; the Ru and Os complexes have the nonbridged structure shown in Figure 7.32. The Fe and Ru complexes in solution are fluxional to $-100°C$, showing only a single resonance in the ^{13}C NMR; the Os complex is rigid at room temperature.[36] The combination of ^{13}C spin-lattice relaxation times, spin–spin relaxation times, and a high-resolution magic angle spinning ^{13}C NMR spectrum allowed evaluation of the barrier to interconversion in $Ru_3(CO)_{12}$.[40] Even at $-138°C$ the axial and equatorial COs could not be resolved, but from broadening of the singlet an activation energy of 4.4 ± 0.5 kcal/mole was obtained.[40]

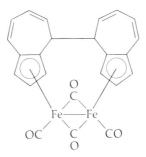

Figure 7.31. The structure of the iron dimer with linked cyclopentadienyls.

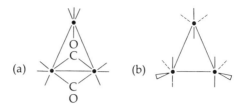

Figure 7.32.   Structures of the trimetaldodecacarbonyls. The bridged structure (a) is for $Fe_3(CO)_{12}$, and the unbridged (b) for $Ru_3(CO)_{12}$ and $Os_3(CO)_{12}$.

The exchange process has been investigated for the trimer $Cp_3Rh_3(CO)_3$, which exists in two isomeric forms (Fig. 7.33).[41] Both isomers undergo fluxional processes without converting into each other. This is surprising since a primary suggestion for fluxional behavior of (b) in Figure 7.33 is termed the merry-go-round for which (a) would be an intermediate.[36] The ^{13}C NMR spectra of (b) at different temperatures were investigated; spectra at two temperatures are shown in Figure 7.34.[41] The quartet at room temperature indicates complete randomization of the three CO groups with the quartet arising from coupling to three equivalent Rh centers. The spectrum recorded at $-125°C$ is consistent with a pairwise bridge-opening mechanism for the exchange process. The activation energy, $\Delta G^{\ddagger} = 8.9$ kcal/mole, was derived from the intermediate-exchange spectra.[41] The limiting slow-exchange spectrum could not be obtained, but the broadening at $-156°C$ led to an estimate of 4.8 kcal/mole for $\Delta G^{\ddagger}$ for the low-temperature process. An additional process occurred at high temperature converting (b) into (a) (Figure 7.33). The scheme in Figure 7.35 was suggested to account for these observations. The essential feature is the conversion of an edge-bridging carbonyl to a face-bridging one. The upper part of the scheme accounts for the low-temperature process in which carbonyl A is always bridging, but B and C are terminal part of the time, explaining the $-125°C$ spectrum. The conversion from X into Y, a face-to-face interconversion through an edge-bridged intermediate, may be an important step in carbonyl scrambling on clusters in general.

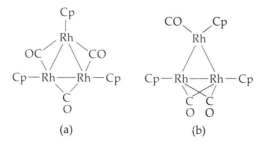

Figure 7.33.   Two isomers of $Cp_3Rh_3(CO)_3$.[41]

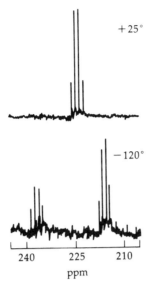

Figure 7.34. ^{13}C NMR spectra of the B isomer of $Cp_3Rh_3(CO)_3$ at room temperature and at $-125°C$. [Reprinted with permission from R. J. Lawson and J. R. Shapley, *Inorg. Chem.* **17**, 772 (1978). Copyright 1978 American Chemical Society.]

7.3.3. Tetranuclear Clusters

Tetranuclear carbonyl clusters, $M_4(CO)_{12}$, M = Co, Rh, Ir, also exhibit nonrigid behavior.[36] These complexes adopt one of the two structures shown in Figure 7.36. Co and Rh have bridged structures; Ir has the unbridged structure. The energy difference between the two structures is probably not large, and the two may be considered intermediates for fluxional behavior of the other structure. The Rh

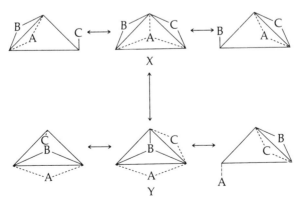

Figure 7.35. Scheme for the scrambling processes in $Cp_3Rh_3(CO)_3$. [Reprinted with permission from R. J. Lawson and J. R. Shapley, *Inorg. Chem.* **17**, 772 (1978). Copyright 1978 American Chemical Society.]

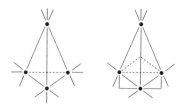

Figure 7.36.   Structures of the tetrametaldodecacarbonyls.

complex $Rh_4(CO)_{12}$ shows only a broad resonance in the ^{13}C NMR at room temperature.[42] Heating to 63°C leads to a quintet as expected for complete equilibration of the carbonyls. The collapse of this spectrum as the temperature is lowered is symmetrical, indicating no site selectivity, and consistent with an interconversion between the bridged and unbridged structures in Figure 7.36.[42] The low-temperature (−65°C) spectrum shows four doublets of equal intensity consistent with the bridged structure for $Rh_4(CO)_{12}$.[43]

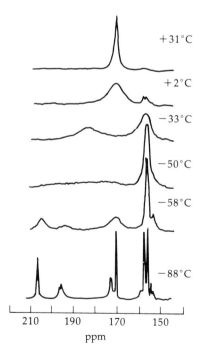

Figure 7.37.   The variable-temperature ^{13}C NMR spectra of $Ir_4(CO)_{11}PPh_2Me$ showing three averaging processes. [Reprinted with permission from G. F. Stuntz and J. R. Shapley, *J. Am. Chem. Soc.* **99**, 607 (1977). Copyright 1977 American Chemical Society.]

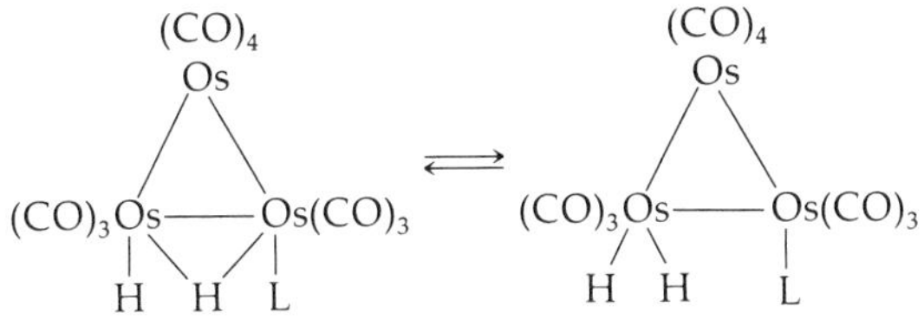

$$\begin{array}{ccc}
(CO)_4 & & (CO)_4 \\
Os & & Os \\
(CO)_3Os\!-\!\!-\!Os(CO)_3 & \rightleftharpoons & (CO)_3Os\!-\!\!-\!Os(CO)_3 \\
H \quad H \quad L & & H \quad H \quad L
\end{array}$$

Figure 7.38. The bridge-terminal hydride exchange in $H_2Os_3(CO)_{11}$ and $H_2Os_3(CO)_{10}PMe_2Ph$,[46] $L = CO, PMe_2Ph$.

The tetrairidium complexes are interesting in that substitution of one carbonyl by a P donor changes the structure from the unbridged isomer to the bridged with the P donor in the basal plane. The fluxional behavior of $Ir_4(CO)_{11}PPh_2Me$ has been investigated.[44] The NMR spectrum indicates three distinct averaging processes as the temperature is raised from $-88°C$ (Fig. 7.37). The low-temperature process equilibrates the bridging carbonyls with the basal, radial carbonyls (defined in Fig. 4.8). The next process (about $-50°C$) equilibrates the two axial carbonyls on the basal Ir with two carbonyls on the apical Ir. The remaining carbonyl is equilibrated in a process above $0°C$. Each of these processes apparently occurs by bridge–terminal exchange with energy differences governed by location of the phosphine.[44]

7.3.4. Hydrido Cluster Compounds

The introduction of a hydride ligand offers further possibilities for nonrigidity of metal clusters.[45] Hydrides are commonly either bridging or terminal, and the exchange between the two is usually facile. The complexes $H_2Os_3(CO)_{11}$ and $H_2Os_3(CO)_{10}PMe_2Ph$ serve as examples of the exchange. The bridging and terminal hydrides have been investigated by proton NMR.[46,47] The reaction is shown in Figure 7.38. The free energies of activation show very small dependence on the

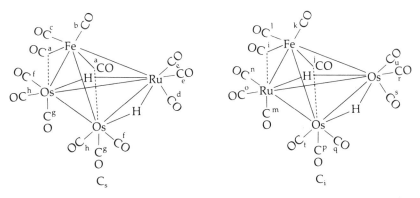

Figure 7.39. Carbonyl labeling scheme for the C_s and C_i isomers of $H_2FeRuOs_2(CO)_{13}$. [Reprinted with permission from W. L. Gladfelter and G. L. Geoffroy, *Inorg. Chem.* **19**, 2579 (1980). Copyright 1980 American Chemical Society.]

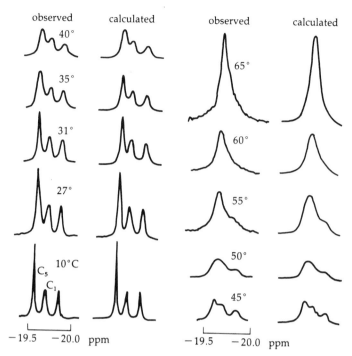

Figure 7.40. Variable-temperature proton NMR spectra for $H_2FeRuOs_2(CO)_{13}$. [Reprinted with permission from W. L. Gladfelter and G. L. Geoffroy, *Inorg. Chem.* **19**, 2579 (1980). Copyright 1980 American Chemical Society.]

ligand L in $H_2Os_3(CO)_{10}L$ for a wide variety of different L groups.[47] A smaller value of $\Delta G^{\ddagger}$ for exchange of bridge and terminal hydrides was observed for smaller or more basic ligands. It has been suggested that the transition state for bridge–terminal exchange involved a double-bridging configuration of one Os–Os bond.[47]

The series of $P(OMe)_3$-substituted derivatives of $H_4Ru_4(CO)_{12}$ are fluxional down to −100°C, indicating the facility of hydride exchange.[48] This facility has led to very little clear mechanistic evidence on hydrogen movement on clusters. Variable-temperature NMR spectra of $[AsPh_4][H_3Ru_4(CO)_{12}]$ showed the exchange process, but the line shapes could not be fit by a simple mechanism.[49]

Three dynamic processes were observed for hydride equilibration on $H_4Ru_4(CO)_{10}(\mu\text{-diphos})$.[50] The lowest energy process (observed between −92 and −131°C) was considered to be movement of a hydride from one edge-bridging position to another. This process was observable because the backbone of the diphos ligand is asymmetric.[50] The second process (−74 to −92°C) was considered to involve conformational changes of the diphos ligand with a free energy of 9.1 kcal/mole. The final process (−40 to −70°C) equilibrated all four hydrides, but the precise mechanism could not be determined.[50]

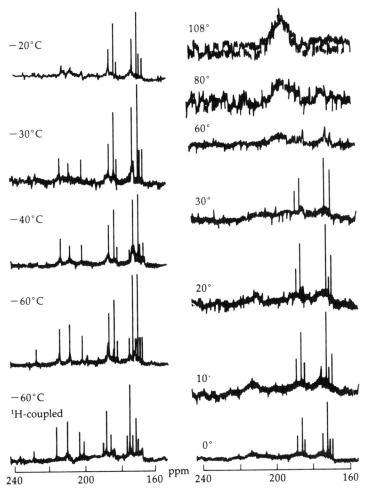

Figure 7.41. Variable-temperature ^{13}C NMR spectra of $H_2FeRuOs_2(CO)_{13}$. [Reprinted with permission from W. L. Gladfelter and G. L. Geoffroy, *Inorg. Chem.* **19**, 2579 (1980). Copyright 1980 American Chemical Society.]

The hydrido mixed-metal clusters $H_2FeRu_3(CO)_{13}$, $H_2FeRu_2Os(CO)_{13}$, and $H_2FeRuOs_2(CO)_{13}$ provide the most thoroughly investigated examples of fluxional behavior in metal-cluster complexes.[51] The structures and labeling scheme for the latter are shown in Figure 7.39. The asymmetry provided by the different metal centers allows considerable mechanistic interpretation from variable temperature proton and ^{13}C NMR spectra. The spectra for $H_2FeRuOs_2(CO)_{13}$ shown in Figures 7.40 and 7.41 indicate several different exchange processes.[51] The carbonyl environments are equilibrated in three distinct exchange processes. The first is exchange of bridge and terminal carbonyls on the Fe, probably by the opening of one carbonyl bridge, a trigonal twist of the $Fe(CO)_3$ unit, and re-formation of the CO bridge. The second carbonyl exchange process equilibrates the COs around the Fe–M–M triangle that contains the bridging COs. A cyclical movement of the COs

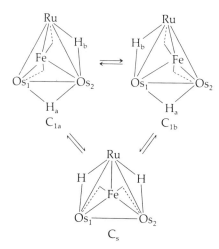

Figure 7.42. Skeletal rearrangement for the fluxional behavior of $H_2FeRuOs_2(CO)_{13}$. [Reprinted with permission from W. L. Gladfelter and G. L. Geoffroy, *Inorg. Chem.* **19,** 2579 (1980). Copyright 1980 American Chemical Society.]

around the triangular face (a type of merry-go-round) has been suggested for this fluxional reaction.[49] As indicated by the proton NMR, the hydrogen atoms are not involved in the two low-temperature exchange processes. The third process is involved in isomerization, H exchange, and total CO scrambling. This was considered a cluster framework rearrangement (Fig. 7.42) with the iron moving away from Os_1 toward Os_2 and H_b moving simultaneously. The combination of framework rearrangement and cyclic CO exchange allows total exchange of carbonyls and hydrogens in $H_2FeRuOs_2(CO)_{13}$. Examination of the three clusters $H_2FeRu_3(CO)_{13}$, $H_2FeRu_2Os(CO)_{13}$, and $H_2FeRuOs_2(CO)_{13}$ showed identical exchange processes, except that the activation barrier for each process increased with the Os content of the cluster.[51]

The ligand scrambling on larger cluster compounds is more complicated to study. Migration of the hydride ligands on $Rh_{13}(CO)_{24}H_3^{3-}$ and $Rh_{13}(CO)_{24}H_2^{2-}$ indicates that the hydrides move rapidly on the inside of the hexagonal close-packed structure, while carbonyl fluxionality occurred by the "normal" bridge-terminal exchange.[52] The movement of hydrogen atoms on the inside of the metal framework is directly applicable to hydrogen diffusion in metals and warrants further investigation.

7.3.5. Metal Rotations Within the Ligand Polyhedron

An alternative postulated mechanism for metal-cluster fluxionality involves reorganization of the metal framework within the polyhedron of ligand atoms. This process has been described in some detail for $Fe_3(CO)_{12}$.[53] In the ground state structure

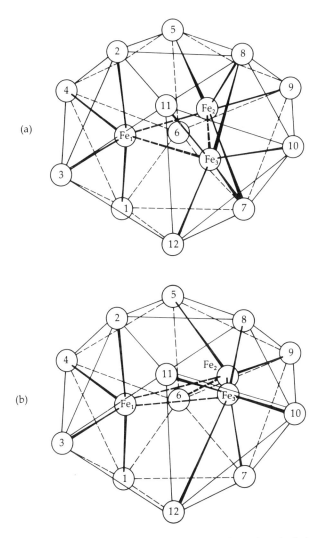

Figure 7.43. Possible orientations of the Fe triangle within an icosahedral arrangement of COs: (a) the observed structure of $Fe_3(CO)_{12}$; (b) the observed structure of $Os_3(CO)_{12}$. [Reprinted with permission from B.F.G. Johnson, *J. Chem. Soc. Chem. Comm.* 703 (1976). Copyright 1976 Royal Society of Chemistry.]

the 12 carbonyls describe an icosahedron, and the Fe atoms lie within this icosahedron, as shown in Figure 7.43a. Rotation of the metal triangle by 30° would lead to the isomer with all COs equivalent and terminal (Figure 7.43b).[54] Similar rotations can lead to other structures. This scheme has the advantage of equilibrating the COs in a single step involving relatively minor atom motion—as required for reactions with very low activation barriers.[53] At this time there is no evidence to differentiate between ligand motion and metal motion for the stereochemical nonrigidity of extremely facile systems in solution. However, the solid-state ^{13}C NMR spectrum of $Fe_3(CO)_{12}$ showed six resonances of equal intensity, which is inconsistent with a static structure and suggesting time averaging of the two disordered molecules observed crystallographically. [Note that on the X-ray crystallographic time scale ($10^{-18}s$) $Fe_3(CO)_{12}$ in the solid state is in the slow-exchange limit, while on the NMR time scale (10^{-1}–$10^{-9}s$) a fast-exchange spectrum is observed.] It has been suggested that the two molecules interconvert by a 60° rotation of the iron triangle, a transition that is very rapid on the NMR time scale.[54]

7.4. Summary

Certain classes of metal complexes can be expected to exhibit stereochemical nonrigidity at moderate conditions without undergoing ligand loss. These include five- and seven-coordinate complexes, unsaturated ring complexes, and metal clusters. In each case a combination of NMR spectra and computer simulation has allowed reasonable mechanistic interpretation of the fluxional behavior.

7.5. References

1. K. F. Purcell and J. C. Kotz, *Inorganic Chemistry* (Philadelphia; W. B. Saunders, 1977).

2. F. A. Cotton and G. Wilkinson, *Advanced Inorganic Chemistry* (New York; Interscience, 1972).

3. E. L. Muetterties, *Tetrahedron* **30**, 1595 (1974).

4. E. L. Muetterties, *J. Am. Chem. Soc.* **91**, 1636 (1969).

5. E. L. Muetterties, *Acc. Chem. Res.* **3**, 266 (1970).

6. L. H. Pignolet, W. DeW. Horrocks, Jr., and R. H. Holm, *J. Am. Chem. Soc.* **92**, 1855 (1970).

7. G. N. LaMar and E. O. Sherman, *J. Am. Chem. Soc.* **92**, 2691 (1970).

8. M. Schumann and H. Elias, *Inorg. Chem.* **24**, 3187 (1985).

9. R. R. Holmes, Sr., R. M. Deiters, and J. A. Golen, *Inorg. Chem.* **8**, 2612 (1969).

10. J. R. Shapley and J. A. Osborn, *Acc. Chem. Res.* **6**, 305 (1973).

11. R. R. Holmes, *Acc. Chem. Res.* **5**, 296 (1972).

12. R. S. Berry, *J. Chem. Phys.* **32**, 933 (1960).

13. C. A. Udovich and R. J. Clark, *J. Am. Chem. Soc.* **91**, 526 (1969).

14. P. Meakin, E. L. Muetterties, and J. P. Jesson, *J. Am. Chem. Soc.* **94**, 5271 (1972).

15. J. P. Jesson and P. Meakin, *J. Am. Chem. Soc.* **96**, 5760 (1974).

16. H. Mahnke, R. J. Clark, R. Rosanske, and R. K. Sheline, *J. Chem. Phys.* **60**, 2997 (1974).

17. J. D. Atwood and T. L. Brown, *J. Am. Chem. Soc.* **97**, 3380 (1975).

18. D. J. Darensbroug and A. H. Graves, *Inorg. Chem.* **18**, 1257 (1979).

19. J. C. Bailar, Jr., *J. Inorg. Nucl. Chem.* **8**, 165 (1958).

20. R. H. Holm, *Acc. Chem. Res.* **2**, 307 (1969).

21. N. Serpone and D. G. Bickley, *Prog. Inorg. Chem.* **17**, 391 (1972) and references therein.

22. J. G. Gordon II and R. H. Holm, *J. Am. Chem. Soc.* **92**, 5319 (1970).

23. A. Y. Girgis and R. C. Fay, *J. Am. Chem. Soc.* **92**, 7061 (1970).

24. P. Meakin, E. L. Muetterties, F. N. Tebbe, and J. P. Jesson, *J. Am. Chem. Soc.* **93**, 4701 (1971).

25. R. Hoffman, B. F. Beier, E. L. Muetterties, and A. R. Rossi, *Inorg. Chem.* **16**, 511 (1977).

26. F. A. Van-Catledge, S. D. Ittel, C. A. Tolman, and J. P. Jesson, *J. Chem. Soc., Chem. Commun.* 254 (1980).

27. P. J. Domaille, R. L. Harlow, and S. S. Wreford, *Organometallics* **1**, 935 (1982).

28. E. L. Muetterties, *Inorg. Chem.* **4**, 769 (1965).

29. L. G. Guggenberger and E. L. Muetterties, *J. Am. Chem. Soc.* **98**, 7221 (1976).

30. F. A. Cotton, *Acc. Chem. Res.* **1**, 257 (1968) and references therein.

31. F. A. Cotton, A. Davison, T. J. Marks, and A. Musco, *J. Am. Chem. Soc.* **91**, 6598 (1969).

32. K. J. Karel, T. A. Albright, and M. Brookhart, *Organometallics* **1**, 419 (1982).

33. D. M. Heinekey and W.A.G. Graham, *J. Am. Chem. Soc.* **104**, 915 (1982).

34. M. Airoldi, G. Deganello, G. Gennaro, M. Moret, and A. Sironi, *Organometallics* **12**, 3964 (1993).

35. R. Ben-Shoshan and R. Pettit, *J. Am. Chem. Soc.* **89**, 2231 (1967).

36. E. Band and E. L. Muetterties, *Chem. Rev.* **78**, 639 (1978) and references therein.

37. R. L. Sweany and T. L. Brown, *Inorg. Chem.* **16**, 415 (1977).

38. R. D. Adams and F. A. Cotton, *J. Am. Chem. Soc.* **95**, 6589 (1973).

39. J. G. Bullitt, F. A. Cotton, and T. J. Marks, *Inorg. Chem.* **11**, 671 (1972).

40. S. Aime, W. Dastrú, R. Gobetta, J. Krause, and L. Milone, *Organometallics* **14**, 4435 (1995).

41. R. J. Lawson and J. R. Shapley, *Inorg. Chem.* **17**, 772 (1978).

42. F. A. Cotton, L. Kruczynski, B. L. Shapiro, and L. F. Johnson, *J. Am. Chem. Soc.* **94**, 6191 (1972).

43. J. Evans, B.F.G. Johnson, J. Lewis, and J. R. Norton, *J. Chem. Soc., Chem. Commun.* 807 (1973).

44. G. F. Stuntz and J. R. Shapley, *J. Am. Chem. Soc.* **99**, 607 (1977).

45. J. Evans, *Adv. Organomet. Chem.* **16**, 319 (1977) and references therein.

46. J. R. Shapley, J. B. Keister, M. R. Churchill, and B. G. DeBoer, *J. Am. Chem. Soc.* **97**, 4145 (1975).

47. J. B. Keister and J. R. Shapley, *Inorg. Chem.* **21**, 3304 (1982).

48. S.A.R. Knox and H. D. Kaesz, *J. Am. Chem. Soc.* **93**, 4594 (1971).

49. J. W. Koepke, J. R. Johnson, S.A.R. Knox, and H. D. Kaesz, *J. Am. Chem. Soc.* **97**, 3947 (1975).

50. M. R. Churchill, R. A. Lashewycz, J. R. Shapley, and S. I. Richter, *Inorg. Chem.* **17**, 1277 (1980).

51. W. L. Gladfelter and G. L. Geoffroy, *Inorg. Chem.* **19**, 2579 (1980).

52. S. Martinengo, B. T. Heaton, R. J. Goodfellow, and P. Chini, *J. Chem. Soc., Chem. Commun.* 39 (1977).

53. B.F.G. Johnson, *J. Chem. Soc., Chem. Commun.* 703 (1976).

54. H. Dorn, B. E. Hanson, and E. Modell, *Inorg. Chim. Acta* **54**, L71 (1981).

7.6. Problems

7.1. Reaction of $NiCl_2$ with excess PEt_3 leads to $Ni(PEt_3)_2Cl_2$, which can be isolated and further reacted with a P ligand, $P(CF_3)F(Cl)$, to yield $Ni[P(CF_3)F(Cl)](PEt_3)Cl_2$ as yellow crystals. Describe this complex in detail given the following data:

  a. Magnetic susceptibility at 25°C is 0.3 B.M.; at 65°, 1.2 B.M.; and at $-100°$, 0.0 B.M.

  b. Molecular weight studies show that no association or dissociation occurs from $-100°C$ to $+100°C$.

  c. The proton NMR spectrum for the complex in C_6D_6 at 20° is a broad resonance displaced substantially from the typical C-H proton region.

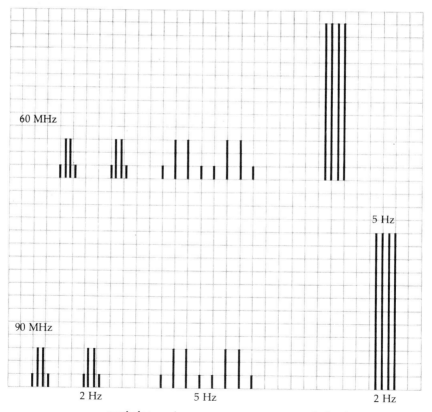

methylene region   methyl region

  d. The proton NMR at $-100°C$ (at both 60 MHz and 90 MHz) is as shown in the preceding graph. Assume rotation about all bonds is infinitely fast at $-100°C$ and that the only magnetic nuclei are ^1H.

7.2. The optically active complex R-cis-Co(en)$_2$Cl$_2^+$ undergoes both racemization and isomerization in CH$_3$OH. Based on the observation that racemization is somewhat more rapid that isomerization, describe the most likely mechanism.

7.3. The analysis of the mechanism of coordination sphere isomerization is generally more complicated than we have presented. The separation of different possibilities is very difficult. For five-coordinate complexes an alternative possibility is termed the *turnstile* mechanism. Consider the theoretical differences between the Berry pseudorotation and the turnstile mechanism; evaluate the possibility of differentiating between the two mechanisms experimentally. [J. I. Musher, *J. Am. Chem. Soc.* **94**, 5662 (1972)].

7.4. Why do five- and seven-coordinate complexes undergo such rapid equilibration?

7.5. The Bailar twist of a *tris*-chelate complex formally involves rotation around the C$_3$ axis. There exist two other unique faces about which rotation can occur. For the three different faces show what isomerization occurs for a *fac*-R-*tris*-unsymmetrical bidentate metal complex.

7.6. What is the order of reactivity toward coordination sphere isomerization for different coordination numbers? Explain.

7.7. The ^{13}C NMR spectra of $(\eta^4$-C$_8$H$_8)$Fe(CO)$_3$ are shown in the graph at different temperatures. Based on these spectra and on the similarity of activation parameters to those for the $(\eta^4$-C$_8$H$_8)$Ru(CO)$_3$ complex, assign the spectra and offer a mechanism for the ring-whizzing phenomenon. [F. A. Cotton and D. L. Hunter, *J. Am. Chem. Soc.* **98**, 1413 (1976).]

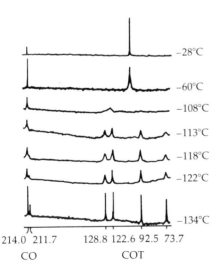

7.8. The fluxional behavior of the complex $(\eta^1\text{-}C_7H_7)Re(CO)_5$ was investigated by the spin saturation transfer technique. Irradiation of the one position (Fig. 7.22) caused the two- and seven-position resonances to decrease in intensity with a $\Delta H^{\ddagger}$ of 18.1 kcal/mole and $\Delta S^{\ddagger} = -5.7$ eu. Interpret this observation mechanistically. [D. M. Heinekey and W.A.G. Graham, *J. Am. Chem. Soc.* **101**, 6115 (1979).]

7.9. Cyclopropenium complexes are another example of ring whizzing. This behavior has been considered in detail for triphenylcyclopropenium complexes of Ni and Pd. Suggest a reasonable mechanism, and illustrate the transition state. [C. Mealli, S. Midollini, S. Moneti, L. Sacconi, J. Silvestre, and T. A. Albright, *J. Am. Chem. Soc.* **104**, 95 (1982).]

7.10. The complex $Ir_4(CO)_{11}(CNCMe_3)$, which has the unbridged structure, undergoes two processes: a low-temperature one that equilibrates 10 carbonyls, and a higher-temperature process that equilibrates the remaining carbonyl with the 10. Describe the processes that are probably occurring. [G. F. Stuntz and J. R. Shapley, *J. Organomet. Chem.* **213**, 389 (1981).]

7.11. The ligands that occupy the equatorial site of a trimetal cluster have an important effect on the merry-go-round process. Why? What ligand characteristics provide for facile equilibration?

7.12. The tetranickel cluster $Ni_4(CNCMe_3)_7$ undergoes a fluxional process that scrambles the isocyanides. The complex contains four terminal and three edge-bridging isocyanides. Suggest a mechanism for the fluxional behavior.

7.13. In the dinuclear indenyl complex $(C_{10}H_8)Fe_2(CO)_4PEt_3$ that contains no bridging carbonyls,

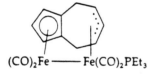

$(CO)_2Fe \text{——} Fe(CO)_2PEt_3$

internuclear CO exchange is observed. Suggest a scheme that incorporates electron redistribution of the rings.

7.14. Solid-state ^1H and ^{13}C NMR studies of $H_4Ru_4(CO)_{12}$ show the carbonyls to be nonfluxional and the hydrides to be fluxional. What mechanisms are indicated? [S. Aime, R. Gobetto, A. Orlandi, C. J. Groombridge, G. E. Hawkes, M. D. Mantle, and K. D. Sales, *Organometallics* **13**, 2375 (1994)].

7.15. Interpret the following variable temperature NMR spectra obtained for $Ru_4(\mu\text{-}H)_4(CO)_{11}P(OEt)_3$. [S. Aime, M. Botta, R. Gobetta, L. Milone, D. Osella, R. Gellert, and E. Rosenberg, *Organometallics* **14**, 3693 (1995).]

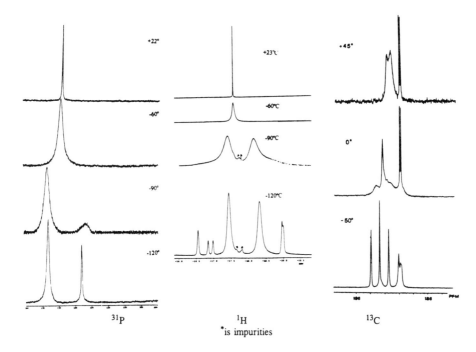

^{31}P ^{1}H ^{13}C

*is impurities

7.16. Four-coordinate *bis* chelate complexes of Ni(II), NiL$_2$ (*bis*(N,N'-dialkyl-2-aminotro-pone iminato)nickel(II)), undergo tetrahedral $\rightleftarrows$ planar equilibria that depend on the alkyl group as follows, R = Me, K = 0.018; R = Et, K = 140; R = Pr, K = 110; R = *i*-Pr, all tetrahedral. These complexes react with pyridine at rates that are strongly dependent on the alkyl group (the planar R = Me complex reacts 10^6 more rapidly than the tetrahedral R = *i*-Pr complex). What implications do these results have for reactions of NiL$_2$ complexes? [M. Schumann and H. Elias, *Inorg. Chem.* **24,** 3187 (1985).]

8

Oxidation-Reduction Reactions—Electron Transfer

Many inorganic reactions involve a change in the oxidation states of metals. The ability of metals to assume a variety of oxidation states (the chromium subgroup has been observed in oxidation states from -4 to $+6$) makes this a common reaction often utilized in synthetic sequences.[1-4] Oxidation-reduction (redox) reactions involve two species that change in oxidation state, with one increasing (losing an electron) and the other decreasing (gaining an electron). The thermodynamics of electron transfer depend on the reduction potentials of the reactants. This topic is treated in general chemistry courses and will not be covered in this chapter. The complexes to be considered here are thermodynamically accessible, but they may have a kinetic barrier to reaction. The kinetic barrier and the mechanism will be our primary focus.

Several examples of electron transfer in inorganic reactions are shown in the following equations:

$$\Lambda\text{-Os(bipy)}_3^{3+} + \Delta\text{-Os(bipy)}_3^{2+} \rightleftharpoons \text{racemization} \tag{8.1}$$

$$k \geq 5 \times 10^{-4} M^{-1}s^{-1}$$

$$\text{Cr(H}_2\text{O)}_6^{2+} + \text{Co(NH}_3)_5\text{Cl}^{2+} \xrightarrow{\text{H}^+} \tag{8.2}$$

$$k = 10^5 M^{-1}s^{-1} \qquad \text{Cr(H}_2\text{O)}_5\text{Cl}^{2+} + \text{Co(H}_2\text{O)}_6^{2+} + 5\text{NH}_4^+$$

$$*\text{Co(NH}_3)_6^{2+} + \text{Co(NH}_3)_6^{3+} \rightleftharpoons *\text{Co(NH}_3)_6^{3+} + \text{Co(NH}_3)_6^{2+} \tag{8.3}$$

$$k = 10^{-8} M^{-1}s^{-1}$$

$$*Fe(o\text{-phen})_3^{2+} + Fe(o\text{-phen})_3^{3+} \rightleftharpoons \tag{8.4}$$

$$k = 10^7 M^{-1}s^{-1} \qquad\qquad *Fe(o\text{-phen})_3^{3+} + Fe(o\text{-phen})_3^{2+}$$

These equations indicate the types of reaction and the diversity of rates observed among electron-transfer reactions in inorganic systems. Reactions such as 8.3 and 8.4, which were investigated in the 1950s as radioisotopes became available, form an important part of electron-transfer chemistry.[1-4] Electron transfer between organometallic complexes has not been as extensively studied. A section at the end of this chapter will summarize the status of organometallic electron-transfer reactions.

The mechanisms for redox reactions fall into two classifications. An *inner sphere electron transfer* is accompanied by marked changes in the coordination sphere of the transition state, usually with a ligand bridging the two metal centers. In an *outer sphere electron-transfer* mechanism the coordination spheres of the reactants remain intact. Differentiating between the two mechanistic types is generally not easy. Reaction rates for exchange that are more rapid than they are for substitution at the metal (as in Eq. 8.4) are good evidence for an outer sphere reaction. In describing the product distribution it is necessary to understand the ligand substitutional lability of the metal centers (Chapter 3). Reaction 8.2 provides an example of the importance of the metal center on the products. The reactant metal complexes involve a Cr^{2+}, which is extremely labile, and Co^{3+}, which is inert; the products are Cr^{3+}, which is inert, and Co^{2+}, which is labile. Thus, the loss of ligands from the Co product is a natural consequence of the lability of the Co^{2+}. Either an inner sphere or an outer sphere mechanism would give the same products.

Inner-sphere:

$$Cr(H_2O)_6^{2+} + Co(NH_3)_5Cl^{2+} \longrightarrow \tag{8.5}$$

$$(H_2O)_5Cr\text{---}Cl\text{---}Co(NH_3)_5^{4+} + H_2O$$

$$(H_2O)_5Cr\text{---}Cl\text{---}Co(NH_3)_5^{4+} \longrightarrow \tag{8.6}$$

$$Cr(H_2O)_5Cl^{2+} + Co(NH_3)_5H_2O^{2+}$$

$$Co(NH_3)_5H_2O^{2+} \xrightarrow{H^+} Co(H_2O)_6^{2+} + 5NH_4^+ \tag{8.7}$$

$$Cr(H_2O)_5Cl^{2+} \rightleftharpoons Cr(H_2O)_6^{3+} + Cl^- \tag{8.8}$$

Outer-sphere:

$$Cr(H_2O)_6^{2+} + Co(NH_3)_5Cl^{2+} \longrightarrow Cr(H_2O)_6^{3+} + Co(NH_3)_5Cl^+ \tag{8.9}$$

$$Co(NH_3)_5Cl^+ \xrightarrow{H^+} Co(H_2O)_6^{2+} + 5NH_4^+ + Cl^- \tag{8.10}$$

$$Cr(H_2O)_6^{3+} + Cl^- \rightleftharpoons Cr(H_2O)_5Cl^{2+} + H_2O \tag{8.11}$$

Thus, simple product distribution is not a good means of distinguishing mechanistic type in electron-transfer reactions. Use of $^{36}Cl^-$ free in solution initially showed no incorporation into the chromium product, indicating an inner sphere mechanism. In this case, because $Cr(H_2O)_5Cl^{2+}$ is inert, isotope labeling allows mechanistic differentiation.

All electron-transfer reactions are first order in oxidant and reductant,

$$rate = k[\text{oxidant}][\text{reductant}] \tag{8.12}$$

allowing no distinction between inner-sphere and outer-sphere processes based on the rate law.[5-10] Activation enthalpies are usually low, with outer-sphere mechanisms a bit lower than inner-sphere ones; both have negative entropies of activation. The standard kinetic parameters are not very useful in differentiating electron-transfer mechanisms. Special considerations are required to distinguish between inner- and outer-sphere mechanisms. In later sections each mechanism for electron transfer will be considered.

8.1. Theoretical Considerations

Electron transfer by either type of mechanism is subject to the restrictions of the Franck-Condon principle. The electron weighs much less than the nuclei; therefore, it moves rapidly compared with nuclei. Thus, during the time of electron transfer the nuclei remain stationary and bond distances cannot change. Before electron transfer from one center to another can take place, the coordination spheres must adjust so that the energy of the system is unaltered on electron transfer. This is most easily considered for self-exchange reactions like those of Equations 8.3 and 8.4. The M–L bond distance is shorter for the higher oxidation state metal. A shorter M–L bond distance for the higher oxidation state metal presumes that π-back bonding is not important in the metal-ligand bond. When pi bonding is important, the lower oxidation state M–L bond may be shorter. An example is $V(CO)_6$ and $V(CO)_6^-$ in which the V–C distances are 2.001 and 1.931 Å, respectively.[11,12] In general the M–L bond distance does change with oxidation state of the metal. In our discussion we shall consider the metal of higher oxidation state to have the shorter M–L bond. Several examples are shown in Table 8.1. Before the self-exchange electron transfer can take place the metal of higher oxidation state must lengthen its bonds, and the one with a lower oxidation state must shorten its bonds. The amount of distortion required to reach a common state is inversely related to the rate of electron transfer. If large changes are required, then the rate of electron transfer may be very slow. Coupled with the principles of electronic structures of transition metal complexes this concept allows an understanding of some of the reactivity changes in electron-transfer reactions. Self-exchange between $Co(NH_3)_6^{3+}$ and $Co(NH_3)_6^{2+}$ is quite slow, as shown in Reaction 8.3. Since Co^{3+} has a low-spin d^6 configuration and Co^{2+} has a high-spin d^7 configuration, the electron transfer involves a spin change

Table 8.1. CHANGES IN M–L BOND DISTANCES
AS THE OXIDATION STATE CHANGES

M	L	M^{II}–L	M^{III}–L	Ref.
Ru	NH_3	2.144(4)	2.104(4)	15
Ru	H_2O	2.122(16)	2.029(7)	16
Co	NH_3	2.114(9)	1.936(15)	15
Fe	bipy	1.97(1)	1.963	17
Co	bipy	2.128(8)	1.93(2)	17

and the reaction would be expected to be very slow. The e_g orbitals are antibonding with respect to the M–L σ-bonding framework, while the t_{2g} orbitals are nonbonding. Changing the number of e_g electrons has a relatively larger effect on the bonding than does changing the number of t_{2g} electrons. Electron transfer into or out of the e_g orbital is typically slower than are transitions involving t_{2g} electrons. Reactions that involve very small M–L changes with oxidation state may undergo very rapid electron transfer. Note that the smallest M–L changes with oxidation state involve differences in t_{2g} electrons (t_{2g}^6 and t_{2g}^5).

Since electron transfer involves rearrangement of the coordination sphere along a normal coordinate, isotope changes should affect the rate. Kinetic isotope effects were measured for redox reactions of $Fe(H_2O)_6^{2+}$, $Fe(D_2O)_6^{2+}$ and $Fe(^{18}OH_2)_6^{2+}$ with $M(bipy)_3^{3+}$ (M = Fe, Ru, Cr).[15,16] The oxygen kinetic isotope effect of k_{16}/k_{18} = 1.08 for Cr was readily predicted as the product of an electronic interaction, and Franck-Condon factors for vibrational modes as given by the effect of mass change on the frequency of the symmetric M–L stretching mode. The substitution of deuterium for hydrogen in the ligand caused a larger kinetic isotope effect (k_H/k_D = 1.3) than could be explained by the change in M–L frequency. It has been suggested that the O–H modes were also involved in the electron transfer.[15]

8.2. Outer-Sphere Electron Transfer

An outer-sphere electron transfer may be considered as the following steps:

$$A + B \rightleftharpoons [A,B] \tag{8.13}$$

$$[A,B] \rightleftharpoons [A,B]^* \tag{8.14}$$

$$[A,B]^* \rightleftharpoons [A^-,B^+] \tag{8.15}$$

$$[A^-,B^+] \rightleftharpoons A^- + B^+ \tag{8.16}$$

where the brackets indicate the reactants are in close proximity (an encounter complex). Reaction 8.14 involves the structural rearrangement necessary for electron transfer; Reaction 8.15 indicates the electron transfer; and the final reaction is the movement of the products apart. Since outer-sphere reactions involve no bond-breaking or bond-making they are amenable to theoretical treatment.[3,7,18] In Marcus's interpretation, for a reaction with $\Delta G^0 = 0$, the free energy of activation is the sum of several contributions.[19]

$$\Delta G^\ddagger = RT \ln \frac{kT}{hZ} + \Delta G_a^\ddagger + \Delta G_i^\ddagger + \Delta G_0^\ddagger \tag{8.17}$$

The first term allows for the loss of translational and rotational free energy on forming the collision complex from the reactants. This term involves constants (Z is the effective collision number) and can be calculated. The $\Delta G_a^\ddagger$ represents the free energy change due to electrostatic interaction between the reactants at the separation distance in the activated complex compared with the interaction at infinite separation. The term $\Delta G_i^\ddagger$ represents the free energy change that is required for rearrangement of the coordination sphere—elongation or contraction of the M–L bonds in the activated complex and perhaps also rearrangement of the ligand. This normally occurs along the vibrational normal modes for the molecule. The $\Delta G_0^\ddagger$ represents solvent sphere rearrangement. The solvent would certainly be more attracted to the higher oxidation state and must rearrange prior to electron transfer. This term is relatively constant for different $M(H_2O)_6$ complexes in the same oxidation state. The $\Delta G_0^\ddagger$ may be quite different for different ligand systems. A few examples for hexaaquo complexes are shown in Table 8.2.[7] In complexes that do not differ greatly in oxidation state or ligands, differences in outer-sphere reactivity cannot arise from $\Delta G_a^\ddagger$ or $\Delta G_0^\ddagger$; therefore, they must arise from changes in $\Delta G_i^\ddagger$. A measure of $\Delta G_i^\ddagger$ is the M–L bond distance changes as shown in Table 8.1 for the different oxidation states. For complexes whose structures are known, the energy barrier for electron transfer can be calculated from the following classical equations:

$$E = 3k_m(r^\ddagger - r_m)^2 + 3k_n(r_n - r^\ddagger)^2 \tag{8.18}$$

$$r^\ddagger = \frac{k_m r_m + k_n r_n}{k_m + k_n} \tag{8.19}$$

where k_m and k_n are force constants, r_m and r_n are the M–L bond distances, and $r^\ddagger$ is the bond distance necessary for electron transfer.[3,7,18] Using these equations for $Co(NH_3)_6^{2+}/Co(NH_3)_6^{3+}$ gives an energy barrier of 6.8 kcal/mole; the value for $Ru(NH_3)_6^{2+}/Ru(NH_3)_6^{3+}$ is negligible since the radius difference is only 0.04 Å.[14] This small radius change illustrates the small effect when a "nonbonding" t_{2g} elec-

Table 8.2.   ENTHALPIES OF SOLVATION FOR HEXAAQUO IONS[7]

$M(H_2O)_6^{2+}$	ΔH (kcal/mole)	$M(H_2O)_6^{3+}$	ΔH (kcal/mole)
V^{2+}	−267	V^{3+}	−610
Cr^{2+}	−274	Cr^{3+}	−662
Mn^{2+}	−259	Mn^{3+}	−655
Fe^{2+}	−282	Fe^{3+}	−629

tron is removed. The difference in self-exchange rates for these two systems is a factor of 10^{15}, and 7 kcal/mole is insufficient to account for such a large difference. The Co complex also must undergo a spin change with a calculated energy difference of 24.6 kcal/mole. Thus, the reorganization energy for the Co–L bond is small compared with the spin-change energy. This simple concept has been considerably expanded in more recent experiments on $Co^{2+,3+}$ reactions.[20-22] Similar considerations for the species $Fe(H_2O)_6^{2+}/Fe(H_2O)_6^{3+}$ and $Ru(H_2O)_6^{2+}/Ru(H_2O)_6^{3+}$ show that the reactivity order

$$k \text{ for } Fe(H_2O)_6^{2+}/Fe(H_2O)_6^{3+} < k \text{ for } Ru(H_2O)_6^{2+}/Ru(H_2O)_6^{3+} < k$$
$$\text{for } Ru(NH_3)_6^{2+}/Ru(NH_3)_6^{3+}$$

is readily accounted for by the reorganization energies during self-exchange.[14]

A useful variation of Marcus's theory called *the relative Marcus theory* utilizes self-exchange reactions to calculate the barrier for cross-reactions.[3,7]

$$Ox_1 + Red_1 \longrightarrow Red_1 + Ox_1 \qquad\qquad k_{11}, \Delta G_{11}^{\ddagger} \qquad\qquad (8.20)$$

$$Ox_2 + Red_2 \longrightarrow Red_2 + Ox_2 \qquad\qquad k_{22}, \Delta G_{22}^{\ddagger} \qquad\qquad (8.21)$$

$$Ox_1 + Red_2 \longrightarrow Red_1 + Ox_2 \qquad\qquad k_{12}, \Delta G_{12}^{\ddagger} \qquad\qquad (8.22)$$

The barrier to the self-exchange reaction is considered to contribute to the cross-reaction as follows:

$$\Delta G_{12}^{\ddagger} = 0.5 \, \Delta G_{11}^{\ddagger} + 0.05 \, \Delta G_{22}^{\ddagger} + 0.05 \, \Delta G_{12}^{0} \qquad\qquad (8.23)$$

The $\Delta G_{12}^{\ddagger}$ is lower than the average of the self-exchange free energies because of the favorable free energy change, ΔG_{12}^{0}.[7] For a series of closely related reactions (a common oxidant or reductant) a plot of $\Delta G_{12}^{\ddagger}$ versus ΔG_{12}^{0} should be linear with a slope of 0.5. This has been demonstrated in oxidation of a series of Fe phenanthroline complexes by Ce(IV), as shown in Figure 8.1.[23] The expression for the rate constant is *(See Cannon p 206)*

$$k_{12} = (k_{11}k_{22}K_{12}f)^{\frac{1}{2}} \qquad\qquad (8.24)$$

where f is defined as

$$\log f = \frac{(\log K_{12})^2}{4 \log(k_{11}k_{22}/Z^2)]} \qquad\qquad (8.25)$$

which becomes significant only if K_{12}, the cross-reaction equilibrium constant, is large. Equation 8.24 is more usefully represented as

$$\log k_{12} = 0.5(\log k_{11} + \log k_{22} + \frac{\Delta E^0}{0.059} + \log f) \qquad\qquad (8.26)$$

at 25°C. Table 8.3 shows the utility of this approach to modeling outer-sphere electron-transfer reactions.[7] It should be emphasized that calculation of rate con-

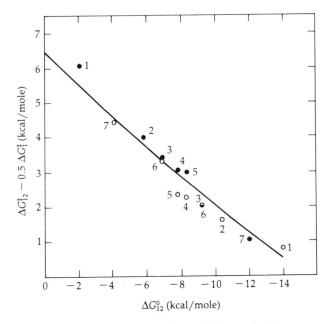

1. tris(3,4,7,8-tetramethyl-1,10-phenanthroline)
2. tris(5,6-dimethyl-1,10-phenanthroline)
3. tris(5-methyl-1, 10-phenanthroline)
4. tris(1,10-phenanthroline)
5. tris(5-phenyl-1, 10-phenanthroline)
6. tris(5-chloro-1, 10-phenanthroline)
7. tris(5-nitro-1,10-phenanthroline)

Figure 8.1. Relation between $\Delta G^{\ddagger}_{12}-0.5\Delta G^{\ddagger}_{1}$ and the standard free energy change of the oxidation-reduction reactions at 25.0°C: open circles, $Fe(phen)^{3+}_{3}$-Ce(IV) reactions in 0.50 F H_2SO_4; closed circles, Fe^{2+}-$Fe(phen)^{3+}_{3}$ reactions in 0.50 F $HClO_4$.

Table 8.3. COMPARISON OF CALCULATED AND OBSERVED RATE CONSTANTS FOR OUTER-SPHERE CROSS-REACTIONS[7]

Reaction	Observed	Calculated
$IrCl^{2-}_{6} + W(CN)^{4-}_{8}$	6.1×10^7	6.1×10^7
$IrCl^{2-}_{6} + Fe(CN)^{4-}_{6}$	3.8×10^5	7×10^5
$Mo(CN)^{3-}_{8} + W(CN)^{4-}_{8}$	5.0×10^6	4.8×10^6
$Fe(CN)^{4-}_{6} + MnO_4^-$	1.3×10^4	5×10^3
$V(H_2O)^{2+}_{6} + Ru(NH_3)^{3+}_{6}$	1.5×10^3	4.2×10^3
$Ru(en)^{2+}_{3} + Fe(H_2O)^{3+}_{6}$	8.4×10^4	4.2×10^5
$Fe(H_2O)^{2+}_{6} + Mn(H_2O)^{3+}_{6}$	1.5×10^4	3×10^4

stants is a difficult task. The success for outer-sphere electron-transfer reactions further emphasizes the special nature of these reactions.

The validity of the Marcus treatment has been demonstrated in so many cases that deviations from the expectation are usually indicative of a mechanistic change.[24] Reaction of $Co(H_2O)_6^{3+}$ with a number of reductants occurs by an outer-sphere mechanism. The parameters from these reactions can be utilized to calculate the $Co(H_2O)_6^{3+,2+}$ self-exchange rates, which are 12 orders of magnitude smaller than the experimental value. This has been used to suggest an inner-sphere mechanism through a water-bridged pathway for the self-exchange.[24]

It has been shown that differences in rates of outer-sphere redox processes (as shown in Table 8.4) do not depend on the free energy gain for reactions where the reorganization energy is small compared with the exothermic contribution.[25] The Marcus theory accommodated the free energy gain completely when quadratic terms were included.[25]

Both charge transfer and intervalence transitions can be used to provide information on thermal electron transfer.[26-28] The relationship between electron transfer and intervalence transitions is shown graphically in Figure 8.3.[26] The intervalence transfer leads to an excited vibrational state through a photochemical reaction. As a long-range, directed electron-transfer process between separated redox sites, intervalence transfer may provide an experimental basis for relating optical and thermal electron transfer. Most studies have centered on Ru^{2+}, Ru^{3+} complexes bridged by aromatic N donor ligands.[26]

$$[(bipy)_2ClRu^{II}(pyz)Ru^{III}Cl(bpy)_2]^{3+} \xrightarrow{h\nu} \qquad (8.27)$$

$$pyz = pyrazine \qquad [(bipy)_2ClRu^{III}(pyz)Ru^{II}Cl(bpy)_2]^{3+}$$

The theories of Marcus and Hush relate the energy of the intervalence transfer to the energy of the electron transfer. Outer-sphere charge transfer transitions have been investigated for a series of mixed-metal ion pairs and used to compare with the enthalpy change associated with the electron transfer. Some data are shown in Table 8.5.[27] Plots of the energy of the optical transition versus the enthalpy change for the charge transfer

$$[M^{II}(CN)_6, Ru^{III}(NH_3)_5L]^- \longrightarrow [M^{III}(CN)_6, Ru^{II}(NH_3)_5L]^- \qquad (8.28)$$

Table 8.4. KINETIC DATA FOR THE REACTION OF BINUCLEAR Co (SUBSTITUTED CARBOXYLATO) WITH SOME REDUCTANTS (STRUCTURE IN FIG. 8.2)[25]

Reductant	CH_3	CH_2F	CHF_2	CF_3	E^0
$Ru(NH_3)_6^{2+}$	0.035	0.12	0.21	0.36	0.10
$Cr(H_2O)_6^{2+}$	1.46×10^{-3}	2.79×10^{-3}	3.97×10^{-3}	6.3×10^{-3}	-0.45
Zn^+ (aq)	5.5×10^8	7.6×10^8	1.0×10^9	1.2×10^7	-2.0

Column header over the four R values: R

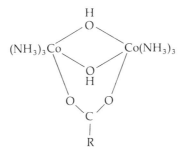

Figure 8.2. Structure of the dinuclear carbonato complexes that are oxidants in the reaction described in Table 8.4.

show linear relationships (Fig. 8.4), indicating the value of charge-transfer transitions to understanding outer-sphere electron-transfer mechanisms. It has been suggested that outer-sphere contact between $M(CN)_6^{4-}$ and an amine face of the $Ru(NH_3)_5L^{3+}$ occurs.[27]

One feature allows the definite assignment of an electron transfer as outer sphere: If the rate of electron transfer is more rapid than substitution at the metal centers, the mechanism can be assigned as outer sphere. This is because inner-sphere electron transfer requires loss of a ligand to open a coordination site, whereas outer sphere needs no such change in coordination. Thus, a reactant such as $Co(NH_3)_6^{3+}$ or $Ru(bipy)_3^{3+}$ that is inert to substitution and does not contain a possible bridging ligand will undergo outer-sphere electron transfer. The examples shown in Table 8.3 further illustrate this feature. Although Cl^- or CN^- can function as a bridging ligand, neither $IrCl_6^{2-}$ with an Ir^{4+} center nor $W(CN)_8^{4-}$ undergoes substitution on the same time scale as the electron transfer. Similar considerations apply to the other examples.

Much of the interest in outer-sphere electron transfer has been in the area of bioinorganic reactions. Cross-reactions between Ni(II) peptides and Ni(III) peptides

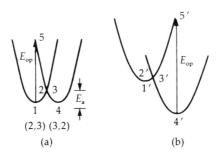

Figure 8.3. Reaction coordinate diagrams for electron transfer showing the optical transitions for (a) $\Delta G^\circ = 0$ and (b) ΔG° negative.

Table 8.5. SPECTRAL AND ENTHALPY DATA FOR OUTER-SPHERE CHARGE TRANSFER IN $[M^{II}(CN)_6, Ru^{III}(NH_3)_5L]^{-}$ [28]

M	L	λ	ΔH
Fe	Py	910	11.0
Fe	4-Mepy	898	11.1
Fe	4-Brpy	932	10.7
Ru	Py	643	10.7
Ru	4-Mepy	627	15.9
Ru	4-Brpy	653	15.3
Os	Py	658	15.3
Os	4-Mepy	626	16.0
Os	4-Brpy	670	14.9

have been used to evaluate the self-exchange rate constants for a number of different complexes.[29] The self-exchange reaction rates varied with peptide structure, but they could be broadly classified according to the protonation, triply deprotonated peptides reacting at 1.2×10^5 $M^{-1}s^{-1}$ and doubly deprotonated peptides reacting at 1.3×10^4 $M^{-1}s^{-1}$. It was suggested that the approach of the complexes was close, probably from an axial direction.[29] Reactions between blue Cu proteins—stellacyanin, plastocyanin, and azurin—and inorganic redox reagents—*bis*(dipicolinate) complexes of Co(III) and Fe(II)—have been examined.[30] The rate data indicate

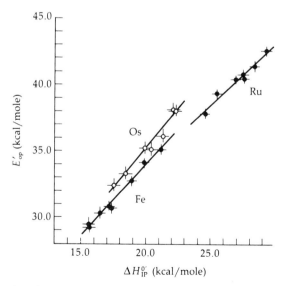

Figure 8.4. A plot of the optical transition versus the enthalpy change for the charge transfer transition (reaction 8.28).

well-behaved reactions (Figure 8.5). The electron-transfer distance of 2–3 Å suggests that the hydrophobic ligand penetrates the hydrophobic region of the Cu–histidine redox unit.

8.3. Inner-Sphere Mechanisms

Inner-sphere mechanisms are characterized by formation of a binuclear transition state/intermediate during the electron-transfer reaction.[1–4,31] A very important facet of the inner sphere process is the bridging ligand that forms part of the coordination spheres of both the oxidizing and the reducing metal ions.[10] The bridging ligand must function as a Lewis base toward both metal centers—it must have two pairs of electrons that can be donated to different metal centers. The electron transfer can be considered as several individual reactions.

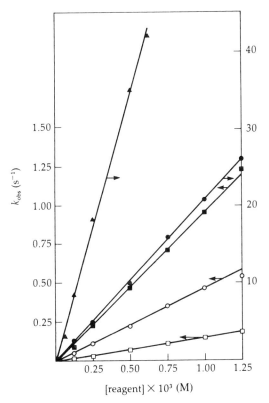

Figure 8.5. Dependences of the observed rate constants on reagent concentration at 25°C. [pH 7.0 (phosphate); μ = 0.2 M]. (■) Fe(dipic)$_2^{2-}$-azurin(II); (□) Co(dipic)$_2^-$-azurin(I); (●) Fe(dipic)$_2^{2-}$-plastocyanin(II); (○) Co(dipic)$_2^-$-plastocyanin(I); (▲) Fe(dipic)$_2^{2-}$-stellacyanin(II).

$$M^{II} + L - N^{III} \longrightarrow M^{II}, L - N^{III} \tag{8.29}$$

$$M^{II}, L - N^{III} \longrightarrow M^{II} - L - N^{III} \tag{8.30}$$

$$M^{II} - L - N^{III} \longrightarrow M^{III} - L - N^{II} \tag{8.31}$$

$$M^{III} - L - N^{II} \longrightarrow M^{III} - L + N^{II} \tag{8.32}$$

The first reaction is the diffusion-controlled formation of a collision complex. The second reaction is the formation of a complex, termed the *precursor complex,* in which the ligand bridges the two metal centers, but in which the electron has not been transferred. Reaction 8.31 is the electron transfer leading to the successor complex. The last reaction is dissociation of the successor complex to products. Any of these reactions may be rate determining as shown by the reaction profiles in Figure 8.6. If Reaction 8.30 or 8.32 is rate determining, then the reaction is substitutionally controlled; however, if the electron transfer is rate determining, as is most commonly observed, then the middle profile is observed, and Reaction 8.31

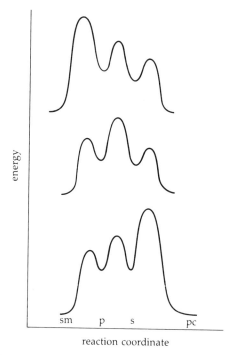

Figure 8.6. Possible energy profiles for inner sphere electron transfer. The relative minima correspond to the starting complexes (sm), the precursor (p), the successor complex (s), and the product complexes (pc). In the top profile the precursor formation is rate determining, in the middle one the electron transfer is rate determining, and in the bottom profile successor decomposition is rate determining.

is rate determining. The scheme shown in Reactions 8.29–8.32 represents an atom transfer in which the bridging ligand is transferred to a different metal center. Atom transfer is not a requirement for an inner-sphere mechanism, but it is a consequence of the kinetic labilities of the two metal centers in the successor complex. The bridging ligand remains with the more inert metal center.

The same theoretical considerations apply for inner-sphere reactions as they do for outer-sphere—bond lengths must adjust—but the same theoretical treatment cannot be accomplished because the reaction also involves bond dissociation. The most important consideration for inner-sphere reactions is the bridging ligand.

It is instructive to consider an example that illustrates an inner sphere reaction— $Cr(H_2O)_6^{2+}$ with $Co(NH_3)_5Cl^{2+}$—for which the substitutional reactivities are known. The Co(III) complex is substitutionally inert in acidic solution, and the Cr(II) complex exchanges extremely rapidly. The substitutional labilities at the two centers are reversed in the products of the electron transfer, with the Co(II) center being very labile and substitution on Cr(III) quite slow.

$$Cr(H_2O)_6^{2+} + Co(NH_3)_5Cl^{2+} \xrightarrow{H^+} \qquad (8.33)$$
$$Cr(H_2O)_5Cl^{2+} + Co(H_2O)_6^{2+} + 5NH_4^+$$

Since the Cr(II) complex is quite labile, the precursor complex can be easily formed. The electron transfer also occurs readily. In the successor complex the Cr(III) center is the most inert, and the Cl–Cr bond remains while the labile Co(II) loses Cl. The absence of $^{36}Cl^-$ incorporation when the reaction is run in the presence of free $^{36}Cl^-$ shows that the Cl is transferred directly.[32] (Use of $^{36}Cl^-$ free in solution initially showed no incorporation into the chromium product, indicating an inner-sphere mechanism.) In this case, because $Cr(H_2O)_5Cl^{2+}$ is inert, isotope labeling allows mechanistic differentiation.

8.3.1. Nature of the Bridging Ligand

The bridging ligand can be very important in inner-sphere reactions. Some data for reduction of $Co(NH_3)_5L^{3+}$ by Cr^{2+} are shown in Table 8.6.[3] The order observed for the halides represents the ability of the ligand to donate to two metal centers. Water has two electron pairs that could be used to bridge metal centers, but an inverse dependence on the $[H^+]$ indicates that OH^- is the actual bridging ligand. The 10^5 increase in rate for OH^- over H_2O adds to the plausibility of this interpretation.[9,10]

When a halide atom functions as the bridging ligand, it is obvious that a single atom is involved. When a ligand such as SCN^- functions as a bridging ligand, there are two possibilities.[10] If the second metal bonds to S, then the reaction is referred to as adjacent attack; if the second metal bonds to N, then it is referred to as remote attack. Attack at both possible sites has been observed for Cr^{2+} as the reductant. Adjacent attack leads to the unstable S-bonded isomer, which rearranges to the more stable N-bonded form.[33]

Table 8.6. RATE CONSTANTS
FOR REDUCTION OF $Co(NH_3)_5L^{3+}$
BY $Cr(H_2O)_6^{2+}$ [3]

L	k
NH_3	8.0×10^{-5}
F^-	2.5×10^5
Cl^-	6.0×10^5
Br^-	1.4×10^6
I^-	3.0×10^6
N_3^-	3.0×10^5
OH^-	1.5×10^6
NCS^-	19
SCN^-	1.9×10^5
H_2O	~ 0.1

$$Co(NH_3)_5SCN^{2+} + Cr(H_2O)_6^{2+} \longrightarrow [(NH_3)_5Co\underset{\underset{N}{|}}{\overset{\overset{C}{|}}{S}}Cr(H_2O)_5^{4+}]$$

$$\text{(8.34)}$$

$$Cr(H_2O)_5(NCS)^{2+} \longleftarrow Cr(H_2O)_5(SCN)^{2+} + Co^{2+} + 5NH_3$$

For remote attack the stable N-bonded isomer forms directly.

$$Co(NH_3)_5SCN^{2+} + Cr(H_2O)_6^{2+} \longrightarrow [(NH_3)_5CoSCNCr(H_2O)_5^{4+}]$$

$$\text{(8.35)}$$

$$Cr(H_2O)_5NCS^{2+} + Co^{2+} + 5NH_3$$

In 1 M H^+, 30% of the electron transfer occurred by adjacent attack.[31,33] It is difficult to prove that remote attack has occurred in some systems. The reaction of

$$\left[(NH_3)_5Co-N \underset{}{\bigcirc}-C \overset{\overset{O}{\diagup}}{\underset{NH_2}{\diagdown}} \right]^{3+}$$

with Cr^{2+} in acid solution leads to the product

$$\left[HN \underset{}{\bigcirc} - C \overset{OCr}{\underset{NH_2}{\diagup}} \right]^{4+}$$

This product cannot arise by interaction of $Cr(H_2O)_6^{3+}$ with the free ligand since the product is unstable with respect to the free amide and $Cr(H_2O)_6^{3+}$, although the reaction is slow enough that it is possible to characterize the intermediate.[9] In a few cases it has been possible to characterize the binuclear species formed in electron-transfer reactions.[31,33–35] For the reaction of 4,4'-bipyridinepentaaminecobalt(III) with aquopentacyanoferrate(II), the rate of formation and dissociation of the precursor were measured.[33]

$$Co(NH_3)_5(4,4'bipy)^{3+} + Fe(CN)_5OH_2^{3-} \rightleftharpoons \qquad (8.36)$$

$$(NH_3)_5Co^{III} - N \underset{}{\bigcirc} - \underset{}{\bigcirc} N - Fe^{II}(CN)_5$$

The rate of dissociation and electron transfer in the precursor were comparable. Electron transfer in binuclear $(NC)_5Fe(II)$-CN-Co(III)(chelate) complexes has been observed by picosecond absorption spectrosocpy. The following scheme has been suggested based on these data.[34]

$$Fe^{II} - CN - Co^{III*} \xrightarrow{h\nu} Fe^{III} - CN - Co^{II} \qquad (<10\ ps) \qquad (8.37)$$

$$Fe^{III} - CN - Co^{II**} \longrightarrow Fe^{III} - CN - Co^{II} \qquad (\sim75\ ps) \qquad (8.38)$$

$$Fe^{III} - CN - Co^{II} \longrightarrow Fe^{II} - CN - Co^{III*} \qquad (\sim95\ ps) \qquad (8.39)$$

$$Fe^{II} - CN - Co^{III*} \longrightarrow Fe^{II} - CN - Co^{III} \qquad (>500\ ps) \qquad (8.40)$$

The first reaction is excitation into the charge transfer band, which leads to Co(II) in a $t_{2g}^6 e_g^1$-excited state. The next reaction is the rearrangement to $t_{2g}^5 e_g^2$. The third reaction is the spin-allowed back electron transfer leading to Co(III) in an excited state $(t_{2g}^5 e_g^1)$ that slowly relaxes to the singlet groundstate.[34]

After the precursor complex forms, there are two possible roles for the bridging ligand in the actual electron transfer.[6] The electron could be transferred to the ligand forming a radical that, in a subsequent step, would transfer the electron to the second metal. The electron might also spend no time on the bridging ligand, which would function merely as a mediator for the electron transfer. Pulse radiolysis of p-nitrobenzoato-pentaamineCo(III) led to formation of a metastable intermediate ascribed to the Co(III) complex of the radical ion, which decayed to the Co(II) complex in a first-order process.[10]

$$(8.41)$$

Because detection of radical ligand species is not usually possible, a rate test has been developed to distinguish the role of the bridging ligand by comparing reductions of Co(III) and Cr(III) with the same reductant and bridging ligand.[10] For the radical mechanism electron transfer takes place from the reductant to the ligand; the nature of the oxidant will therefore have a small effect on the rate. When the ligand is merely mediating the electron transfer, the nature of the oxidant should be much more important. Data in Table 8.7 demonstrate this procedure for distinguishing the role of the bridging ligand in inner-sphere electron transfer.[10] For the ligands that would not be expected to form a radical, the ratio $k(Co)/k(Cr)$ is around 10^6, which indicates an important effect on the nature of the oxidant. Fumarate and maleate have a much smaller dependence on the nature of the oxidant, and it has been concluded that these reductions occur through a ligand-centered radical.[10]

Table 8.7. RATES OF REACTION OF Cr(II) WITH Co(III) AND Cr(III) COMPLEXES, $Co(NH_3)_5L^{2+}$ OR $Cr(H_2O)_5L^{2+}$ [a]

L	$k(Co)/k(Cr)$
NCS$^-$	1.4×10^5
F$^-$	3.4×10^7
OH$^-$	2.3×10^6
$\begin{matrix}-O\\ \diagdown\\ C-CH=CHCO_2H(trans)\\ O\diagup\end{matrix}$	0.4
$\begin{matrix}-O\\ \diagdown\\ C-CH=CHCO_2H(cis)\\ O\diagup\end{matrix}$	50

[a] L = bridging ligand [10]

The ability of a ligand to mediate an electron transfer has been ascribed to a matching of the symmetry of metal and ligand orbitals.[10] When the symmetry of the orbitals of the metal ions that donate and accept the electron are the same, ligands with orbitals of matching symmetry may provide a lower energy pathway for electron transfer. Thus, if the reductant donates an e_g electron to an e_g orbital, then a ligand such as chloride (a σ carrier) would be a better bridging ligand than would azide or acetate. Electron transfer involving a t_{2g} orbital would be better accommodated by an azide or an acetate, presumably because of better overlap of the t_{2g} orbital with the π system of azide or acetate. Existing data are in agreement with these suggestions, but they are insufficient to test the concepts adequately.

It is difficult to prepare complexes in which the effect of the nature of the bridging ligand on the rate of electron transfer could be investigated, usually because one of these metal centers is labile and the rate of formation of the precursor is difficult to separate from the rate of electron transfer. An ingenious approach to circumvent this problem involved the preparation of binuclear Ru(III)–Co(III) complexes, which were then reacted with a rapid one-electron reducing agent.[36] This produced the binuclear precursor complex,

$$Co(III)-L-Ru(II)$$

which could be used to measure the rate of electron transfer between the Ru(II) and the Co(III) centers. The results for bridging pyridine carboxylate ligands are shown in Table 8.8.[10] These results show an inhibiting effect of the CH_2 group, and the importance of conjugation in the bridging ligand.

Table 8.8. RATE CONSTANTS FOR INTRAMOLECULAR ELECTRON TRANSFER IN $(NH_3)_5Co^{III}-L-Ru^{II}(NH_3)_4(H_2O)^{4+}$ WHERE L = PYRIDINECARBOXYLATE[10]

L	k
	1×10^2
	1.6×10^{-2}
	1.6×10^{-3}

8.3.2. Other Considerations for Inner-Sphere Reactions

The role of the bridging ligand is the most important feature of inner sphere reactions, and it has dominated research on reactivity toward electron transfer by an inner-sphere mechanism. The *cis* ligand has a very small effect if the data in Table 8.9 are indicative of inner sphere processes in general.[7] The electronic requirements of inner-sphere electron transfer for macrocyclic ligands have been discussed in terms of a three-center transition state bonding scheme.[37]

Discussions of inner-sphere reactions are complicated by the fact that any of the three components to an inner sphere reaction (Reactions 8.30–8.32) may be rate determining. The possibilities in the free energy profiles are shown in Figure 8.6. In the first case the rate-determining step is formation of the precursor complex (Eq. 8.30). In the second case the rate-determining step is the electron transfer. In the third case decomposition of the successor complex is rate determining. Each type has been observed, but the second is the most common, involving the rearrangement of the coordination sphere as discussed earlier.

8.4. Long-Range Electron Transfer

One of the most active research areas in electron transfer involves long-range electron transfer, transfers in which the distance between the oxidant and the reductant may be up to 25 Å.[38] The primary impetus for the research lies in a wish to understand electron transfer in biological systems where electron transfer often takes place over considerable distances.[38] The capability for rapid electron transfer between organic donor and acceptor sites separated by rigid saturated hydrocarbon groups has been investigated.[39] The types of complexes are shown in Figure 8.7. These complexes were treated with solvated electrons, and the approach to equilibrium between the two ends was investigated. As shown in Figure 8.7 the rates for electron transfer from one end to the other range from half-lives of <0.5 ns to 1 μs.[40] This fast electron transfer is surprising considering the long distances (~ 15 Å) and insulating nature of the organic backbone, but it certainly makes the electron transfer between metal centers in biological molecules more understandable. The

Table 8.9. RATES OF ELECTRON
TRANSFER BETWEEN Cr^{2+} AND
cis-$Co(en)_2LCl^{2+}$ [7]

L	$k(M^{-1}s^{-1})$
NH_3	2.3×10^5
H_2O	4.2×10^5
Py	9.4×10^5
Cl^-	7.7×10^5
F^-	9×10^5

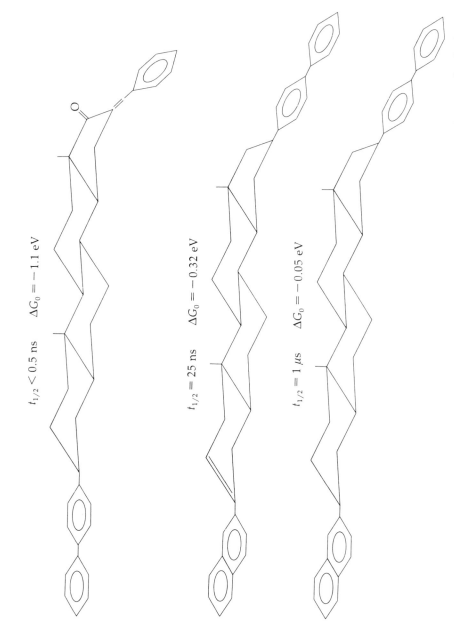

$t_{1/2} < 0.5$ ns $\quad \Delta G_0 = -1.1$ eV

$t_{1/2} = 25$ ns $\quad \Delta G_0 = -0.32$ eV

$t_{1/2} = 1$ μs $\quad \Delta G_0 = -0.05$ eV

Figure 8.7. Systems in which long-range electron transfer has been studied. The half-lives and free energies for intramolecular electron transfer are shown above each di-ended steroid.

distance from the redox site to the surface in several metalloproteins was estimated from reactions with Fe(EDTA)$^{2-}$.[40] The calculations indicated electron-transfer distances ranging from 3 to 10 Å, depending on the protein and inorganic redox reagent.[40] Electron transfer has been observed in a Zn(II), Fe(III) hybrid hemoglobin at 25 Å.[38] The electron transfer was initiated by flash photoexcitation forming Zn(III), which is a good reductant for Fe(III) in the long-range process. The distance is known from crystallography to be 25 Å in a rigid structure, and the rate constant was 60 ± 25 s^{-1}.[41] Similar experiments on pentaammineruthenium(III)(histidine-33)-ferricytochrome C have shown electron transfer between the metal centers 15 Å apart at a rate of 20 s^{-1}.[42]

There is considerable theoretical and experimental interest in these long-range electron transfers, but at this point a full explanation is not available.[38] The application to biological systems will certainly keep this an active research area.

8.5. Multi-electron Transfer

A number of electron-transfer reactions yield products that have changed by more than one electron.[43]

$$U^{IV}(aq) + Tl^{III}(aq) \longrightarrow U^{VI}(aq) + Tl^{I}(aq) \tag{8.42}$$

$$Cr^{II}(aq) + Tl^{III}(aq) \longrightarrow Cr^{IV}(aq) + Tl^{I}(aq) \tag{8.43}$$

$$Sn^{II}(aq) + V^{V}(aq) \longrightarrow Sn^{IV}(aq) + V^{III}(aq) \tag{8.44}$$

It is usually difficult to determine whether a two-electron transfer occurs simultaneously or in steps. The absence of large amounts of Cr(III)(aq) in Reaction 8.43 is evidence that two electrons may be transferred. A more recent report has suggested that a four-electron "simultaneous" transfer occurs from a chelated Cr(V) complex.[44] The reaction is shown in Figure 8.8. The suggested scheme involved formation of a hydroxylamine precursor followed by a four-electron transfer.

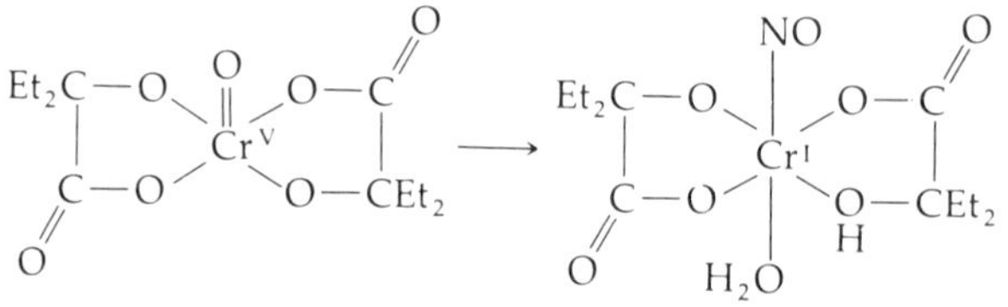

Figure 8.8. A reaction that has been suggested to proceed by a four-electron transfer.

8.6. Organometallic Electron-Transfer Reactions

Electron-transfer reactions of 18-electron organometallic complexes have received growing attention. The field of electron-transfer catalysis where an organometallic complex is activated by oxidation or reduction provides the impetus for such studies. Organometallic complexes lie on the border between the fields of inorganic and organic chemistry. The terminology for electron-transfer reactions of organometallic complexes demonstrates the influence of organic (single-electron transfer, SET, and nucleophilic reactions) and inorganic (outer- and inner-sphere electron transfer) nomenclature. For the metal carbonyl anions whose reactions dominate the electron-transfer reactions of 18-electron organometallic complexes, SET and outer-sphere terms are equivalent, and nucleophilic and inner-sphere reactions are frequently equivalent.

Electron transfer has frequently been observed in catalytic and synthetic reactions of organometallic complexes. Simple reduction

$$Fe(CO)_5 + 2Na(Hg) \longrightarrow Na_2Fe(CO)_4 + CO \tag{8.45}$$

$$Cp_2Fe_2(CO)_4 + Mg \longrightarrow Mg[CpFe(CO)_2]_2 \tag{8.46}$$

and oxidation

$$2Ni(CO)_4 + CS_2 \longrightarrow 2NiS + C + 8CO \tag{8.47}$$

$$Fe(CO)_5 + 2CuCl_2 \longrightarrow Cu_2Cl_2 + FeCl_2 + 5CO \tag{8.48}$$

reactions are commonly observed. Disproportionation reactions provide another example of electron transfer.[45]

$$Co_2(CO)_8 + 2L \longrightarrow [Co_2(CO)_3L_2][Co(CO)_4] + CO \tag{8.49}$$
$$L = phosphine\ ligand$$

$$3V(CO)_6 + 6L \longrightarrow [VL_6][V(CO)_6]_2 + 6CO \tag{8.50}$$
$$L = oxygen\ or\ nitrogen\ base$$

Reactions between cationic and anionic carbonyls to form M–M bonds also involve a formal electron transfer.[46,47]

$$[Mn(CO)_6]^+ + [Mn(CO)_5]^- \longrightarrow Mn_2(CO)_{10} + CO \tag{8.51}$$

$$[Co(CO)_3(PPh_3)_2][Co(CO)_4] \xrightarrow{\Delta} Co_2(CO)_6(PPh_3)_2 + CO \tag{8.52}$$

The primary method of synthesis of heterobimetallic complexes may involve electron transfer between metal halides and metal carbonyl anions.[48–51]

$$CpFe(CO)_2I + NaMn(CO)_5 \longrightarrow Cp(CO)_2FeMn(CO)_5 + NaI \tag{8.53}$$

$$Re(CO)_5Cl + NaMn(CO)_5 \longrightarrow MnRe(CO)_{10} + NaCl \tag{8.54}$$

$$Mn(CO)_5Cl + NaCo(CO)_4 \longrightarrow MnCo(CO)_9 + NaCl \qquad (8.55)$$

$$Pt(Py)_2Cl_2 + 2NaCo(CO)_4 \longrightarrow Pt(Py)_2[Co(CO)_4]_2 + 2NaCl \qquad (8.56)$$

This type of reaction sometimes leads to formation of homobimetallic complexes. Indeed, a rather simple change from Reaction 8.53

$$2[CpFe(CO)_2]^- + 2Mn(CO)_5Br \longrightarrow Cp_2Fe_2(CO)_4 + Mn_2(CO)_{10} \quad (8.57)$$

gives a different product distribution.[52] An interesting example of electron transfer from $Mn(CO)_5^-$ to $Cr(\eta^6\text{-}C_6H_6)_2^+$ is as follows[53]:

$$2[Cr(\eta^6\text{-}C_6H_6)_2][Mn(CO)_5] \longrightarrow 2Cr(\eta^6\text{-}C_6H_6)_2 + Mn_2(CO)_{10} \qquad (8.58)$$

This reaction occurs at reasonable rates just above ambient temperatures.

8.6.1. Metal Carbonyl Anions

Quantitative treatment of electron-transfer reactions of metal carbonyl anions is limited by irreversibility of oxidation of the anions and reduction of the related dimers. Very rapid reactions of the resulting odd-electron complexes cause electrochemical irreversibility. These problems have been solved by kinetic corrections to the electrode potential,[54] very rapid electrochemical measurements,[55] and assumptions regarding the M–M bond strengths.[56] The resulting half-reaction potentials for some metal carbonyl anions are provided in Table 8.10. The other parameter that is important for reactions of the metal carbonyl anions is the nucleophilicity of the metal carbonyl anion, defined as the logarithm of the second-order rate constant for displacement of iodide from MeI:

$$M^- + MeI \longrightarrow MeM + I^- \qquad (8.59)$$

These values are also given in Table 8.10.[57]

Metal carbonyl anions react by outer-sphere mechanisms, where $M^- \rightarrow M\bullet + e^-$ is important, and by nucleophilic displacement mechanisms (inner sphere with atom transfer).

8.6.2. Outer-Sphere Reactions

Gas phase kinetics have been used to evaluate self-exchange between the metallocenes of Mg, Fe, Co, and Ru.[58,59] These self-exchange reactions occur by outer-sphere mechanisms. The much lower efficiency for manganocene indicates a substantial barrier from internal rearrangement, as expected from a d^5 high-spin complex.

The kinetics of oxidation of $Cp_2Fe_2(CO)_4$ by $Ru(bipy)_2Cl_2^+$ and by $[CpFe(CO)]_4^+$ have been reported.[60]

$$Cp_2Fe_2(CO)_4 + 2Ru(bipy)_2Cl_2^+ \xrightarrow{\text{MeCN}} \qquad (8.60)$$

$$2[CpFe(CO)_2(NCMe_3)]^+ + 2Ru(bipy)_2Cl_2$$

Table 8.10. HALF-REACTION REDUCTION POTENTIALS FOR ONE-ELECTRON AND TWO-ELECTRON TRANSITIONS OF METAL CARBONYL SPECIES[54-56]

Metal Carbonyl, M_2	$M_2 + 2e^- \rightarrow$ $2M^-$ (V)	$M\bullet + e^- \rightarrow$ M^- (V)	Nucleophilicity $(\ln k_2)$ $M^{-2} + MeI \rightarrow$ $MeM + I^-$
$Co_2(CO)_8$	−0.123	−0.15	−4.7
$Cp_2Cr_2(CO)_6$	−0.639	−0.70	−2.9
$Cp_2Mo_2(CO)_6$	−0.336 (−0.56[a])	−0.79	−0.94
$Cp_2W_2(CO)_6$	−0.330	~−0.8	−0.67
$Mn_2(CO)_{10}$	−0.400 (−0.69[a])	−0.97	0.41
$Re_2(CO)_{10}$	−0.523	−1.2	4.3
$Cp_2Fe_2(CO)_4$	−1.186 (−1.44[a])	−1.7	Large

[a]Values from Pugh and Meyer,[55] which are probably more accurate and are systematically 0.25 V more negative.

$$Cp_2Fe_2(CO)_4 + 2[CpFe(CO)]_4^+ \xrightarrow{\text{MeCN}} \qquad (8.61)$$

$$2[CpFe(CO)_2(NCMe)]^+ + 2[CpFe(CO)]_4$$

As usual for oxidation-reduction reactions, the rate is first order in $Cp_2Fe_2(CO)_4$ and the oxidant. The rates are rapid ($k = 5 \times 10^4$ M^{-1}s^{-1}), requiring stopped-flow techniques. Very low activation enthalpies (2–6 kcal mol^{-1}) and negative entropies of activation (−25 cal K^{-1} mol^{-1}) are found.[60] The initial step is suggested as transfer of an electron from the Fe–Fe bond to the oxidant. For both oxidants the mechanism involved outer-sphere electron transfer.

To ascertain the characteristics expected for outer-sphere electron-transfer reactions of the metal carbonyl anions, reactions have been examined with Co(o-phen)$_3^{3+}$ and with 3-acetoxy-N-methylpyridinium.[61]

$$2M^- + 2[Co(o\text{-phen})_3]^{3+} \longrightarrow M_2 + 2[Co(o\text{-phen})_3]^{2+} \qquad (8.62)$$

These reactions are first order in [M$^-$] and [Co(o-phen)$_3^{3+}$] or [py$^+$], as expected for electron transfer. The iron complex, Fe(o-phen)$_3^{3+}$, reacts too rapidly for kinetic examination. For the metal carbonyl anions reacting with Co(o-phen)$_3^{3+}$ the order is as expected from measured potentials [i.e., CpFe(CO)$_2^-$ > Re(CO)$_5^-$ > Mn(CO)$_5^-$ > CpMo(CO)$_3^-$], but the rates are within a factor of 3.[61] The 3-acetoxy-N-methylpyridinium shows the same order with the rates spanning a factor of 30. For the same anions, reactions with MeI (used to define the nucleophilicity) span rates of 10^6. Thus, the outer-sphere reaction is much less dependent on the nature of M$^-$. The effect of the ligand in Mn(CO)$_4$L$^-$ (L = CO, PPh$_3$, PEt$_3$) on electron transfer is also reversed for reaction with Co(o-phen)$_3^{3+}$ (rate decreases by a factor of 6, L = CO > PEt$_3$ > PPh$_3$) in comparison to reaction with MeI (rate increases by factor of 50, L = PEt$_3$ > PPh$_3$ > CO).

Reactions of mononuclear metal carbonyl anions, M^- ($M^- = Co(CO)_4^-$, $CpFe(CO)_2^-$, $Re(CO)_5^-$, $Mn(CO)_4L^-$ [L = CO, PPh_3, PBu_3, $P(OPh)_3$], and $CpM''(CO)_3^-$ [M'' = Cr, Mo, W]), with metal carbonyl dimers, $M_2' = Co_2(CO)_8$, $Cp_2Fe_2(CO)_4$, $Re_2(CO)_{10}$, $Mn_2(CO)_{10}$, $Cp_2M_2''(CO)_6$ [M'' = Cr, Mo, W], and $Cp_2Ru_2(CO)_4$), also show outer-sphere reaction mechanisms[61]:

$$2M^- + M_2' \longrightarrow M_2 + 2M'^- \qquad (8.63)$$

The direction of these reactions can be predicted from the two-electron reduction potentials for the dimers ($M_2 + 2e^- \rightarrow 2M^-$) (see Table 8.10). The estimated potentials correctly predict product formation when Equation 8.63 occurs. The anion of any dimer with a more negative two-electron potential will react by Reaction 8.63. Thus, for instance, $CpFe(CO)_2^-$ reacts with all dimers by Equation 8.63. The kinetics show the rate to have a first-order dependence on $[M^-]$ and $[M_2']$. For a given M_2', the nature of M^- has a relatively small effect on the rate of the reaction, which is consistent with an outer-sphere process.

The reaction of $Fe(CO)_4^{2-}$ with $Mn_2(CO)_{10}$ emphasizes that $Fe(CO)_4^{2-}$, which has been termed a *super nucleophile*, may also react by an outer-sphere electron transfer.[62]

$$2[Fe(CO)_4]^{2-} + Mn_2(CO)_{10} \longrightarrow [Fe_2(CO)_8]^{2-} + 2[Mn(CO)_5]^- \qquad (8.64)$$

The $Fe_2(CO)_8^{2-}$ could form from a reaction of $Fe(CO)_4^{2-}$ with $Fe(CO)_5$ or by dimerization of $[Fe(CO)_4]^{\bullet-}$. However, reaction of $Fe(CO)_4^{2-}$ with $Fe(CO)_5$ occurs two orders of magnitude more slowly, implicating $[Fe(CO)_4]^{\bullet-}$ in the reaction. Slow addition of $Fe(CO)_4^{2-}$ prohibits the generation of sufficient quantities of $[Fe(CO)_4]^{\bullet-}$ to dimerize, and $MnFe(CO)_9^-$ becomes a significant product from coupling of $[Fe(CO)_4]^{\bullet-}$ and $[Mn(CO)_5]^{\bullet}$.[59] Product and rate studies provide strong support for an outer-sphere electron transfer mechanism for reaction of $Fe(CO)_4^{2-}$ with $Mn_2(CO)_{10}$.

8.6.3. Inner-Sphere Reactions

A more nucleophilic metal carbonyl anion may abstract an electrophilic group from a less nucleophilic metal carbonyl anion[63,64]:

$$M^- + M'X \longrightarrow M\text{-}X + M'^- \qquad (8.65)$$

M^- is a more nucleophilic metal carbonyl anion than M'^-
$X = H^+, Me^+, Br^+, CO^{2+}$

Proton exchange reactions have been measured between the Group 6 anions.[63]

$$[CpW(CO)_3]^- + HMoCp(CO)_3 \longrightarrow \qquad (8.66)$$
$$HWCp(CO)_3 + [CpMo(CO)_3]^-$$

By measuring the self-exchange rates and using the pK_a values for the metal carbonyl anions, these proton transfer reactions were shown to follow relative Marcus-type relationships. Similar transfer reactions of methyl groups

$$[CpFe(CO)_2]^- + MeMoCp(CO)_3 \longrightarrow \tag{8.67}$$
$$MeFeCp(CO)_2 + [CpMo(CO)_3]^-$$

may follow Marcus-type relationships.[64] The dependence on the organic group transferred, $H^+ > CH_2Ph^+ > Me^+ > Et^+ > Ph^+$, is consistent with nucleophilic attack. The absence of metal carbonyl dimers indicates that outer-sphere processes are not operative for these reactions.

Very interesting examples of inner-sphere reactions are the CO^{2+} transfers observed between metal carbonyl anions and metal carbonyl cations[62,65]:

$$[Re(CO)_5]^- + [Mn(CO)_6]^+ \longrightarrow [Re(CO)_6]^+ + [Mn(CO)_5]^- \tag{8.68}$$

$$[Fe(CO)_4]^{2-} + [Re(CO)_6]^+ \longrightarrow Fe(CO)_5 + [Re(CO)_5]^- \tag{8.69}$$

$$[Mn(CO)_3(PPh_3)_2]^- + [Mn(CO)_6]^+ \longrightarrow \tag{8.70}$$
$$[Mn(CO)_4(PPh_3)_2]^+ + [Mn(CO)_5]^-$$

Isotopic-labeling studies show that one CO is transferred from the cation to the anion.[65] In addition no ^{13}CO appears in the product cation or anion when the reaction is accomplished under a ^{13}CO atmosphere. These observations are not consistent with a single-electron transfer to 17-electron complexes. Reactions 8.68–8.70 provide more examples of single step, two-electron processes. These reactions may be considered as two-electron transfer from the reactant anion to the reactant cation and back transfer of a CO or as CO^{2+} transfer reactions between two metal carbonyl anions. Consideration as CO^{2+} transfer facilitates comparison to Reactions 8.66 and 8.67.

Each of these inner-sphere reactions may be described as nucleophilic attack of the metal carbonyl anion on the group to be transferred:

$$M^- + X\text{-}M' \longrightarrow [M\text{-}X\text{-}M']^- \tag{8.71}$$

The more nucleophilic metal carbonyl anion then retains the electrophilic group X. The dependence of reaction rate on the metal carbonyl anion is consistent with the nucleophilicity, as shown in Table 8.10.

8.6.4. Seventeen-Electron Complexes

Seventeen-electron complexes are inherently unstable. However, the huge increase in reactivity for 17-electron complexes offers possibilities for activation of 18-electron complexes through a process called *electron-transfer catalysis*. Seventeen-electron complexes are formed by oxidation of 18-electron complexes, by reduction of 18-electron complexes accompanied by ligand dissociation, and by homolytic cleavage of the M–M bond of dimers. Only two types of reactions for 17-electron complexes have been studied: ligand substitution and electron transfer (including atom transfer).

A competing reaction for many 17-electron complexes (especially those generated by photochemical cleavage of metal dimers) is dimerization (see Chapter 4).

$$2[Mn(CO)_5]^\bullet \longrightarrow Mn_2(CO)_{10} \tag{8.72}$$

The dimerization reactions occur at rates approaching the diffusion-controlled limit. For example, dimerization of $[Mn(CO)_5]\bullet$ occurs with a rate constant of 9×10^8 $M^{-1}s^{-1}$, and dimerization of $[Re(CO)_5]\bullet$ occurs with a rate constant of 3×10^9 $M^{-1}s^{-1}$.[66]

Seventeen-electron complexes undergo facile oxidation, reduction, and atom-transfer reactions. The rapidity of these reactions has made study difficult, but the importance in radical-chain processes warrants discussion. The speed of atom-transfer reactions can be illustrated by the reactions of 17-electron metal carbonyl complexes with halide complexes.[66] A few reactions with CCl_4 are shown in Table 8.11.[66]

$$M\bullet + CCl_4 \longrightarrow M\text{-}Cl + [CCl_3]\bullet \tag{8.73}$$

Hydrogen-transfer reactions have also been reported.

Seventeen-electron complexes may be oxidized or reduced:

$$M\bullet + e^- \longrightarrow M^- \tag{8.74}$$

$$M\bullet \xrightarrow{S} M^+\text{-}S + e^- \tag{8.75}$$

S = solvent

$M\bullet$ = a 17-electron complex

Burke and Brown examined reactions of $[Re(CO)_4L]\bullet$, L = PMe_3, $P(O\text{-}i\text{-}Pr)_3$, with N-methylpyridinium salts.[67] These reactions occur with electron-transfer rate constants in the 10^6–10^9 $M^{-1}s^{-1}$ range. From Marcus-type analysis, an intrinsic barrier of 3 kcal mol^{-1} was obtained for the 17-electron Re complexes.[67] The facile oxidation and reduction of 17-electron complexes makes them quite susceptible to disproportionation reactions:

$$3V(CO)_6 \xrightarrow{\text{acetone}} [V(acetone)_6][V(CO)_6]_2 \tag{8.76}$$

Such reactions are probably quite important in the disproportionation of dimers. Disproportionation of $V(CO)_6$ was interpreted as an inner-sphere reaction with transfer of an electron through an isocarbonyl.[68] The observation that $V(CO)_5PBu_3$ disproportionates 10^5 times more slowly than does $V(CO)_6$ suggests that large ligand effects are possible for such reactions.

Table 8.11. RATE CONSTANTS FOR REACTION 8.73[66]

M•	$k(M^{-1}s^{-1})$
$Mn(CO)_5$	1×10^6
$Re(CO)_5$	3×10^7
$Re(CO)_4PMe_3$	2×10^9
$CpMo(CO)_3$	2×10^4
$CpW(CO)_3$	1×10^4

The $CpM(CO)_3\bullet$, M = Mo, W, complexes provide examples of the electron-transfer reactions of 17-electron species.[69,70] The W complex [formed by flash photolysis of $Cp_2W_2(CO)_6$] reduces ferricenium ions and oxidizes decamethylfer-rocene, $Cp_2^*Fe.$[69]

$$CpW(CO)_3\bullet + Cp_2Fe^+ \longrightarrow CpW(CO)_3^+ + Cp_2Fe \qquad (8.77)$$

$$k = (1.89 \pm 0.04) \times 10^7 M^{-1}s^{-1}$$

$$CpW(CO)_3\bullet + Cp_2^*Fe \longrightarrow CpW(CO)_3^- + Cp_2^*Fe^+ \qquad (8.78)$$

$$k = (2.23 \pm 0.07) \times 10^8 M^{-1}s^{-1}$$

These reactions indicate the facility of oxidation and reduction of 17-electron complexes. The oxidation (Reaction 8.77) was greatly accelerated by the presence of PPh_3 ($k = 3 \times 10^9\ M^{-1}s^{-1}$), which suggests that the 19-electron complex, $CpW(CO)_3PPh_3\bullet$, is a stronger reducing agent.[69]

Atom-transfer and electron-transfer reactions of 17-electron complexes are very important in radical-chain processes. An example is shown for the radical-initiated substitution of $HRe(CO)_5$ (see Chapter 4).[63]

$$HRe(CO)_5 + R\bullet \longrightarrow RH + [Re(CO)_5]\bullet \qquad (8.79)$$

$$[Re(CO)_5]\bullet + L \longrightarrow [Re(CO)_4L]\bullet + CO \qquad (8.80)$$

$$[Re(CO)_4L]\bullet + HRe(CO)_5 \longrightarrow HRe(CO)_4L + [Re(CO)_5]\bullet \qquad (8.81)$$

etc.

Chain reactions take advantage of the very rapid atom transfer (in other cases electron transfer) and ligand substitution of 17-electron complexes.

8.7. Summary

Electron transfer is one of the primary reactions of transition metal complexes. Its rate depends on the concentration of the oxidant and the reductant. Electron-transfer mechanisms are of two basic types: An outer-sphere mechanism between metal centers with intact coordination spheres and an inner-sphere mechanism in which a ligand bridges the two metal centers. In either case a primary consideration in the rate is the changes in the bond lengths and angles necessary before electron transfer can occur. Outer-sphere reactions can be adequately described by theoretical methods. Conjugation in a bridging ligand aids electron transfer, although electron transfer can occur over very long distances, even in the absence of conjugation.

8.8. References

1. K. F. Purcell and J. C. Kotz, *Inorganic Chemistry* (Philadelphia: W. B. Saunders, 1977).

2. F. A. Cotton and G. Wilkinson, *Advanced Inorganic Chemistry* (New York: Interscience, 1972).

3. R. G. Wilkins, *The Study of Kinetics and Mechanism of Reactions of Transition Metal Complexes* (Boston: Allyn and Bacon, 1974).

4. F. Basolo and R. G. Pearson, *Mechanisms of Inorganic Reactions* (New York: John Wiley and Co., 1968).

5. H. Taube, *J. Chem. Ed.* **45**, 452 (1968).

6. L. E. Bennett, *Prog. Inorg. Chem.* **18**, 2 (1973).

7. J. E. Earley, *Prog. Inorg. Chem.* **13**, 243 (1970).

8. N. Sutin, *Acc. Chem. Res.* **1**, 225 (1968).

9. H. Taube and E. S. Gould, *Acc. Chem. Res.* **2**, 321 (1969).

10. A. Haim, *Acc. Chem. Res.* **8**, 264 (1975).

11. S. Bellard, K. A. Rubinson, and G. M. Sheldrick, *Acta Cryst.* **B35**, 271 (1979).

12. R. D. Wilson and R. Bau, *J. Am. Chem. Soc.* **96**, 7601 (1974).

13. H. C. Stymes and J. A. Ibers, *Inorg. Chem.* **10**, 2304 (1971).

14. P. Bernhard, H.-B. Burgi, J. Hauser, H. Lehmann, and A. Ludi, *Inorg. Chem.* **21**, 3936 (1982).

15. J. Guarr, E. Buhks, and G. McLendon, *J. Am. Chem. Soc.* **105**, 3764 (1983).

16. E. Buhks, M. Bixon, and J. Jortner, *J. Phys. Chem.* **85**, 3763 (1981).

17. F. Moattar, J. R. Walton, and L. E. Bennett, *Inorg. Chem.* **22**, 550 (1983).

18. N. Sutin, *Prog. Inorg. Chem.* **36**, 441 (1983).

19. R. A. Marcus, *Ann. Rev. Phys. Chem.* **15**, 155 (1964).

20. J. F. Endicott and T. Ramasami, *J. Phys. Chem.* **90**, 3740 (1986).

21. J. F. Endicott, D. R. Stranks, and T. W. Swaddle, *Inorg. Chem.* **29**, 385 (1990).

22. R.M.L. Warren, A. G. Lappin, B. D. Mehta, and H. M. Neumann, *Inorg. Chem.* **29**, 4185 (1990).

23. G. Dulz and N. Sutin, *Inorg. Chem.* **2**, 917 (1963).

24. J. F. Endicott, B. Durham, and K. Kumar, *Inorg. Chem.* **21**, 2437 (1982).

25. H. Cohen, S. Efrima, D. Meyerstein, M. Nutkovich, and K. Wieghardt, *Inorg. Chem.* **22**, 688 (1983).

26. T. J. Meyer, *Acc. Chem. Res.* **11**, 94 (1978).

27. J. C. Curtis and T. J. Meyer, *Inorg. Chem.* **21**, 1562 (1982).

28. T. J. Meyer, *Prog. Inorg. Chem.* **30**, 389 (1983).

29. C. K. Murray and D. W. Margerum, *Inorg. Chem.* **22**, 463 (1983).

30. A. G. Mauk, E. Bordignon, and H. B. Gray, *J. Am. Chem. Soc.* **104**, 7654 (1982).

31. A. Haim, *Prog. Inorg. Chem.* **30**, 273 (1983).

32. C. Shea and A. Haim, *J. Am. Chem. Soc.* **93**, 3055 (1971).

33. D. Gaswick and A. Haim, *J. Am. Chem. Soc.* **96**, 7845 (1974).

34. B. T. Reagor, D. F. Kelley, D. H. Huchital, and P. M. Rentzepis, *J. Am. Chem. Soc.* **104**, 7400 (1982).

35. R. W. Craft and R. G. Gaunder, *Inorg. Chem.* **14**, 1283 (1975).

36. S. S. Isied and H. Taube, *J. Am. Chem. Soc.* **95**, 8198 (1973).

37. F. P. Rotzinger, K. Kumar, and J. F. Endicott, *Inorg. Chem.* **21**, 4111 (1982).

38. D. Devault, *Quart. Rev. Biophy.* **13**, 390 (1980).

39. L. T. Calcaterra, G. L. Closs, and J. R. Miller, *J. Am. Chem. Soc.* **105**, 670 (1983).

40. A. G. Mauk, R. A. Scott, and H. B. Gray, *J. Am. Chem. Soc.* **102**, 4360 (1980).

41. J. L. McGourty, N. V. Blough, and B. M. Hoffman, *J. Am. Chem. Soc.* **105**, 4470 (1983).

42. J. R. Winkler, D. G. Nocera, K. M. Yocom, E. Bordignon, and H. B. Gray, *J. Am. Chem. Soc.* **104**, 5798 (1982).

43. A. G. Sykes, *Kinetics of Inorganic Reactions* (Oxford: Pergamon, 1966) Chapter 8.

44. N. Rajasekar, R. Subramaniam, and E. S. Gould, *Inorg. Chem.* **21**, 4110 (1982).

45. J. E. Ellis, *J. Organomet. Chem.* **86**, 1 (1975).

46. T. Kruck and M. Hoffer, *Chem. Ber.* **97**, 2289 (1964).

47. M. Absi-Halabi, J. D. Atwood, N. P. Forbus, and T. L. Brown, *J. Am. Chem. Soc.* **102**, 6248 (1980).

48. R. B. King, P. M. Treichel, and F.G.A. Stone, *Chem. Ind. (London)*, 747 (1961).

49. N. Flitcroft, D. K. Huggins, and H. D. Kaesz, *Inorg. Chem.* **3**, 1123 (1964).

50. G. Sbrignaldello, *Inorg. Chim. Acta*, **48**, 237 (1981).

51. P. Chini, G. Longoni, and V. G. Albano, *Adv. Organomet. Chem.* **14**, 285 (1976).

52. D. J. Maltbie and J. D. Atwood, unpublished results.

53. I. Wender and P. Pino, *Metal Carbonyls in Organic Synthesis* (Wiley, New York, 1968), Vol. 1.

54. M. Tilset and V. D. Parker, *J. Am. Chem. Soc.* **111**, 6711 (1989).

55. (a) J. R. Pugh and T. J. Meyer, *J. Am. Chem. Soc.* **114**, 3784 (1992). (b) J. R. Pugh and T. J. Meyer, *J. Am. Chem. Soc.* **110**, 8245 (1988).

56. M. S. Corraine and J. D. Atwood, *Organometallics* **10**, 2315 (1991).

57. C.-K. Lai, W. G. Feighery, Y. Zhen, and J. D. Atwood, *Inorg. Chem.* **28**, 3929 (1989).

58. J. R. Eyler and D. E. Richardson, *J. Am. Chem. Soc.* **107**, 6130 (1985).

59. (a) R. M. Nielson, M. N. Golovin, G. E. McManis, and M. J. Weaver, *J. Am. Chem. Soc.* **110**, 1745 (1988). (b) G. E. McManis, R. M. Nielson, A. Gochev, and M. J. Weaver, *J. Am. Chem. Soc.* **111**, 5533 (1989).

60. J. N. Braddock and T. J. Meyer, *Inorg. Chem.* **12**, 723 (1973).

61. (a) M. S. Corraine, C.-K. Lai, Y. Zhen, M. R. Churchill, L. A. Buttrey, J. W. Ziller, and J. D. Atwood, *Organometallics* **11**, 35 (1992). (b) M. S. Corraine and J. D. Atwood, *Organometallics* **10**, 2315 (1991). (c) M. S. Corraine and J. D. Atwood, *Organometallics* **10**, 2647 (1991).

62. Y. Zhen and J. D. Atwood, *Organometallics* **10**, 2778 (1991).

63. (a) S. S. Kritijánsdóttir and J. R. Norton, *J. Am. Chem. Soc.* **113**, 4366 (1991). (b) C. Creutz and N. Sutin, *J. Am. Chem. Soc.* **110**, 2418 (1988).

64. P. Wang and J. D. Atwood, *J. Am. Chem. Soc.* **114**, 6424 (1992).

65. (a) Y. Zhen, W. G. Feighery, C.-K. Lai, and J. D. Atwood, *J. Am. Chem. Soc.* **111**, 7832 (1989).
(b) Y. Zhen and J. D. Atwood, *J. Coord. Chem.* **25**, 229 (1992).

66. T. L. Brown, in *Organometallic Radical Processes*, ed. W. C. Trogler (Elsevier, Amsterdam, 1990), p. 67.

67. M. R. Burke and T. L. Brown, *J. Am. Chem. Soc.* **111**, 5185 (1989).

68. T. G. Richmond, Q. Z. Shi, W. C. Trogler, and F. Basolo, *J. Am. Chem. Soc.* **106**, 76 (1984).

69. S. L. Scott, J. H. Espenson, and W.-J. Chen, *Organometallics* **12**, 4077 (1993).

70. T.-J. Wan and J. H. Espenson, *Organometallics* **14**, 4275 (1995).

8.9. Problems

8.1. Write inner- and outer-sphere mechanisms for the reaction of $V(H_2O)_6^{2+}$ with $Co(NH_3)_5N_3^{2+}$. Suggest data that could be used to differentiate between the two.

8.2. Use the following data for the isotopic exchange reactions to calculate the rate constants for the cross-reactions:

	E_0	k
$*Ce(IV)(aq) + Ce(III)(aq) \rightleftarrows Ce(IV)(aq) + *Ce(III)(aq)$	1.44	4.6
$*Fe(CN)_6^{3-} + Fe(CN)_6^{4-} \rightleftarrows Fe(CN)_6^{3-} + *Fe(CN)_6^{4-}$	0.68	300
$*MnO_4^- + MnO_4^{2-} \rightleftarrows MnO_4^- + *MnO_4^{2-}$	0.56	3.6×10^3

8.3. Use the data in Table 8.1 to estimate the rate of self-exchange for $Ru(H_2O)_6^{2+,3+}$. [P. Bernhard, H.-B. Burgi, J. Hauser, H. Lehmann, and A. Ludi, *Inorg. Chem.* **21**, 3936 (1982).]

8.4. Consider the following data:

SECOND-ORDER RATE CONSTANTS FOR THE REDUCTIONS OF
COMPLEXES I–IV BY Cr(II) AND V(II) AT 25°C, $\mu = 1.0$ M $(LiClO_4)$

Complex	$k_{Cr} \times 10^3 (M^{-1}s^{-1})$	$k_v (M^{-1}s^{-1})$	k_{Cr}/k_v
I	1.16 ± 0.09	0.056 ± 0.005	0.021
II	1.15 ± 0.03	0.065 ± 0.005	0.018
III	1.66 ± 0.02	0.085 ± 0.003	0.020
IV	1.89 ± 0.09	0.096 ± 0.005	0.020

Suggest mechanisms based on these data. [M. Hery and K. Wieghardt, *Inorg. Chem.* **15**, 2315 (1976).]

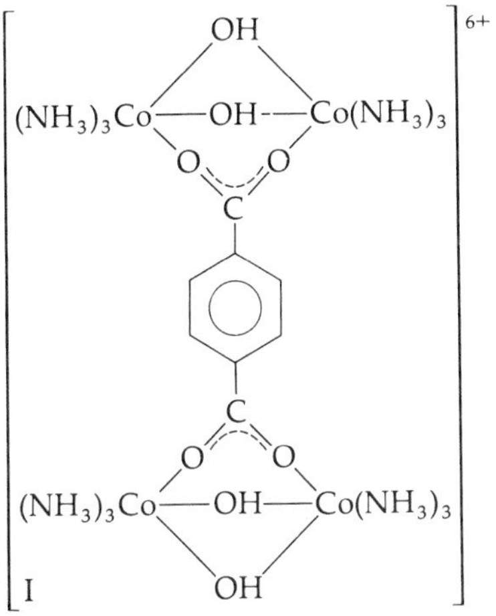

8.5. The transfer of electrons between centers in biological molecules has been modeled by attaching an oxidizing agent to a protein containing Fe(III). Briefly describe the nature of the experiment. [S. S. Isied, G. Worosila, and S. J. Atherton, *J. Am. Chem. Soc.* **104**, 7659 (1982).]

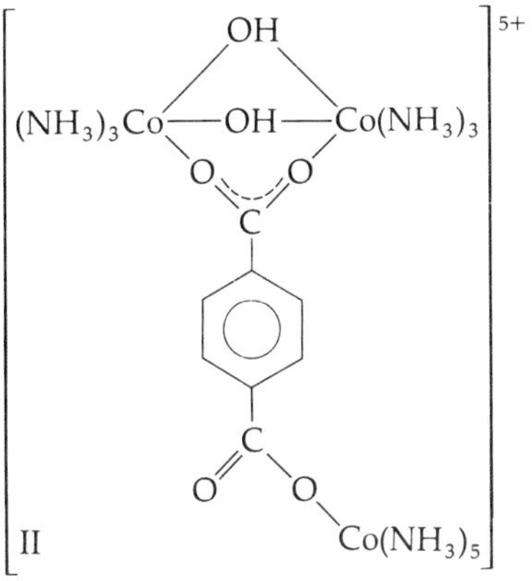

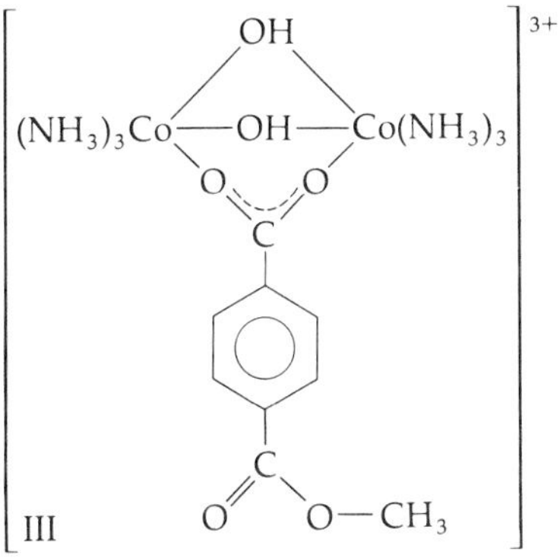

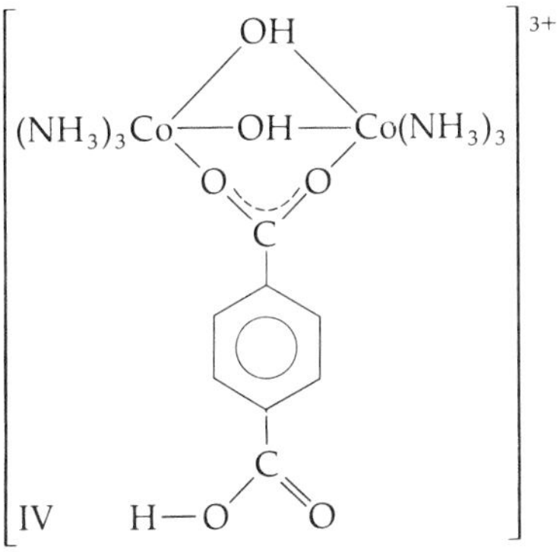

8.6. For reactions of $Cp_2Fe_2(CO)_4$ with oxidants

$$Cp_2Fe_2(CO)_4 + 2Ox \xrightarrow{CH_3CN} 2CpFe(CO)_2(NCCH_3)^+ + 2Red$$

$$Ox = Ru(bipy)_2Cl_2^+ \text{ or } [CpFe(CO)]_4^+$$

the rate law is

$$\frac{-d[Cp_2Fe_2(CO)_4]}{dt} = k[Ox][Cp_2Fe_2(CO)_4]$$

and activation parameters,

Ox	$\Delta H^{\ddagger}$	$\Delta S^{\ddagger}$
$Ru(bipy)_2Cl_2^+$	6.2	−17.0
$[CpFe(CO)]_4^+$	1.8	−31.0

Suggest a mechanism [J. N. Braddock and T. J. Meyer, *Inorg. Chem.* **12**, 723 (1973)].

8.7. The activation enthalpy for reduction of cis-$Co(en)_2Cl_2^+$ by Cr^{2+} is -6 ± 4 kcal/mole. Explain the negative value. [R. C. Patel, R. E. Ball, J. F. Endicott, and R. G. Hughes, *Inorg. Chem.* **9**, 23 (1970).]

8.8. Explain the following two reactions:

$$cis\text{-}Cr(H_2O)_4(N_3)_2^+ + \text{*}Cr(H_2O)_6^{2+} \rightarrow \text{*}Cr(H_2O)_5N_3^{2+} + Cr(H_2O)_6^{2+} + N_3^-$$
$$k = 1.9 \text{ M}^{-1}\text{s}^{-1}$$

$$cis\text{-}Cr(H_2O)_4(N_3)_2^+ + \text{*}Cr(H_2O)_6^{2+} \rightarrow cis\text{-}\text{*}Cr(H_2O)_4(N_3)_2^+ + Cr(H_2O)_6^{2+}$$
$$k = 60 \text{ M}^{-1}\text{s}^{-1}$$

[R. Snellgrove and E. L. King, *J. Am. Chem. Soc.* **84**, 4609 (1962).]

8.9. Why is $Co(CN)_5^{3-}$ a common choice for preparation of a bridged redox species?

8.10. The reaction of $Co(NH_3)_6^{3+}$ with $Co(CN)_5^{3-}$ occurs at approximately the same rate as that of $Co(NH_3)_5F^{2+}$ with $Co(CN)_5^{3-}$. What does this indicate about the mechanism?

8.11. In the reaction of $Co(NH_3)_5SCN^{2+}$ with $Fe(H_2O)_6^{2+}$ the intermediate $Fe(H_2O)_5(NCS)^{2+}$ can be detected. Suggest a mechanism.

8.12. The reduction of $Co(NH_3)_5H_2O^{3+}$ by $Cr(H_2O)_6^{2+}$ occurs at a rate of approximately 0.1; $Co(NH_3)_5OH^{2+}$ is reduced by $Cr(H_2O)_6^{2+}$ at a rate of 1.5×10^6 (all rates expressed in $M^{-1}s^{-1}$). Reduction of these two oxidants by $Ru(NH_3)_6^{2+}$ led to rate constants of 3.0 and 0.04 for the aquo and hydroxo complexes, respectively. What do these data suggest for the mechanisms of these redox reactions?

8.13. Calculate the self-exchange rate constant for $Mn(H_2O)_6^{2+,3+}$ given the following data:

$$Co(H_2O)_6^{3+} + Mn(H_2O)_6^{2+} \xrightarrow{k=48\,s^{-1}M^{-1}} Co(H_2O)_6^{2+} + Mn(H_2O)_6^{3+}$$
$$Co(H_2O)_6^{3+} + Co(H_2O)_6^{2+} \xrightarrow{k=13.5\,s^{-1}M^{-1}} \text{exchange}$$
$$Co(H_2O)_6^{3+}\ e^- \rightarrow Co(H_2O)_6^{2+} \qquad E = 1.86 \text{ V}$$

[D. H. Macartney and N. Sutin, *Inorg. Chem.* **24**, 3403 (1985).]

8.14. $Os(CN)_6^{4-}$ has a ^{13}C resonance at 142.9 ppm with a line width of 4.5 Hz. As one adds $Os(CN)_6^{3-}$ the line width increases. The following data were generated at ionic strength of 0.5 ($NaClO_4$)

$[Os(CN)_6^{3-}]$	Line Width of $Os(CN)_6^{4-}$ (Hz)
0.4×10^{-3} M	9
0.7×10^{-3} M	11
1.0×10^{-3} M	16
1.3×10^{-3} M	18
1.6×10^{-3} M	20
1.9×10^{-3} M	22
2.1×10^{-3} M	27

Calculate the self-exchange rate constant for $Os(CN)_6^{3-,4-}$. [D. H. Macartney, *Inorg. Chem.* **30**, 3337 (1991).]

8.15. The electron-transfer reaction between $Co(NH_3)_5H_2O^{3+}$ and $Fe(CN)_6^{4-}$ does not show the typical first-order dependence on oxidant and reductant, but shows rate-limiting behavior when k_{obs} is plotted versus $[Fe(CN)_6^{4-}]$. Explain this behavior. [I. Krack and R. Van Eldik, *Inorg. Chem.* **28**, 851 (1989).]

8.16. Two parallel paths are observed in the self-exchange reactions of $Fe(H_2O)_6^{2+,3+}$. What are these paths? [W. H. Jolley, D. R. Stranks, and T. W. Swaddle, *Inorg. Chem.* **29**, 1948 (1990).]

Index